Construction Project Management

An integrated approach

Peter Fewings

Taylor & Francis
Taylor & Francis Group

LONDON AND NEW YORK

First published 2005
by Taylor & Francis
2 Park Square, Milton Park, Abingdon, Oxon OX14 4RN

Simultaneously published in the USA and Canada
by Taylor & Francis
270 Madison Ave, New York, NY 10016, USA

Reprinted 2007, (twice), 2008

Taylor & Francis is an imprint of the Taylor & Francis Group, an informa business

© 2005 Peter Fewings

Typeset in Sabon by
Newgen Imaging Systems (P) Ltd, Chennai, India
Printed and bound in Great Britain by
MPG Books Ltd, Bodmin, Cornwall

British Library Cataloguing in Publication Data
A catalogue record for this book is available
from the British Library

Library of Congress Cataloging in Publication Data
A catalog record for this book has been requested

ISBN 10: 0–415–35905–8 (hbk)
ISBN 10: 0–415–35906–6 (pbk)

ISBN 13: 978–0–415–35905–4 (hbk)
ISBN 13: 978–0–415–35906–1 (pbk)

This book is dedicated to my wife Lin for her love and wholehearted support

Contents

Figures

Tables

Case studies

Foreword

Professional management of construction projects is crucial to improved performance and the reputation of the industry. There are many books written on project management and they are written from different perspectives. This book is specific to the art and science of the management process as it applies to construction projects.

The author recognises a holistic integrated approach where the techniques have taken second place to the skills of bringing together people from different cultures as a conductor brings together an orchestra. However, our construction project teams need to combine their many skills in producing quite complex products, from sonatas to symphonies, with very little opportunity for rehearsal. They are often required to assemble new teams in new places with new materials or new combinations of materials which need to work well over many years and, like music, they are on public display. The project manager is required to make great music by forward planning and co-ordinating the many different activities seamlessly to strict cost, time and quality targets which reflect fully the clients business requirements. A successful project should be profitable for all parties and makes everyone happy.

Construction Project Management: An Integrated Approach examines the key issues such as business need, clear project definition, strategy, the relationship of risk and value, quality improvement and organisational culture, providing the theory behind some of the best practice mantras and challenges a thoughtful consideration of where we should be. The case study material provides backing for some of the developments that are taking place and the wide ranging nature of the book is helpful in seeing the whole picture.

Recently we have seen parts of the construction industry stir itself to a higher level of awareness of the perennial problems which have dogged it and the causes of them. What is needed now is for more customer focused, highly trained project teams to grab the initiative. We need to measure our performance throughout the whole supply chain in order that we can establish the value of the innovations that occur and only in that way will we be able to blow our trumpets. This book provides some of the insight that we need.

Alan Crane CBE
Executive Director of The Movement for Innovation
Chair of The Movement for Innovation and
Rethinking Construction

Acknowledgements

The author would like to thank all those who have contributed to the content of this book. My chapter contributor Martyn Jones for Chapter 12 and contributions to Chapters 6 and 14, Tony Westcott for his contribution to Chapter 8 and for Carol Graham's tireless work in reading the text through on my behalf and pointing out numerous inconsistencies and text improvements (have I got the apostrophe right?). For so many contributors to the case studies, who gave of their time to patiently explain processes and products. Although I have done my best to get it right any remaining mistakes are mine alone. I would particularly like to thank Tony Brown who has allowed me to use his material on the Andover North prime contract and spending valuable time to give me assistance. I would also like to mention Paul Harding, Jerry Hayes, Paul Dempster, Grant Findlay, Bruce McDonald, Sem Kyambula, Mike Howlett and Saleem Ackram who were instrumental in inspiring me and giving of their time.

As the case studies are mainly anonymous, I would like to thank those who have made the material possible. As well as live case studies, we have been able to draw on excellent web material that has been duly acknowledged in the text and used with permission. Mansell have particularly allowed me to make use of material for a case study.

We would also like to thank the University of the West of England, Faculty of the Built Environment for the support they have given to this project. We are grateful to the University for allowing us to access module material, which has helped in the preparation of this book.

Peter Fewings
Martyn Jones
Tony Westcott

Introduction

Many books have been written on project management and there are two approaches to it. One deals mainly with the tools and techniques of project management and provides instruction on what they are and how to use them. The other approach takes a managerial viewpoint and is concerned more with the context and the way in which decisions are made and the tools which are most appropriate in that situation. This book is more allied to the managerial approach, analysing how techniques have been applied in traditional and best practice and synthesising additional guidance on evaluating contextual factors, which make the projects unique.

In construction in particular, there is a long history of project management and standard systems have been set up which have become comfortable, but have not always produced the best value for the client. Every project is different and has at least a unique location, and due to the time and budget constraints the final product is an untested prototype, which has been subject to continuing design variations. Therefore, at first, it is a particular challenge to an industry that has not standardised its products. The industry is also quite fragmented with inexperienced clients and separate design and construction organisations. The supply chain can often be quite long with some detailed design provided at a second and third tier contract level, with little direct labour provided by the main contractor. This presents additional challenges to the construction project manager who needs to co-ordinate the design and construction sides and make decisions based on the promises of others.

In addition to this the profitability of contractors has been low; there is a high turnover of construction business and many consultants are on low remuneration. It is a particular challenge to deliver to the tight time, cost and quality targets that are set by most of the clients. In response, construction has had to adopt a much more client-orientated view. This view allows alternative procurement strategies, where design and construction are much more integrated and opportunities for the development of the brief to take account of project constraints and ongoing business opportunities are recognised. Risk management (RM) is about managing uncertainty and there is a chapter dealing with this concept in the context of increasing value.

It is important to look at both the function of the product (effectiveness) and the efficiency of the process and this means a closer relationship between the client and the whole project team including construction. Project management has an important role to play here, by understanding the client's business as well as the project objectives and incorporating these in the design, method and sequence of the finished construction. Chapters on business development, value management (VM), design management and supply chain management (SCM) will cover this issue of increasing value. VM in its simplest sense is a way of reducing waste in order to maximise productive output. This waste comes in all forms and reports such as Latham [1], Atkins [2], Egan [3, 4] have particularly recognised inefficiencies in the construction process and set targets for improvement. Toyota, who are well known for pioneering lean manufacturing techniques, have recognised that there is as much as 60% of unproductive activity in their system, so there is always room to improve.

Project management is at least 50% people management. This book emphasises the importance of people skills, considering such things as collaboration, leadership, organisation, culture, communication, delegation, motivation and negotiation. Interpersonal skills are gained from experience so the purpose of discussion here is to raise awareness of the importance of these skills and the ways in which they might be developed in different situations. It will cover the issues which arise in managing a supply network and communicating across a broad inter-professional team. It will support the concept of leadership and innovation and look at ways of developing competence in negotiation, delegation and developing teamwork in the specific context of construction and engineering projects. There is an instinctive role in managing people and lately a more formal role in initiatives such as Investment in People and Respect for People, which recognises the increasingly key role of developing people. Both of these do have potential for development in order to really improve teamwork and interpersonal skills in project management where there is such diversity. The development of synergistic teamwork is of particular interest to this text and there is also a particular chapter on engineering the psycho-productive environment.

The structure of the book is based on the life cycle of a construction project as portrayed in the *Code of Practice for Project Management for Construction and Development* [5]. This is so that, it will be easy to cross refer the material in this book with The Code of Practice. The Chartered Institute of Building (CIOB) have welcomed this, but this book is designed to stand alone in offering guidance to best practice for construction professionals and offers illustrative case study material and our views are not necessarily those held by the Working Group for the Code of Practice.

The text should be of great use to those who are completing their studies in construction and project management. To support this aim, key concepts have been introduced and developed at the beginning of the chapter. This will also give a proper foundation for the development of innovative practice which will be discussed at the end of relevant chapters, drawing on current research and appraising the way forward. Although much reference is made to documentation and case studies in the UK and European context, the research carried out suggests that the practical applications and challenges for an integrated project management approach apply to many countries, so it is hoped that a wider readership will be able to use the book and any comments will be welcome.

The managerial approach focuses on decision making. In order to support this, the authors have referenced the Gateway Review™ process published originally in the Treasury Procurement Guidance documents [6], which has been updated slightly by the Office of Government Commerce (OGC) and reference to this can be made through their website. This provides a generic framework of five decision 'gateways' allied to the life cycle of a capital expenditure project and identifies the necessary activity to achieve these from a client perspective. Detailed notes specify the relevant government procedures, but the model is useful to the project manager in identifying the information and its timing to elicit important client and project decisions for any type of project. The approach will be discussed in Chapter 2.

The chapters are roughly in life cycle order, but some chapters have also been added to develop specific themes such as RM and VM, supply chain management, leading and managing teams. These thematic chapters will particularly take the structure discussed earlier.

The key features which together will differentiate the book from other texts will be:

- An integrated approach to hard and soft management issues at all stages of the construction project cycle.
- An open systems approach emphasising the key role of managing external factors in gaining project success.
- The key role of new forms of procurement such as Private Finance Initiative (PFI), as a system to encourage the culture of partnering and which can be broadly adopted in integrating the life cycle of capital provision and maintenance.
- Recognition of the need to engineer the psycho-productive environment to improve the efficiency and delivery of a project.

Definitions

A *project* is defined in BS 6079:2000 *Guide to Project Management* [7] as

> A unique set of co-ordinated activities with definite starting and finishing points, undertaken by an individual or organisation to meet *specific objectives* within defined schedule, cost and performance parameters.

It is important to understand how the definition of a project can apply to lots of types of activities and not just construction. If you understand this, then you will see the differences between project management and general management and why project management is used in construction.

Features of a project include non-repetitive, goal orientated, holding a particular set of constraints, measurable output and changes something through the project being carried out [8].

Project management may be defined as

> The overall planning, co-ordination and control of a project from inception to completion, aimed at meeting a client's requirements in order to produce a functionally and financially viable project that will be completed on time, within authorised cost and to the required quality standards.
>
> (CIOB [5])

Programme management is the management of several related parallel or sequential projects. When concepts such as strategic partnering and supply chain management are discussed there is a need to look at issues which are 'inter' projects, as value is often built on carrying forward knowledge and experience between projects. Strictly these are business strategies, but they are recognised as being project driven as they are based on business with a single client. An example of programme management would be Bristol Harbourside Development where the developer is Crest Nicholson, who has an overall programme based on reaching commitments to several users, yet they have let out the design to several architects and contractors to cover infrastructure and various buildings or building types.

Project definition is the development of the project brief and scope up to the planning application stage, so that the risks have been identified and the value for money (VFM) has been optimised to suit the business needs.

Body of Knowledge

The two well-known institutions supporting and certifying the area of project management have produced supportive documentation defining generic competencies of a project manager.

The Association for Project Management (APM) [9] Body of Knowledge breaks down these competencies into seven groups. Project management (covers various stages in the PM process), organisation and people (skills for people management), techniques and procedures (tools and techniques used and developed specifically in connection with project management) and general management (competencies required of management in general). This has now been rehashed into seven. It now extracts a group of commercial competencies, people competencies distinguished from the organisational ones, strategic competencies that recognise the project planning processes of risk, value, quality and health and safety, and control competencies related to time, cost, change and information. This appears to emphasise the importance of distinguishing the strategic, technical and people management functions.

The American Project Management Institute (PMI) [10] talks about five groups of processes which are initiating, planning, executive, controlling and closing processes which are clearly linked to the life cycle of a project mentioned earlier. They differentiate between

- project management processes which are applicable to all projects – these are discussed in some detail;
- the product-orientated processes which vary according to the application of project management, for example, construction, particularly because the life cycle stages can differ between different applications.

The project management processes are integration management, scope management, time management, cost management, quality management, human resource management, communications management, RM and procurement management.

A typical product-orientated process in construction would be the commissioning of building air conditioning, which makes use of a commissioning engineer to test and adjust environmental parameters of the system in the context of the building envelope. The main output here is user comfort. This would be quite different from the commissioning of say, a computer software project, where the main output is user satisfaction and efficiency.

The competencies try to provide some common ground between projects. They are generic but as construction is going through a relearning curve, they potentially offer a basis for revisiting the processes and skills which provide effective project outcomes. There is no good reason for remaining insular in construction. In particular, there is a definite need to consider the skills for managing stakeholders. The Code of Practice has a similar life cycle approach.

It is assumed that the reader has some knowledge of construction terms, but a glossary has been compiled to make these clearer. Key definitions and concepts have been covered in the text and a particular feature of the text is to relate construction project management to management theory, which is relevant for greater understanding of the context, for example, organisation theory.

References

1 Latham M. (1994) *Constructing the Team*. Final Report of the Government Industry Review of Procurement and Contractual Arrangements in the UK Construction Industry. HMSO, London.

2 Atkins W.S. (1994) 'Secteur strategic study on the construction sector final report'. *In Association with the Centre for Strategic Studies in Construction, University of Reading; Euroquip; Dorsch Consult; Prometeia; Finco Ltd; Agrotech Inc and ESB International*. F2347.050/SDR/ECC.6A. For the ECC 111/4173/93.

3 Egan J. (1998) *Rethinking Construction: The Report of the Constrution Task Force*. DETR July.

4 Egan J. (2002) *Accelerating Change*. Strategic Forum for Construction/DTI. HMSO, London.

5 CIOB (2002) *Code of Practice for Project Management for Construction and Development*. 3rd Edition. Blackwell Publishing, Oxford.

6 Treasury (2000) *Guidance No. 2 Value for Money*. HMSO, London.

7 British Standards Institution (2000) *Guide to Project Management*. BS 6079. BSI, UK.

8 Maylor H. (2000) *Project Management*. Financial Times/Pitman Publishing, London.

9 Dixon M. (ed.) (2000) *Project Management Body of Knowledge*. The Association for Project Management. High Wycombe, London.

10 Project Management Institute (2000) *A Guide to the Project Management Body of Knowledge*. PMI, Upper Darby, PA.

Chapter 1

Project life cycle and success

Project management is not a new concept, but it has emerged since the Second World War as a methodology that can be applied to intensive periods of work with a specific objective, which can be isolated from general management so that expenditure can be ring fenced and the synergy of a team is engaged. However not all managers are able to cope with the dynamic nature of projects, where decisions have to be made fast and planning and control have to be very tight. Large projects such as the NASA space programme, the Channel Tunnel and the Polaris submarine programme have developed techniques for project management that have set a pattern for subsequent ones. These projects have also had to develop specific roles and create management structures to suit and satisfy various interests, both within the project and contract and outside. Construction work particularly lends itself to project management because of the temporary and unique nature of the work.

This chapter will look at the project as a whole from inception to completion. It will look at how projects are managed and try to understand how they are led and what the critical factors of success are:

- Defining the construction project life cycle.
- Distinguishing specific project management activities and allocating roles throughout the life cycle.
- Investigating factors which affect the way that projects are managed.
- Determining the critical factors for project success.

Project life cycle

The life cycle of a project from a client's point of view really starts when there is a formal recognition of project objectives, generally termed the inception, and through to the delivery of these objectives – generally called the completion or project delivery. Activities relating to the conception of a project may take place over an extended period before formal recognition. Related activities commissioning the project also take place in the period

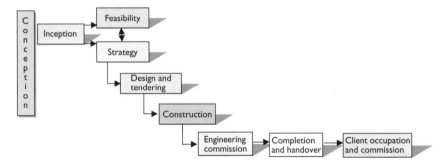

Figure 1.1 Life cycle of construction projects.
Source: Based on the CIOB (Figure 3.1) [1].

after formal completion. In construction projects, this inception is generally associated with the commissioning of external consultants and a life cycle indicating the main elements of its life cycle as shown in Figure 1.1.

There are many different ways of looking at projects depending upon the viewpoint of the participant. Different parts of a life cycle are often managed by different people and not all organisations are involved in the project all the way through from inception to completion of a building project. For example, a main contractor gets involved from tendering through to handover of the building for the client's fitting out – just two parts of the client's project life cycle, but for them it is a complete project with an inception and a completion. Different industries have different terminology and put emphasis on different parts of the life cycle.

There is a need to be flexible in the view of the construction life cycle, as there are now many different forms of procurement, which put a different emphasis on different stages of the cycle such as Design, Build, Finance and Operate (DBFO), which has a strong contractor provider involvement in the inception, feasibility and operation phases, which is certainly not the case in traditional procurement. This broader, more integrated viewpoint helps to recognise the client's position as a developer or a user of an asset, which is essential to the success of any project. The project life cycle model in Figure 1.1 is robust enough to cover a wide variety of procurement approaches and allows for the development of innovative approaches to meet client requirements best. The development cycle is wider and tracks the building or structure to a change of use or to demolition and recycling for the next development opportunity. At this stage another building project may emerge. The end of a construction project is the handover to a facilities management team, who maintain the structure and services.

The life cycle of construction projects starts at *inception*, at the stage where a client's business case for a building or a refurbishment is communicated to a professional team to develop the constraints. Outline planning consent may have been achieved, but only if a site has already been chosen. The inception process may be an extended period. In order to proceed, a client has to test the *feasibility* of the business case they have. The fundamental go ahead will have been reviewed by the beginning of the inception stage and the main focus for the project manager at this stage is to define options for project feasibility and their financial viability or benefits. The feasibility stage therefore, can include investigation of alternative site locations, funding options, design option appraisal, value enhancement, comparative estimates and life cycle costing. It considers the associated project constraints and marketing implications and is closely tied up with the strategy for the project. At the end of this period, a feasibility study needs to test affordability (fit within an outline budget and cost plan) and meet constraints (an option to achieve planning permission and give best value for the client). The client needs to make critical decisions which suit the needs of their business. A key part of this is to identify and allocate risks and to carry out a functional analysis to optimise VFM. The user or facilities management groups may be involved at this stage.

Strategy deals with how a project is carried out and controlled, such as the procurement route that is chosen, the cost, the programme, the control systems, the quality management and the methodology for construction. Strategy is a partly parallel activity to feasibility as the viability is often dependent on the strategy. For example, the funding of the project is tied up with the programme time and the cash flow availability. Strategy also needs to identify the right procurement method and determine the organisation structure of the project. A key part of this stage is to produce a project execution plan (PEP) that fully analyses and allocates the risk issues. It also specifies how the project is going to be planned and organised through the subsequent stages of the project life cycle. The brief needs to be developed to ensure a full understanding of the client's requirements and the design and construction strategies need to be co-ordinated within the project constraints. Outputs include a PEP. If a construction manager can be brought in at this stage more reliable information on construction planning and methodology is available.

Pre-construction (design and tendering) appoints the full design and construction team and includes the full development of the design scheme, detailed drawings, tendering and mobilisation of resources for construction. There is a clear responsibility to manage design and procurement and to identify a start date for construction which is related to the handover and occupation of the building. Risk and value factors continue to be managed so that the client gets best value. Outputs include further statutory permissions such as building regulations, integrated design drawings, tender documents, contractor appointment, an agreed contract price and a contract programme and a pre-tender health and safety plan. It is notoriously difficult to control diverse design activity to meet the deadlines, anticipate the timing and nature of statutory consents or even appeals and to predict a market price which will comply with the budget constraints.

The construction phase is self explanatory, but it has a particular emphasis on the control of time, quality and cost and the management of many other issues such as supply chain, health and safety planning, the environment and change. Outputs here will include construction stage programmes, construction health and safety plans, method statements, cash flow forecasts, quality assurance schemes and change orders. In taking on a contractor, there is a risk of conflict if information is not available, if things get changed a lot, or the project is delayed. Conflict management, leadership and team building skills are used a lot in this stage.

Engineering commissioning comes at the latter end of construction. It is distinctive as its outputs should include the efficient functioning of the building. The management of the process includes the signing off of various regulatory requirements such as building regulations, fire and water certificates, gas and electrical tests.

Practical completion is certified by the project manager for the formal *handover* of the fabric and systems to the facilities team. Liability is not limited by occupation and there is a responsibility to put defects right if and when they occur after handover. Documentation and a health and safety file (HSF) are handed over for the safe and efficient use of the building's systems and maintenance schedules. Handover is sometimes called *close out*, because it suggests a focused period of preparation to ensure the project and the documentation are in order and the facilities team and users are properly briefed and inducted. The main output is not documentation, but to pass on the knowledge for running the building safely and efficiently.

The client's occupational fit out follows full or sectional completion of the contractor work and may well involve a new project team. This period often has intensive collaboration with user groups and FM teams. During this period there is a need to commission equipment, move personnel and induct occupiers in the use of the building and in its emergency procedures. Outputs include fitting out and the production of health and safety policies, user manuals and training programmes.

Post-project appraisal and review is the final stage. The objective is to evaluate success in meeting the objectives as set out in the business case/ project brief and to look at lessons learnt and to carry forward improvements, where relevant, to the next project or phase. Outputs from this stage are client satisfaction surveys, production incentives and project process reports, which may inform projects for the future.

Project management

It is important to distinguish project management responsibilities and there are a number of definitions such as:

- The planning, monitoring and control of all aspects of a project and the motivation of all those involved to achieve the project objectives on time and to cost, quality and performance.

  (BS6079 Guide for Project Management)
- The art of directing and co-ordinating human and material resources through the life of a project by using modern management techniques to achieve predetermined goals of scope, cost, time, quality and participant satisfaction.

  (The Project Management Institute)

The common elements of project management in these definitions and the CIOB definition mentioned in the introduction are time, cost and quality management and these can be viewed as a triangle as shown in Figure 1.2.

These three dimensions of control – time, cost and quality – represent the specific project efficiency factors. They are managed for the satisfaction of the client's requirements, but in themselves are secondary to the client's business needs, which are likely to be determined by the market. For example, programme control is a subset of finishing in time for the Christmas sales period when dealing with a retail client. Quality is not absolute, but related to the need for a building's cladding to efficiently keep out the weather and still look good for 10 years before the next refurbishment. It is likely that the client will prioritise one dimension more than another.

The project manager is the leader of the team and acts on behalf of the client as well as trying to maintain an efficient project team. Another important aspect of the PMI definition is the management and motivation of the 'human resource' and achieving 'participant satisfaction'. Understanding how participants from different organisations work together and having the

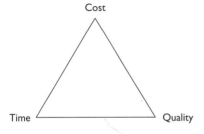

Figure 1.2 The three dimensions of project management control.

skill to direct them towards common project goals, without upsetting them, is a very important part of project management.

Project management is about balancing the needs of all the participants in the project. Forcing the pace of design for the sake of construction or vice versa may mean inefficiencies later. The phrase 'sustainable profitability' comes to mind when trying to integrate the business needs of the contractors and the consultants with the requirements of the client. However, the project manager's prime concern is to try and get value for the client and on occasions conflict may arise within the project team. On these occasions conflict needs to be fairly managed so as not to de-motivate the project team. This in itself is counterproductive.

Project team roles

It is acknowledged earlier that the leadership of the construction project may change during the project life cycle under some types of procurement. For example, in UK traditional procurement it is most likely that the architect or the engineer will take the lead in the inception and design stages and will act on behalf of the client. During construction the main contractor will have a leading role. For large or complex projects the client appoints an executive project manager with a direct leadership of the project team through whom the client communicates.

In this case, there will be a single point of contact for the client co-ordinating the design, construction and other professional roles shown in Figure 1.3. This defines the communication channels, but the actual contract is signed by the client with individual organisations and so there are differing contractual links and communication links.

The supply chain is managed day-to-day through the communication links and things can go badly wrong if the client is allowed to pursue direct contact

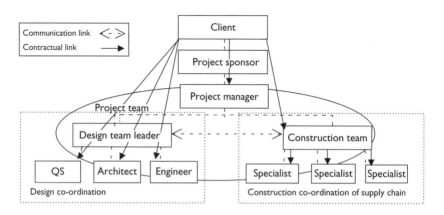

Figure 1.3 Project structure diagram – executive project management model.

through their contractual links, thus undermining the co-ordination role of their project manager and the design and construction managers. The construction and design sides need to co-ordinate their operations right through the life cycle and this model of an executive project manager feeding through to respective design and construction co-ordinators ensures that this happens. In the traditional system, the brief taker tends to take this role by default and hands on the batten to the main contractor during the post tender stage. This creates problems when the design is changed and the impacts on construction time, quality, cost and health and safety are not reassessed. On a complex project the extra cost of a project manager is well worth it.

The design manager's role is to co-ordinate the various design functions and if necessary specialist design expertise as and when needed. This role is also traditional. The main contractor or construction manager's role is to tender specialist packages, set up site procedures and integrate the construction programme and the interfaces between specialist packages. More about that would be discussed in Chapter 12. They have accountability for meeting the contract programme and budget and working with the design team to implement a quality product. This role is also traditional.

Where there is a lack of trust and respect between the designer and the contractor, or the contractor and the client, the culture can become one of 'them and us' and costs and time constraints are often at risk to the disadvantage of the client. Conciliation is a key role for the project manager to manage disputes, discrepancies, omissions and changes so that the team works smoothly.

It is normal in construction for roles to be assigned to different members of an inter-organisational team, for example, quantity surveyor, engineer/ architect, construction manager and client. The project manager is the leader of the project team with a single point of contact with the client.

It is important to try standardising some definitions in order to understand project management as people use terms to mean a lot of different things. We need to consider the roles in the context of the *client's* total view of a project that is, a finished building or facility, not *'our view'*, which covers just the length of a participant's involvement. For example, a contractor and a specialist both call the person they appoint to run the contract for them, the project manager, and the former in a traditional contract only spans the construction phase and the latter only covers the length of time the specialist installation is on the site. The Book of knowledge and two key documents 'Code of Practice for Construction Project Management' [1] (CPPM) and 'The BS6079 [2] Guide to Project Management' give us good guidance in this respect. Some of the most important of these definitions are given in the glossary at the end of the book.

Table 1.1 illustrates how different members of the team play leading roles at activities which are predominant in each of the life cycle stages assuming a project structure using an executive project manager. This is not meant to

Table 1.1 Roles of project team at each stage of implementation

Stage	Role	Client approvals	Leader
Inception	Client business objectives interpreted to strategic brief		**Client advisor**
	Professional interpretation and development of brief		PM
	to determine value and performance		
	Outline planning	CA1	Architect
Feasibility	Test for viability and/or option appraisal		**PM**/QS
	Project risks assessed		
	Outline design and costplan		**Architect/QS**
	Funding and location		PM
	Client go ahead on scheme	CA2	
	Appoint professional design team		**PM**/Architect
Strategy	Decide on procurement route, RM,		**PM**/(CM)
	cost control and quality management	CA2a	
Scheme design	A scheme design and planning application	CA3	PM/**Design team leader**
	Cost plan and cost checks – iterative with client		QS
	Build ability testing. Building regulations approval		PM/CM
	Health and safety co-ordination		**Planning supervisor**
Tender	Prepare detail design and bill of quantities		**Arch**/QS
	Tender documents	CA4	
	Pre-tender health and safety plan	CA4a	**PS**
Construction	Appoint contractor(s)	CA5	**CM**/QS/Architect
	Mobilisation of construction process, tender		
	sub contractors, health and safety plan		**CM/**
	Time, quality and cost control		**QS**
Commissioning	Test and snag all systems		**CM**/Architect
	Ensure equipment compliance and efficiency to		
	meet client objectives		
Post-project review	Feed back into future projects		
	Lessons for client	CA6	Client
	Lessons for PM		**PM**
Occupation	Recheck 'underuse' conditions		Client/User
	Manuals and training		

Abbreviations: PM = project manager, CM = construction manager, QS = quantity surveyor, PS = planning supervisor, CA = client approval.

be an exhaustive coverage of the activities which the team performs. However, it does give you an idea of the main connections between the different players. It is based on the Royal Institute of British Architects (RIBA) Plan of Work stages.

The client has a continuing interaction with the project team and may even take a lead role in project management. Their essential role is to sign off each stage so that work may proceed to the next stage. These are called client approval points or gateway reviews. For a more specific reference to the public procurement gateway review system there is more information in Chapter 2 on project definition. The final gateway review is the post project appraisal.

Choosing the project manager: skills and functions

Choosing the project manager is important to the culture of the project and will be the responsibility of the client with guidance from advisors. It is quite common for legal or other professions to provide early advice in these areas. So what is required of the project manager?

The APM [3] lists eight characteristics of a certified project manager. These are an open, positive, 'can do' attitude, common sense, open mindedness, adaptability, inventiveness, prudent risk taker (weighing up the risk), fairness and commitment. Many of these skills are associated with managing people and themes like partnership, negotiated tenders and building the team have been a feature of periodic industry-wide reports.

From a client's point of view these qualities will make the project manager easier to work with, reliable and realistic. However, they might want to add 'a willingness to see those things which are important to their business' and to adapt the system and parameters of the project to suit. They would also need to have assurance of general competence and experience especially in dealing with projects that have similar characteristics as their own. From a project team point of view being open minded and taking the lead in fairness and commitment are expected of someone who has high expectations of their performance and has to deserve team loyalty.

The project manager's role will be related to the procurement route chosen by the client. In the case of the executive structure, the project manager manages all aspects of the project from inception to completion. In a traditional procurement the project manager changes in different stages of the project life cycle. Either way there are four key tasks for the project manager acting on behalf of the client:

1   Guide and advise the client.
2   Manage the resources to carry out project activities.
3   Build the project team.
4   Ensure customer requirements are met.

In traditional procurement systems where there is a sharing of project leadership, there is potential for tasks (1), (3) and (4) to be dealt with in a piecemeal way. The appointment of a dedicated project manager should help to provide customer focus and to draw the client into the team. This gives more certainty to the client approval process.

Meeting customer requirements means gaining a knowledge of the customer's business, sharing problems at an early stage, so that trust is built up and reviewing project goals at regular intervals to make sure that the developing brief meets expectations. Guiding and advising the client is of particular importance, referring back to the need to reconcile the client's brief with construction constraints. Building the project team is about developing relationships between professionals and ensuring efficient communications between diverse and numerous participating professionals and contractors.

The resources managed will have a different focus at different stages of the life cycle. For example, at the inception stage managing funding is critical; at the design stage it is a matter of managing information; at the construction stage physical resources such as materials, plant and labour become important.

So how important is it that a project manager has a technical background? Can a construction project be managed by a generic project manager? Are there certain types of people or personalities which are better at managing construction projects?

Certainly, tasks (3) and (4) could be fulfilled by any experienced project manager and production management experience will help with task (2). Task (1) is likely to require industry knowledge of how procedures work and what the economic, the technological and the legal constraints are. Here it is probable that a competent technical training is not sufficient for reading economic situations or providing broad enough advice to a client who wants to push the technological frontiers. In this sense technical knowledge is important, but very much part of a range of skills and experience which will apply to construction projects as much as they will to any other project. It is useful at this point to remember the APM Body of Knowledge which deconstructs to 38 different areas of competence.

Ethics in project management

Meredith *et al.* [4] mention credibility (technical and administrative), sensitivity, leadership and the ability to handle stress, as four important attributes for a project manager. These are often connected with the professional status of a manager to give client confidentiality and updating knowledge. In respect to leadership, there is a strong point to be made for a code of ethics. However a fuller code is suggested by Ireland *et al.* [5], who suggest accountability in maintaining high standards, in conduct and leadership of the team, in relations with employers and clients and accountability to the wider community and the reputation of the profession. Meredith and Mantel see that the ethical position for a project manager is in the middle, representing the interests of all parties fairly to one another.

Project complexity

The degree of complexity is not directly dependent on value. The overall rating has been shown to be useful for assessing the experience of staff required, the workload generated and the degree of systemisation and formality required. Maylor [6] speaks of organisational resource and technical complexity. He makes an overall rating by multiplying them together for comparison of project complexity.

Organisational complexity increases with the number of organisations and stakeholders that are involved and the degree of integration of their work – high for a construction project even of a small size. Six organisations, which is quite basic for construction, theoretically according to Graicunas's formula [7] give 222 inter-relationships between the different supervisors. Building contracts are also quite complex.

Resource complexity increases with the value of the project and the significance and the range of the resources. In terms of construction, this ranges from very small to extremely large. A site manager will be assigned full time to a project with a value which exceeds £1–2m.

Technical complexity increases with the use of non-standard technology, the building constraints which exist and the complexity of the technology, for example, the repair of a historic building's roof compared with the use of prefabricated trusses. The combination of high ratings for at least two areas makes a lot of difference. See Case study 1.1.

Determining the critical factors for success

If we take success as the delivery of a product that meets the expectations of the client at the same time giving profitable business to the provider, the facts are clear that the construction industry has a flawed record in the delivery of projects. Major projects have been subject to cost and time overruns and the performance of the product has often fallen short of the criteria of the client let alone the expectations. In addition to this, the profitability of the industry has been in question and major investors are reconsidering their exposure to certain sectors of it.

How do you define a project? What is it that makes a project a success for the various parties involved? Are there inherent factors common to all projects or is it in the nature of the project ? Are the factors within the control of the project team or are they, as in the case of political factors, to be responded to? Do you select projects which have built in success and by definition avoid others?

Case study 1.1 Construction complexity

Table 1.2 Comparison of building complexity

Case study description	Complexity factor			
	Organisational	Resource	Technical	Combined index
Construction of a primary care doctor's surgery/pharmacy. Features atrium, low energy usage and novel design. Fast programme on new site	There is a full range of trade organisations working closely together on a lot of different spaces. There is storage and access on the car park, where offices can be accommodated Rating 4/10	Value £1.5m Fast programme Good quality Wide range of resources Rating 6/10	The innovative design has taken a long design development and went through planning with some qualifications for redesign. The building has modern materials and prefabricated components. Redesign to improve Building Research Establishment Environmental Assessment Method (BREEAM) rating Rating 5/10	$4 \times 6 \times 5 = 120$
Refurbishment of historic church hall. Features new roof, repair of stonework, demolition and creation of new basement area under the floor	There are a limited number of subcontractors say 12 involved as there is a limited range of finishes Space is very limited and the site is cramped with no outside storage in the city centre Rating 6/10	Value £600 000 Range of resources is quite low Programme affects use of the building so time is tight High quality Rating 4/10	The design of the roof is complex, uses large non-standard timbers and made to measure stonework with advanced craft techniques. Innovative technology for forming the basement. Long gestation period, but planning acceptable Rating 9/10	$6 \times 4 \times 9 = 216$

If we compare two smaller projects we can illustrate the complexity index which emerges using Maylor's [8] overall complexity measure in Table 1.2.

The first project has demanding requirements, but is not unusual for most contractors. The rating will go up with the combination of greater speed and innovation and quality. The complexity factors are increased for the second case study because of the organisational and technical factors and these are despite the size of the project. Difficult technical solutions make a lot of difference to construction projects as resources – craftsmen and materials in particular, are geared up for standard building construction. Organisational complexity is very dependent on the unique features of the site and often quite small projects present almost insuperable problems as illustrated. The fragmentation between design and construction often means that insufficient buildability has been achieved at design stage, exacerbating the execution stages with later design adjustments that can create conflict. On a conservation project this is usual, but targets are harder to achieve and need more experience and supervision for the size of the job.

Industry-wide studies

The factors of success have been reviewed in the context of many different reports in recent years. Many of them are industry-wide. The results of these reports have often depended on the viewpoint of the party who has commissioned them. Many wider consensus reports have been heralded in principle, but have failed when there has been an attempt to implement the details and interpretation of the reports, which has been harder to harmonise between different parties to the contract. They are summarised in Table 1.3.

The reports connect success to productivity, reduction or successful management of conflict, greater efforts to be client focused, reduced defects/waste and strategic management to respond to external factors.

Pinto's and Slevin [9] 10 factors of success, as determined from a survey of project managers, are shown in Figure 1.4 and were ranked. The 10 factors were listed under the categories:

- Strategic factors critical in the early stages of the life cycle.
- Tactical factors (shaded) which became most important in the latter stages.

Top management support, detailed programmes and budgets for control were the strategic factors and these remained critical and most important throughout the project. Interestingly, the next most important issue was the involvement of the client in the project team. Troubleshooting was ranked lowest.

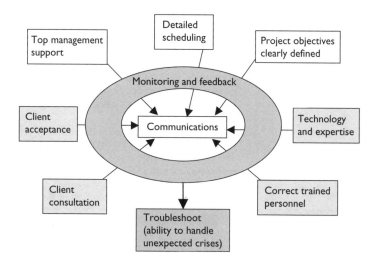

Figure 1.4 Factors of success.

Source: Adapted from Pinto and Slevin [9].

Table 1.3 Some reports indicating the need to improve construction success

Report and title	Broad findings
Morris and Hough (1987), Anatomy of major projects	A review of 5 major project case studies and the issues which caused success or failure indicate planning for external factors
Slevin and Pinto (1986–8), Several reports on determination of critical success factors	A 50 item instrument has been developed to measure a project's score on each of 10 factors on over 400 projects. Report concludes that strategic issues are important throughout the whole life cycle with tactical issues only being equal in the later stages
Reading Building 2000 Report (1990) Reading centre for strategic studies in construction	Identifies the major issues to address by the year 2000 to keep UK industry competitive worldwide. This was eclipsed by the Latham and Egan Reports
The Linden Bovis Report (1992)	Unfavourably comparing productivity between the industries of the UK and the USA
The Latham Report (1994), Constructing the team	Suggested that a productivity improvement of 30% was needed and possible by re-engineering the construction process to eliminate confrontational contracts and relationships and introduce partnering and contract change in the Housing, Construction and Regeneration Act
The Egan Report (1998), Rethinking construction	Supports value and process improvements including introducing selection on value to reduce defects, increase safety and to introduce benchmarking to make it possible to track continuous improvement. Introduces 5 drivers for success and the enhancement of client VFM. Targets set similar to Latham
The Construction Clients Forum Report (1998), Constructing improvement	Calls for a pact to be made between the client and supply side designed to improve VFM and the profitability of the supply side
Strategic Forum for Construction (2002), Accelerating construction (Egan update report)	Measures improvements over the intervening 4 years since 1998, indicating that demonstration projects where Egan principles have been practised have reached the targets set and far exceeded industry averages
Strategic Forum for Construction (2003)	Sets up a toolkit for the integration of the project team in the better delivery of a project in partnership. The parameters that it sets are very wide, but it represents a methodology to integrate the client and the team in a continuous improvement culture in pursuance of Egan and Clients Forum Reports

However Baker [10] has said that factors are different for different industries and also qualified that failure may be defined in different ways. Pinto and Mantel [11] say that there are fundamental causes across industries for failure such as poor planning, insufficient senior management support and getting the wrong project manager.

Table 1.4 Project success factors from different viewpoints

Users' criteria		Project manager's criteria	
Successful projects should	*(%)*	*Successful projects should*	*(%)*
Meet user requirements	96	Meet user requirements	86
Happy users	71	Commercial success	71
Meet budget	71	Meet quality	67
Meet time	67	Meet budget	62
Achieve purpose	57	Achieve purpose	62

Source: Adapted from Wateridge [13].

Applying this to construction it is clear that the brief and its interpretation, effective scheduling and control, together with continued client involvement at the highest levels, are critical. The idea has been broached that success can be seen from different stakeholders' point of view, the main ones being the client/user and the project management team. Morris and Hough [12] cite the Thames Barrier project as a successful project even though it took twice the planned time and cost four times the budget, because it provided profit for most contractors.

Wateridge [13] in his research on measuring success compares the major criteria as seen from the viewpoint of the users and the project manager. Table 1.4 indicates a subtle difference between the two points of view for the top five criteria.

The table indicates that although meeting the user requirements is not in contention between the users and the project team, the idea of making them happy takes a different status in the two points of view and is replaced in the project team's mind by commercial success (making the project team happy!) It is interesting to note that budget and time are *more* important to users. Perhaps quality has been defined in a different way by the users as it is not otherwise mentioned in the top five.

Conclusion

Project management has been around for a long time in construction, but has not always delivered the value that clients have been promised. The unique nature of the product needs to be properly planned for success, understanding the client business and not ignoring the key role of external events. The move is for large clients seeking to enhance value and not to tolerate under-performance. Integrating the events in the project life cycle more fully (especially design and construction), makes it move away from traditional procurement methods so it is important to look at innovative ways of delivering projects. Clients need to be clear about the impact of the approvals and decisions they are making. The later chapters connect

RM with VM and also look more closely at design management and supply chain and how procurement can be managed in a better way.

The professional roles in projects are changing, but there is a need to manage a fragmented process seamlessly and to this end there is a special chapter on the development of DBFO and PFI schemes. The development of a strategic plan is important for all areas of project management indicating the organisational structure, culture and leadership, the strategy for more sustainable building and the promotion of effective teams. The credibility of a project manager depends not just on their technical ability, but on their ability to provide an ethically sensitive service to all the parties to the contract and to deliver client satisfaction and advice.

The research carried out on project success has not been specifically picked up by the recent industry reports, but it is clear that success depends upon the degree of strategic planning, the ability to integrate and getting the right person to match the job, in order to achieve the process-focused improvements which are being recommended by the reports.

References

1 CIOB (2002) *Code of Practice for Project Management for Construction and Development*. 3rd Edition. Blackwell Publishing, Oxford.
2 BSI (2000) *Guide to Project Management*. BS 6079. BSI, UK.
3 Dixon M. (ed.) (2000) *Project Management Body of Knowledge*. The Association for Project Management. High Wycombe, London.
4 Meredith J.R., Mantel S.J. Jr, Shafer S. and Sutton M. (1995) *Project Management: A Managerial Approach*. 3rd Edition. John Wiley & Sons, New York.
5 Ireland L.R., Pike W.J. and Shock J.L. (1982) 'Ethics for project managers'. *Proceedings of the 1982 PMI Seminar/Symposium on Project Management*. Toronto, Canada.
6 Maylor H. (2003) *Project Management*. 3rd Edition. Financial Times/Prentice Hall, UK. p. 37.
7 Graicunas V.A. (1937) 'Relationship in organisation'. In *Papers on the Science of Administration*. University of Columbia. As quoted in L.J. Mullins (1993) *Management and organisational Behaviour*. 3rd Edition. Pitman Publishing, London.
8 Maylor H. (2003) *Project Management*. 3rd Edition. Financial Times/Prentice Hall, Edinburgh. Table 2.5.
9 Pinto J.K. and Slevin D.P. (1987) 'Critical factors in successful project implementation'. *IEEE Transactions on Engineering Management*. February. EM-34: 22–7.
10 Baker N.R., Bean A.S., Green S.G., Blank W. and Tadisina S.K. (1983) 'Sources of first suggestion and project failure/success in industrial research'. *Conference on the Management of Technological Innovation*. Washington, DC.
11 Pinto J.K. and Mantel S.J. (1990) 'The causes of project failure'. *IEEE Transactions on Engineering Management*. November 1990.
12 Morris P.W.G. and Hough G.H. (1991) 'The anatomy of major projects: a study of reality of project management', John Wiley & Sons, Chichester, UK.
13 Wateridge J. (1998) 'How can IS/IT projects be measured for success?'. In *International Journal of Project Management*. Elsevier Science Ltd, Amsterdam. 16(1): 62.

Chapter 2

Building the client business case

This chapter looks at the concept of using project management to fulfil a client's business objectives. Clients may be classified into public and private, profit and non-profit making. The term business case implies the need to justify a need. In this sense, all categories of client will have a business case and this will define the client objectives in the context of the project. Construction projects are complex and expensive and need to justify the expenditure and the need.

The objectives of this chapter are:

- Client objectives as an outcome of client values and type.
- Project constraints and objectives as they constrain the project.
- The principles of presenting a business case.
- To present the context of decision making, using stage gate reviews.
- The concept of benchmarking and building value in the business case.
- Identifying the project stakeholders.

These issues are supplemented with some case studies as a way of illustrating the theory and presenting the current practice in commissioning a building work. Business planning advice has been well documented by many agencies and the Gateway Review™ programme and the Government procurement guides available through the Office of Government Commerce website [1] are a good example of best practice. For private inexperienced clients, publications like the Construction Clients Confederation 'starter' charter [2] will help focus the advice. A balanced approach is needed in the context of client type, building uses, project's size, public versus private clients and the unique stakeholders of each. There is also a need to view business planning as an open system that is heavily affected by project and environmental constraints. The business case is a starting point or a benchmark for the level of performance required. It is very easy to erode the value of the business case, but the integrated approach allows for working together with the project team to preserve and improve that value.

According to Egan [3], value improvements require that performance measurement is carried out. Maylor [4] reminds us that it is easy to measure the wrong things and that real improvements are made by long-term measurements across projects so that the supplier behaviour changes are permanent and not just reactive. The process of feasibility and funding appraisal is dealt with, in Chapter 3.

Project constraints and client objectives

Table 2.1 differentiates objectives. The job of the project manager is to understand the client objectives and to ascertain the priorities. It is also to provide a professional service which not only develops the business case by applying the right tests to the assumptions made, but can also advise on the specific project constraints. These project constraints come in the form of external circumstances which constrain the design and construction process.

Project constraints can be classified as:

- Economic factors, which affects funding and market prices. Market prices for the tendering of construction work vary significantly according to supply and demand. Positively, local authorities may recognise employment opportunities and provide tax breaks for certain locations.
- Ethical and environmental choice.
- Resource availability.
- Time constraints.
- Technical/design issues, cost versus quality balance, life cycle.
- Planning constraints which exist to the type of developments noted, making some locations easier than others to gain permission or apply

Table 2.1 The difference between client and project objectives

Project objectives	Client objectives
Efficiency, to given time, cost and quality levels	Statement of need
Teamwork	Functional facility
Technical task	Financially viable

Figure 2.1 Balancing constraints and objectives.

conditions. For example, building height restrictions on certain sites. Local councils, through the planning system, may seek to impose section 106 agreements (planning gain) on developers. These are designed to contribute towards community and transport issues created by new development, in return for the benefit gained to the client business. Highway authorities may require planning gain to provide additional road widening, improvement of junctions or the provision of traffic signalling.

- Physical site constraints. Site access might cause expense, restrictions or limitations on the positioning of the building. Ground conditions determine foundations.
- Neighbour concerns.
- Health and safety issues.
- Legal requirements such as durability, contamination and sustainability covered by the Building Regulations and various Environmental Acts.

Some of these arise out of a technical knowledge of the construction process so, it will appear strange to the client. The main role of the project manager is to make known these constraints to the client and guide them to best value, whilst maintaining the real essence of the client's objectives as in Figure 2.1.

Client objectives

Company goals are a framework and provide some guidance for the direction of the business, for example, the type and the location of markets to be in, the investment needs and the growth rates of the business. Strategic planning does not define the building project, but should give some justification for a project. The client needs to be sure of testing key assumptions, giving clear requirements, having a good management commitment, sufficient skilled resources and flexibility of contractual arrangements to cope with change.

Figure 2.2 shows the steps in the process in moving from the client objectives to the project brief.

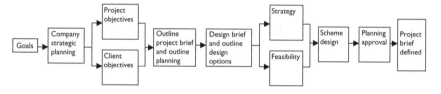

Figure 2.2 Process of developing brief from objectives.

Client objectives for a new building project define the business case for it and lead to the development of a project brief. SMART is a well-known mnemonic that can be applied overhauling a client's objectives to ensure effective implementation.

Specific	they need to identify the outcomes clearly (e.g. business volumes, returns and markets).
Measurable	so it is feasible to monitor attainment, for example, a budget or a functional space (e.g. so many beds, spaces or car parking spaces).
Achievable/agreed to	check that different objectives like space and budget or complexity and time scale do not clash. Ownership of the objective also helps commitment to it.
Realistic	to the available resources of funding, time scale, materials, labour etc.
Time bound	with a programme for delivery of the objectives. This will be a high-level programme indicating key dates for attainment of the objectives, such as decanting or occupation. They should be in the context of the business need.

Client objectives are the starting point for defining the project, but may need to be clarified by the project team. Value can be enhanced by optimising solutions to meet the essential requirements and to separate out the desirables as a 'wish list'. Client objectives recognise the external constraints and will inform the commissioning of a specific project. For example, Dyson have an objective to expand their business and to do so they have decided to move away from the current site in Wiltshire to Malaysia, because they recognise that production is cheaper there and because the conditions for planning permission in expanding on their current site are considered more onerous. Building in Malaysia is cheaper and they have calculated that even with export costs, they can undercut their competitors in European markets.

Different types of clients have different types of objectives. The main client types are private, public and developer. But it is also important to 'get into the client's shoes' in order to understand their business and the basis of the client brief. Below are a few simplistic examples of generic objectives for specific client types:

1   A *manufacturer* needs functional efficiency and VFM, to meet performance criteria and to start production as soon as possible. They will be a secondary client as the building is a means to an end.
2   A *developer* needs a cheap, quick and attractive building. This will be a primary use as the building is being used as a commodity in itself.

3   *A public body* needs a building, that lasts a long time, is efficient use of tax payers' money, is within yearly budget and is low cost to run. This is again a secondary use of the building.

Masterman and Gameson [5] have classified clients on a two dimensional grid as experienced or inexperienced and primary (wanting buildings because buildings themselves are their business) and secondary (those who use buildings to house their business).

The experience of the client will determine the degree of involvement that can be expected from the client in developing objectives. It will also determine their understanding of the project constraints. A small manufacturer is unlikely to have built very often and so, is inexperienced. British Airports Authority (BAA) on the other hand is experienced and has strong influence on the procurement of its built assets. This is quite simplistic and Green [6] believes that the consumer led market has led to a more organic iterative approach to arriving at the brief.

Project objectives

These objectives are associated with the efficiency and effectiveness of the project process. The project manager has particular responsibility to meet these as well as to help the client to meet their own business objectives. They are traditionally to do with project budget, quality and programme (time), but there are also other aspects of project objectives which are important to the success of the project and these will be considered later.

The time–cost–quality triangle in Figures 2.3–2.5 indicates the need to understand the balance between each of the parameters in agreement with the

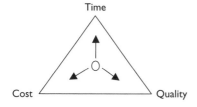

Figure 2.3 Project objectives time, cost and quality.

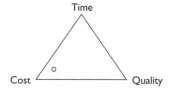

Figure 2.4 A single priority for the LA.

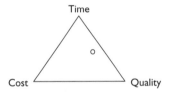

Figure 2.5 A double priority for the National Olympic Committee.

client requirements. A single priority is shown by 'pushing the ball' into one corner. A double priority (quite normal) is shown by 'pushing the ball' to the middle of one side. It becomes much more difficult to manage if a triple priority is given. The first two do not imply that the third factor is unimportant.

Time, cost and quality are all important to a client, but one may be more important than another. For example, a local authority is almost always tied to the lowest price. So once the budget is set for a school it is embarrassing for the budget to be exceeded, because it means going back to the central government, providing less school places or capturing money from another scheme.

Alternatively an Olympic stadium must be ready in time for the event and the quality is important to the ambassadorial role that is so important to a government electing to hold the Olympics. This means that a budget will be secondary to the quality and the programme as shown in Figure 2.5.

Project objectives also include *managing people* and creating synergy from the team so that there is greater productivity. An open and blame free culture, a conducive working environment, good communication with regular short meetings and agreed document/information distribution all help to create this type of win–win environment.

Poor management of people produces the 'them and us' approach. This approach may occur between the contractor and the client or between the design and the construction teams. In both cases the effect is unproductive use of time. For example, a failure to look at ways of getting over problems to mutual advantage and time spent trying to apportion blame and making claims leads to one or both parties out of pocket.

At tender stage the 'them and us' approach produces a minimum bid and claims for extras. This produces bad feelings and a 'win–lose' or 'lose–lose' environment between the client and the contractor. This can be avoided by an open attitude on both sides and a commitment to collaboration.

The project manager may also agree to some *task orientated objectives* such as zero defects, planning to substantially improve health and safety risks and shared bonuses for reducing costs below budget. This particular agenda would relate to the improvements which are being recommended by the Egan Reports (1998 and 2002)[1] and others.

Best practice

The Construction Client Forum (CCF) (2002)[2] recommend seven steps for the client to take when considering a construction investment and these are:

- Providing client leadership, both for the improvement in procurement processes and for the supply side to develop and innovate to meet clients' requirements.
- Providing and setting clearly defined, and where possible, quantifiable objectives and realistic targets for achieving these.
- Fostering trust throughout the supply chain by treating suppliers fairly and ensuring a fair payment regime.
- Promoting a team-based, non-adversarial approach amongst clients, advisors and the supply chain.
- Adopting a partnering approach wherever possible.
- Identifying risk and how best to manage it.
- Collecting and interpreting data on the performance in use of their construction solutions, for purposes of feedback.

The Clients' Construction Group (2004),[3] which succeeded the Forum describes it now, as optimising built asset value throughout the cycle of verifying business needs, assessment of options, development and implementation of procurement strategy, project delivery and post project review.

Presenting a business case

A business case is presented at the inception stage of a construction project and will confirm the expected need for a new construction works to take place. In addition it will contain constraints such as budget limits, date when required and some performance requirements.

Business budget constraints are different from project constraints and refer to at least making a profit margin which covers the risks of the investment and a return to the shareholders, if it is a profit making venture. Public ventures need to objectively value benefits against the costs, if they are to make the justification. Non-profit ventures do not have to cover shareholder expectations, but still need to cover the risk of not breaking even and costs of new investment.

Time constraints are often indirectly connected with returns. A new supermarket can afford greater capital costs if it is able to start trading earlier to a ready market. Every day saved will give a profit bonus.

Quality constraints are based on balancing the durability of higher cost materials and best quality workmanship with the reduced maintenance costs. Cheaper materials may be used where less durability is required or there is a shorter term interest in the asset or where capital funds are scarce. The additional benefits of quality are that it projects an attractive image to the customer.

According to the CIOB [7] a sound business case prepared for presentation at a projection inception will

- be driven by needs;
- be based on sound information and reasonable estimation;
- contain rational processes;
- be aware of associated risks;
- have flexibility;
- maximise the scope obtaining the best value from resources;
- utilise previous experiences.

We shall now look at Case study 2.1 to test the strength of the business case.

Case study 2.1 Student accommodation feasibility

The business case for the provision of a new University student accommodation is to replace the outdated facilities and possibly enhance the location of the residences by reducing distance from the University and making them useful for alternative use out of term time. One problem the University has, is the unavailability of borrowed capital credits and so it has to generate capital by selling assets or it has to commission the provision on the basis of a revenue charge, for example, leasing or PFI. It also has to consider that because UK students contribute fully to their accommodation and to some amount of their fees that accommodation has to be attractive to draw students to the University. The University has investigated the alternative routes which are open to them without capital finance and compared them with the cost of doing nothing. The option to proceed with a leaseback arrangement with a housing association was eventually adopted.

The objectives of the University are to provide 400 student places in a safe, secure and reliable environment under the overall control of the University. It proposes single or grouped en suite rooms with the provision for catering services to widen the use out of term time for conferences and the provision of laundry and common rooms for communal purposes. The issues which are covered in the business case are:

- The opportunities for sites – some existing sites have restrictive covenants on redevelopment so newer sites may be looked at.
- The performance of the service will be in accordance with the University design standards and outside provision of Facilities Management (FM) services will be judged against the University's own FM standards.

- Initial costs and rents must be affordable to students and competitive in the private market.
- Management of the flats by others, but not on matters of discipline, pastoral care and leadership will involve the University.
- Risks judged to be significant were the construction cost, the timing overruns and the subsequent defects in construction, the funding and ownership of assets, the standard of management of the service, the long-term maintenance and equipment replacement, the levels of student occupation and the out of term use *and* the bad debts from unpaid rents.
- Transfer and retention of risk – to transfer construction problems, the level of availability and the standards of service and the out of term use fully to the provider by delegating FM services as well as property ownership in return for a fixed service charge. In addition, transfer the risk for below 80% term time occupancy to the provider. To retain risk for student discipline, pastoral care, rent collection and provision of wardens. To provide affordable rents for the students.

In order to keep the service charge down, they also considered the permanent transfer of obsolete land to be used for independent development by the provider.

Analysis

If we compare this case study with the criteria for a sound business case indicated by the CIOB Project Management Code of Practice then we can make some favourable comparisons with the points made:

- Driven by needs: the University was commissioning new residences because of the changing demands of more market orientated students and the rising costs of maintaining a catered service and shared rooms. The desire for renewing stock alone is not a sufficient driver to attract students, who also have changing needs.
- Flexibility: the danger is that the fashion for single flatted rooms may change again, for example, students might prefer to pay less and share or have catering facilities. Conference use does give some flexibility.
- Sound information: reasonable estimation has been used to assess the market, but the general trends have been quoted and not the specific market research for the project.
- Rational processes: a written business case and the use of option analysis using discounted cash flow for comparison with each other and doing

nothing is rationale as long as the disadvantages of doing nothing are also quantified and the risk transfer is quantified objectively. A tendering process, which is competitive or negotiated from a strong position is also important. In this case, there was some concern that only one housing association was prepared to bid.

- Awareness of associated risk: a risk register associated with a probability index means that the comparative impact of risk can be assessed.
- Scope and best value of resources: as the University did not favour rate capital credits or borrowing rights it was considering a number of options which provided provision other than immediate ownership. The release of land to a developer kept rents down, but was it a short-term view and did they get a VFM for it?
- Previous experience: this was based on some favourable reviews of a development in similar circumstances.

It would be fair to say that there is on balance, a sound business case. The University has a strong sense of its requirements and responsibilities.

Developing value in construction: the Gateway [8] framework for decision making

The Gateway Review™ is a project life cycle procurement guide for public contracts developed initially by the UK Treasury, in order to ensure consistency and VFM in government and other public contracts. It is used here as a generic model to underline the key client decision points. There are six decision points called gateways as shown in Figure 2.6:

- Gateway 0 establishes whether it is a business need.
- Gateway 1 assesses the high level business case and budget and makes way for expenditure on the outside consultants.
- Gateway 2 assesses the feasibility study and the proposed procurement strategy and gives the *critical* go ahead to proceed with the contract documentation for the implementation of the project. A proper PEP should be in place to proceed.
- Gateway 3 assesses the contractor tender bids report and provides the go ahead for design detail and construction. (In this integrated model it is assumed that the scheme design is done by the contractor, but a separate design contract could be procured prior to tender competitions by the contractor.) There are two additional decision points at the completion of planning application and the detail design stages.
- Gateway 4 is the acceptance of the finished project either for a separate client fit out contract or for occupation.
- Gateway 5 is after occupation when a benefits evaluation takes place.

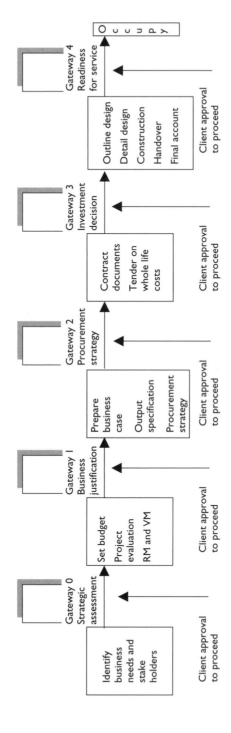

Figure 2.6 Diagrammatic of the gateway procurement life cycle: integrated process.

Source: Based on OGC [8].

These gateways are the basis for getting approval to proceed to the next phase of the project cycle and making sure that there is adequate information available for client decision making for the viability of the project. This system has several points to assess value, but note the early and later VM opportunity for gateways 1 and 3.

This system has been set up for the client, but the project manager is appointed after gateway 1 and at this stage a feasibility study is commissioned with an outline design and a risk assessment and here a more general VM opportunity with the project team appointed. A PEP is prepared which covers the strategy for the job and considers the programme, cash flow, procurement strategy, project organisation, health and safety plan, environmental impact, design management etc.

The plan establishes how the project is going to be carried out whilst the feasibility establishes how the project can be delivered viably and within budget time and quality constraints. A project manager is the key person in providing good advice on the most suitable procurement strategy and in developing a master plan for the delivery of the project to the customer's needs and for the best value.

The system above which has been promoted by the UK government in promoting better procurement efficiency would suit prime contracting, construction management or design and management, but may have to be adjusted for PFI, design and build (D&B) and traditional procurement. These more complex forms of procurement are not necessarily the cheapest options and this is another aspect of matching the procurement type to the specific project imperatives.

VFM is increasingly involving an appraisal of whole life cycle costing (WLC), which analyses the cost between capital cost (CAPEX) and operating and maintenance costs (OPEX). This puts into question the acceptance of the lowest capital price tender and forces the clients to think through the impact of a building's design, the energy costs, the location and the financial benefits as a single entity with the capital value of the building created. The true consideration depends on reliable figures being available and the value of WLC to the client will vary according to their post contract interest in the building. Some procurement options such as prime contracting and DBFO put the CAPEX and OPEX risks with the provider and this is another driver for WLC appraisal. OPEX costs are also recognised as needing to be sustainable in a climate where scarce resources are being increasingly recognised.

The process of VM and life cycle costing are further discussed in Chapter 8.

Project stakeholders

Project stakeholders are those who have an interest in the project process or the outcome. The obvious parties to the contract have an interest in the *outcome* of the project. The client wants to get a building which meets expectations, therefore generating expected returns, the member organisations of the project team want to make good returns, gain experience and to build a reputation to get further work and users/employees want a resource, which is comfortable and convenient. Those interested in the *process* might be community based and the employees of the organisation. The community want a minimal disruption for neighbours, courteousness by project workers and employees want an interesting project.

As a project manager working on behalf of the client and responsible for the project dynamics both the client's stakeholders and the project's stakeholders become important to manage. The client's stakeholders, which are established at verification of the need stage, need to be made known to the project manager and will influence the project management approach indirectly. We will mainly deal here with management of the client's stakeholders, but some of the stakeholders are related to the execution of the project and not to the business need.

Newcombe [9] suggests that there are different requirements for the management of stakeholders depending on their degree of influence (power) and significance. There is a need often to deal with the conflicting requirements of the stakeholders – so who do you satisfy and who do you disappoint? A project manager needs to satisfy the client's ultimate interests, which means making decisions to ensure the project's efficiency objectives don't obliterate the business effectiveness objectives.

Power   may issue from a stakeholder's ability to take action which would be helpful or detrimental to the project outcome. A good example is the Planning Authority who can hold up or stop the project and also any legally enforceable action such as compliance to fire regulations. These people may not have excessive interaction, but may have significant adverse impacts.

Significance   issues from the effect on working relationships and the amount of interaction that there is between the stakeholder and the project team. These can have a slowing down and souring effect which affects the productivity of the project. An example of this type of stakeholder would be a key specialist contractor who provides less than full resource requirements.

The community   is also a stakeholder and may be able to have considerable effect where a group is able to obstruct progress. A client may feel this pressure and make instructions for change as a powerful stakeholder. A project team will be affected by delays.

Managing these phenomenon is important and consists of a mixture of selling the project objectives, satisfying the customer and compensating the stakeholders who have become dissatisfied. Making the logic of the project clear (selling) is a longer-term solution as it attempts to draw stakeholders behind the project and saves abortive resources by being preventative rather than reactive. Compensating stakeholders may be expensive, but may bring a better project solution if other benefits are given for lost amenity.

Stakeholder satisfaction

Maister's [10] law of customer service is stated as

satisfaction = perception − expectation

It is a good benchmark for managing stakeholders, as it has an effect on the initial planning. This law distinguishes between what the stakeholder sees as a product outcome or a level of service with what they were expecting from the project. In achieving stakeholder satisfaction, it is not enough to perform satisfactorily as this will simply neutralise expectation and provide zero satisfaction. Stakeholder management is about making the sum positive by exceeding expectations. In practice, this means advertising the core objectives that will be the main outcomes and what must be achieved *and* setting about achieving some extras.

From a contractor's point of view competitive tendering leaves no room for optional extras for the client, which leaves the area of service as the most powerful way of exceeding expectations. It is also clear from the second Egan Report (2002)[4] that the exceeding expectations principle is clearly enshrined in their vision statement so that customer value is optimised. Negotiated tenders (also espoused by Egan) leave more room for the client to arrange a 'win–win' situation.

Certainly, the Maister's first law of service [11] is clear that the way in which a service is provided and the importance of being able to do what you promise, is most important in keeping stakeholders on your side. Continuing service improvement is also important in keeping the stakeholders satisfied. Case study 2.2 refers to the efforts of the client to satisfy the internal and the external stakeholders in moving their headquarters (HQ).

Case study 2.2 Internal stakeholder, relocate HQ

A major insurance company makes a strategic decision to move its office from the city centre to a peri-urban position to gain more efficient running expenses, pull all personnel together into the same building and make efficiency savings and finally to take tax breaks being offered by the (LA) to gain employment opportunities. The customers are not affected as most business is done remotely, the business becomes more efficient and so produces better returns to shareholders, but the existing employees are mainly disappointed because many of them have to travel further and do not have access to the benefits of city centre shopping and amenities in their lunch breaks and after hours. The local community is residential and is extremely worried that a large new facility in the vicinity will cause even worse traffic congestion and parking problems and loss of visual amenity and green fields. What does the company do?

In their new building plans they offer a gymnasium and sports amenities and ample car parking space and an interesting 'street' environment to 'buy off' the employees and they develop their grounds to provide a lake, ample planting with walks and picnic sites, that will provide alternative lunch time activities and screen and encourage wildlife to stay in the area. This will both partly satisfy the local community and satisfy the Local Planning Authority, who have qualified the planning permission to meet local objections in the planning appeal. The shareholders have suffered no long-term loss to their returns and the solution is integrated to partly alleviate loss to the other stakeholders.

Case study 2.3 looks at the controlling factors for a public client and the need to manage the various aspects, so the satisfaction and project constraints are kept in balance.

Case study 2.3 External stakeholders for a new swimming pool

A well publicised objective for an LA to supply a 25 m swimming pool during 2003 to supplement sports facilities in the area may have helped them to be elected, but the programme slipped to 2004 because of the budget availability. In this case the stakeholders are the council tax paying public, neighbours and the LA.

When they open the swimming pool in the summer of 2004 they decide that they will compensate by fitting out a new gym suite for fitness training to offset the disappointment of being late. This came out of earlier local consultation during the planning stages which was not promised in the final scheme. Stakeholder satisfaction has been achieved by exceeding outcomes and offering people more ownership by consulting them. A basic 25 m swimming pool in autumn 2003 might have got them re-elected. A basic pool in 2004 would have caused frustration and possibly dissatisfaction because people would have made plans and have had to alter them.

The LA budget is a major constraint to the project. They wanted to include the fitness gym suite, preferably within the same budget. Budget savings may be made by considering one or more of the following:

- a VM exercise to question other requirements in the brief such as the amount of car parking and the orientation of the building, which may reduce the expense of the access road; or
- a life cycle approach would take the opposite view to achieve reduced running costs, increased component life and use this as an argument for an increased capital budget; or

- a later cost cutting exercise would, in contrast, reduce the roof specification, the thickness of the tarmac and generally reduce the quality and increase the running costs of the building.

The first two now becomes a specific project design issue and must be levered into the brief development stage. In terms of managing stake-holders, the client needs to be directed to make these decisions early.

Conclusion

The business planning process begins before the inception stage of the project and informs the client about the feasibility of the project in outline terms. The briefing and ongoing development of the brief takes place in the next phase of feasibility testing when a solution is engineered by clearly communicating the project objectives and reconciling them with the project constraints. This stage optimises the value and reviews the effect on the stakeholders to mitigate the conflicts which may arise.

Egan and the Construction Clients' Group have issued reports which indicated that the integration of the client into the construction process is a necessary and not an optional development that must be made if value is to be built into the process. This has brought about the wider use of negotiation during the procurement process in order to make the client's objectives more accessible to the whole supply side and build in value. The integrated project team is seen as the maintenance of supply chains from project to project so that the team is not always learning on the job. This depends a lot on the greater involvement of the client in choosing limited partners for repeat work and naming suppliers and a lot of work needs to be done to convince one-off clients to be more involved. It also means making a strategic move away from the single stage competitive tendering with selection of the lowest price. This system ignores other aspects which bring value to the business case such as earlier finish, guaranteed fixed prices, flexibility, sustainability and lesser life cycle costs. More strategic long-term relationships are possible by adopting forms of the contract that allow more direct contact between the contractors and the client such as prime contracting, construction management, design and build and PFI, where appropriate.

Demonstration projects have been set up on £500m of projects through the Movement for Innovation (M4i) [12]. Public clients are specifying these forms more and taking up the challenge to be best practice clients by giving training to their procurers. Some clients are committing themselves to the Construction Clients' Charter [13] which commits them to continuous improvement by training for client leadership, working in integrated teams, whole life costs (WLC) and respect for people. The industry as a whole is being encouraged to participate through what is currently known as the Constructing Excellence programme which is building up a growing body of best practice material for client and project team and encouraging self improvement clusters of like-minded construction industry players. Bottom line results indicate that on demonstration projects there is increased profits for contractors of 2% and that costs are being reduced by 4% compared with the average achieved in construction. This suggests that the better partnership of clients with the supply chain is sustainable where there is a win–win situation.

Notes

1 Strategic Forum for Constructions (2002) *Accelerating Change*. Department of Trade and Industry (DTI), London. This is a progress update on the Egan targets set in 1998.
2 Construction Client Forum (CCF) (1998) Constructing Improvement. This report is one in a series of UK construction client reports proposing major change in traditional procurement and tendering procedures in order to promote a pact between the client and their supply team and promote better VFM by reducing process and product waste.
3 Construction Clients' Group (2004) *Process Map*. Developed as guidance to clients for the steps of gaining leadership and optimising built asset value.
4 Strategic Forum for Construction (2002) *Accelerating Change*. DTI, London. A document which reviews and updates the progress since the Construction Taskforce Egan Report (1998) Rethinking Construction. DTI, London.

References

1 The OGC (2004) *Gateway Review*^{TM} *and Achieving Excellence Guides*. http://www.ogc.gov.uk
2 Construction Clients Confederation (2002) *Starter Charter*. www.clientsuccess. org.uk
3 Egan J. (2002) *Accelerating Change*. Strategic Forum for Construction, UK.
4 Maylor H. (2003) *Project Management*. 4th Edition. Pearson Education, Edinburgh.
5 Masterman J.W.E. and Gameson R.N. (1996) Client Characteristics and Needs in Relation to their Selection of Building Procurement Systems. *Proceedings of CIB92 Symposium East Meets West: Procurement Systems*. In A. Walker (ed.), *Project Management in Construction*. 4th Edition. Blackwell Science, Oxford.
6 Green S.D. (1999) 'Partnering: the propaganda of corporatism'. *Journal of Construction Procurement*, 5(2).
7 CIOB (2002) *Code of Practice for Project Management Construction and Development*. 3rd Edition. Blackwell Publishing, Oxford.
8 The OGC (2004) *Achieving Excellence Procurement Guide*, in Project Procurement Lifecycle: The Integrated Process, UK. http://www.ogc.gov.uk
9 Newcombe R. (1999) 'Stakeholder management'. *Keynote Address to the CIB65 Conference on Customer Satisfaction*. Cape Town University, Melbourne, Australia.
10 Maister D.H. (1993) *Managing the Professional Service Firm*. Free Press, New York.
11 Office of Fair Trading (1999) *Customer Satisfaction Survey*. HMSO, London.
12 Movement for Innovation (M4i) (1998) http://www.m4i.org.uk
13 Construction Clients' Group (2004) Construction Clients' Charter, http://www.clientsuccess.org.uk (accessed September 2004).

Chapter 3

Project definition

Every construction project needs to have a clear briefing which is realistic and derives from the client's objectives and satisfying the business case. The initial brief at inception is developed in the feasibility and strategy stages. The concept of *project definition* is discussed in the context of project management, as a managed stage trying to reach an agreed design scheme and methodology. The main objectives of the chapter are:

- Determining the elements of project definition.
- Mapping the construction process.
- Assessing project feasibility and affordability.
- Managing the project scope.
- Dealing with external factors.
- Balancing risk and value and allocating risk.
- Project evaluation techniques.

Determining the elements of project definition

Project definition is carried out in the period from receiving the performance specification during the inception stages up until the receipt of full planning permission.

The RIBA Plan of Work lists work stages to reach the planning application stage when the clients' approval allows a scheme to be submitted. These stages are A: inception, B: feasibility, C: outline proposals and D: scheme design. However, these are strongly related to the traditional forms of contract and procurement and assume a two-way relationship between the client and their designers. In Figure 3.1 the RIBA plan has been linked to the relevant project life cycle stages [1]:

A   Inception is the appraisal identification of the clients' requirements and possible constraints on development. It considers a probable procurement route.

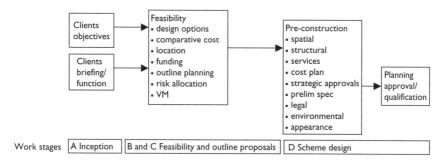

Figure 3.1 Development of brief showing early RIBA Plan of Work stages.

B   Strategic briefing preparation by or on behalf of the client. It also identifies organisational structure and the range of consultants which might be appointed.

C   Outline proposals, taking account of any feasibility studies which have been produced. Estimate of cost. Review of the procurement route.

D   Detailed proposals and completion of the detail brief. Application for full development control approval.

The RIBA work stages provide a managed process with a greater emphasis on the design process. Traditionally the architect has led the management of the project definition stage, which may differ from other different procurement systems. It is important also to consider the cross over of the strategy stage with feasibility and to understand that the programme, the client funding cash flow, the risk and value assessments and the organisation of the project may have important impacts on the design and feasibility. The normal way to proceed with the development is to include option appraisal or an iterative process to develop the design based on the client and the site constraints considered.

A fuller picture

The CIOB Code of Practice [2] therefore talks about an outline brief moving towards a full definition and this can include an extensive development on complex or controversial construction projects as shown in Figure 3.2.

The outline brief should cover the clients' objectives, functional requirements, project outline scope and constraints. The detailed brief provides the resource requirements, a scheme design and a feasibility and strategy appraisal. The development of the brief is a critical stage at which the scope

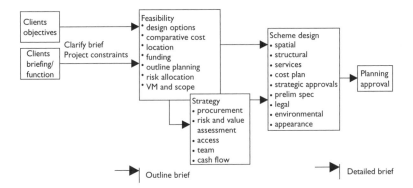

Figure 3.2 Project definition process.

of the project is properly established and verified by the project constraints (Chapter 2), which ultimately leads to the planning permission to proceed. This stage is iterative and requires the involvement of the client with the designer and ideally, if procurement method allows the involvement of the construction manager. At the end of the project definition stage the detailed project brief should have determined scope as fully as possible, because beyond this stage changes to scope become much more expensive. A system of cost checking is vital for control as it informs the designer of the impact of design change.

The crossover between strategy and feasibility provides an axis for checking between the process and the product. It means that:

- Design development can be cost checked.
- Project organisation and communications are properly considered.
- Procurement method can be evaluated to suit the client's unique requirements.
- The impact of design on construction method and construction time can be assessed.
- A master plan for the key time constraint of programme and cash flow is in place.
- Risk is identified and allocated.
- Best value is managed.

The project manager has a key responsibility to co-ordinate the whole process and to ensure that the strategic/feasibility issues are properly integrated with the development of the design proposals. This is also recognised as a separate skill by the RIBA. The following are managed in relation to the external environment:

- Clarification of the brief.
- Feasibility and affordability.
- Scope management, change and contingency.
- Risk assessment of external factors.
- Funding and location.
- Design management.
- Stakeholder management.
- Organisation and culture.

Some of these will be considered in the pages that follow.

In the Gateway Review™, system project definition comes between gateways (1) and (2) and identifies the production of a feasibility study, a procurement strategy and an output based (performance) specification.

The design brief

Gray [3] talks about three distinct types of knowledge controlling design. These originate from

- the client in the early stages of inception;
- the individual designers in the concept and scheme design stage;
- the design/construction manager in the detail design, specialist and construction stages.

The first two types of knowledge are particularly relevant to project definition, but by no means exclude the third type.

Hellard [4] sees that there are four possibly conflicting elements to the brief

Function	Technical and physical requirements to meet the business case.
Aesthetic	Satisfaction of human subjective aspects.
Cost	Both capital and running costs.
Time	The logistic requirements for commercial completion and occupation.

The client may wish to determine some or all of these elements in the outline brief, depending on the degree of innovation and flexibility the client wishes to give to the team. This is the right stage for a VM workshop. Only then can a working brief be established.

Process mapping

Process mapping is used to model and test the system that is intended. In projects it is used to make sure that all operations carried out in the project life cycle are planned and covered. Formally this can be drawn as a flow chart indicating the activities (in rectangles) and key decision points (in diamonds). It is often connected with a series of different pro forma documents. Electronically these are accessed as links across the intranet or the project extranet, making it a good communication tool and a knowledge base across all members of the project team. Problems which arise are related to the compatibility of systems and the shortfalls which arise in the degree of information accessible. Sophisticated systems may access comprehensive external sources through knowledge portals. More integrated internal systems may be linked to a knowledge management intranet, which identifies a wide range of experience and resources within the organisation.

The advantages of the formal system are that it works well in quality assurance, the disadvantage is that the system may become inflexible and discourage innovation and improvement, though it is possible to build in review and improvement stages. Case study 3.1 refers to an FM organisation carrying out maintenance, and smaller projects not exceeding £300 000 for a client who has several complex facilities at different locations.

Case study 3.1 Managing the estate

An FM company operates a three year rolling contract extendable on the basis of performance. The value of work on one of their sites is £5.5m per year on which they are paid a fee. The amount of work may vary each year and the fee adjusted on a pro rata basis. Their work is broken down into maintenance and projects. The proportion by value is approximately 50% for projects and 25% each for responsive and planned maintenance. There are approximately eight managers and technical officers employed to supervise projects. There are 33 documents, which are used to progress the project from inception to completion and also statutory forms to cover building regulations and Construction Design and Management Regulations (CDM) procedures. Contract information has been grouped together in a process map which covers the areas of quality assurance, health and safety, programmed work, planned maintenance, emergency callout, non-programmed work, financial management, works management stores and resource management.

The system is linked directly then to each of the relevant forms in order to ensure complete use of the system. Looking specifically at the projects the process map could be constructed as follows. There are four key players and two audit agencies. The process is mapped in Figure 3.3.

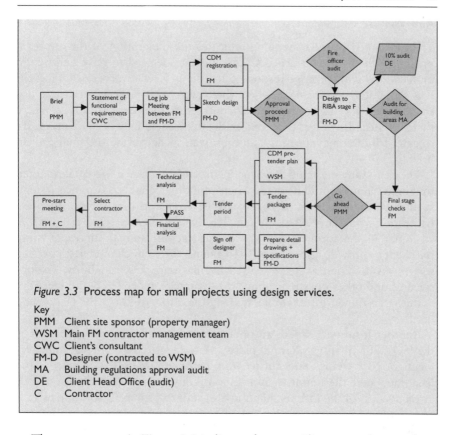

Figure 3.3 Process map for small projects using design services.

Key
PMM   Client site sponsor (property manager)
WSM   Main FM contractor management team
CWC   Client's consultant
FM-D   Designer (contracted to WSM)
MA   Building regulations approval audit
DE   Client Head Office (audit)
C   Contractor

The process map in Figure 3.3 indicates four specific approval stages from initial client request up to the pre-start meeting and indicates also the responsibilities. This map could also be linked to the specific contract pro formas that are relevant at each of these stages so that relevant checklists are processed and approvals signed up. There should also be an opportunity for a review of this process. For example, when the job is logged there is a checklist of 11 items for the extent of the design services covering such items as:

- Does the brief need amplification with the client?
- Does programme need amplification with the client?
- Is contact with the utilities required?
- Is a policy on hazardous material required?
- Has a planning supervisor been appointed?

There is also a sign off for the procurement manager to consider before the subcontractor starts work. This is called the initial quality check, which covers, for example, safety checks, statutory requirements and use and protection of existing services.

Analysis

The case study process mapping indicates the integration of the approval process with the design process. This indicates a need to have good communications with prompt responses for the project to proceed efficiently. It is a useful system to have where there are repeat projects and there is a single designer and some regular subcontractors, as in the case study. Where there is a broader range, then a more flexible system may become necessary to cater to different reporting and approvals in delivering the final project brief.

The case study mapping has benefits for ensuring that a system is adhered to, but decisions should be made when adopting a quality assurance system, so that bottom-up improvements may be built into the system otherwise frustration will set in. This system has two feedbacks. One is from the customer in the form of a customer satisfaction annual survey, inviting the client to provide a scoring of 1–10 against different characteristics of the service given. These were measured against tangibles, reliability, responsiveness and other questions about the assurance engendered and empathy with the customer, for example, making the effort to know the customers' needs.

Another form was called 'an opportunity for improvement' referring to the ISO 9002 quality system in place and to identifying technical improvements or cost savings relevant to other contracts. Members of staff at the 'coal face' take the initiative and responsibility for filling in the form. The attractiveness of the form is that it invites staff to propose a solution as well as identifying a problem or query. To promote use, responses should be made to the initiator.

Feasibility and viability

Looking at it proactively, feasibility seeks a solution that is possible within the applicable constraints, whilst meeting the client objectives. Investigations are carried out that will give an overall picture of the costs and constraints of the project and whether it is viable.

There are two different types of feasibility. At the outline business case stage the client, before inception, considers the viability of the project – whether they need the project and whether there is a suitable return to make an investment in bricks and mortar. This will often take the form of a development appraisal where the estimated costs of the development are weighed against the income and benefit of the investment and over what period they will break even and make a return to the business. It will assess whether the investors' minimum profit margin has been generated. Profits must justify the risk and the additional effort of investment.

At project feasibility stage, the information originates from the outline project brief and the business case. Essentially it is a second stage of feasibility and looks to either optimise value within the parameters which have been set or it will provide an option appraisal of different designs, locations, funding and methods. Viable alternatives can now be assessed against client values.

Client value is an important concept in feasibility and refers to the underlying beliefs of the organisation and also may be emphasised in the individual beliefs of senior managers responsible for the project. Client values determine the priorities and the underlying rationale for the decisions taken. Typically the balance between image and aesthetics and function/utility will determine design acceptance. Other issues are the degree of weight that is afforded to environmental issues and sustainability. Political and stakeholder values will also have an influence on the design and choice.

Public view – cost–benefit analysis

Cost–benefit analysis is a way of taking into account factors other than income, that are included in the wider term of benefits. Non-financial costs should also be recognised such as, loss of environmental facility and homes blighted by additional noise. A cost–benefit analysis is more relevant in the investment appraisal of a public project, that offers social facilities. At inception stage the main consideration is that there is an acceptable business case.

Cost–benefit analysis is typically viewed as a soft management approach to the issue of feasibility, but in public projects, which are not driven by commercial concerns, then the benefits have to be justified. If comparable projects are to be assessed, then it is important that factors included are formally validated as the same for all projects and these benefits have a standard valuation in comparable projects. Different benefits may cross over and care is needed not to double count. The system is categorised and valued as in Figure 3.4. If the costs are as the matrix, then the net benefit = £20 000.

For flood defence work there is a Government guide, identifying value for benefits such as the number of properties which would be saved from

Direct costs £120 000	Direct benefits £90 000
Indirect costs £10 000	Indirect benefits £60 000

Figure 3.4 Cost–benefit matrix.

Table 3.1 Comparative examples of the elements of cost–benefit analysis

	Benefit	Cost
Tangible (measurable in financial accounts)	For example, homes protected from flooding	Construction costs, land, fees compulsory purchase etc., maintenance
Intangible (measurement of financial worth requires discretion and may vary according to client values)	Better wildlife habitats	Loss of developable land and visual amenity

flooding by major proposed flood defence works, with a weighting put against factories and offices. Currently, agricultural land gets a very low weighting. There are also some measurable social benefits, which may be relevant to certain schemes in order to make sure that urban schemes are not the only schemes that get the go ahead on the basis of a priority list and scarce resources. Third, party costs are those that the Environmental Agency incur in dealing with floods.

The example in Table 3.1 indicates the way in which costs are balanced against benefits for a flood defence scheme. Direct costs and benefits are those which are incurred from the building and direct revenue for the service. Indirect benefits are those accrued to third parties such as the benefits for owners now being able to insure or to sell their properties. In practice, a lot of weight is put upon well-populated areas with less urgency and therefore benefit is put on scattered populations.

Cost–benefit analysis and value

At feasibility stage a public body may carry out a cost–benefit where there is a shortfall between the cost and the income. This should be a rigorous and realistic assessment of the financial worth of the benefit and the cost of intangibles. The Millennium Dome is an example where a grossly over optimistic assessment of the visitor figures was made and there was a justified sharp criticism of the justification made. On the other hand, the Sydney Opera House went badly over budget to a factor of 10 and has paid for itself over and over again as a perennial visitor attraction and an income earner.

The difference between the public and the private developer is the worth of the benefit which is used to weigh the balance. In the case of a sports centre for a private developer the income from fees will offset the building costs, fees and land. A private developer will pay substantial premiums for residential land near water. Here, there is a direct benefit from the saleable price. An indirect benefit is the regeneration of contaminated land around old docks and recreational access without public cost. A public body can price the benefit of providing the community with a social amenity which meets the council objectives and reduces crime amongst young people. In both cases, the project is designed at minimum cost which will allow the appropriate level of performance and levels of safety and meet the objectives which are different.

In the case of social cost this is often transferred to the private developer by the legal system. Under the planning system, a new housing estate may have to provide a proportion of affordable housing, contributions towards a new school, a better road junction or open grassed areas. Under environmental law, a new owner is responsible for clearing up site contamination and may have to provide acoustic barriers, holding ponds and flood defences and actually improve visual amenity.

In many cases it makes sense to set up a public private partnership (PPP) for developments which are more socially beneficial. Here a developer is offered incentives such as tax breaks for development in unemployment black spots. Or public money is provided for infrastructure improvements that make the area newly accessible and clears up major contamination. The public contribution is based on the final value to the private party – a host of regeneration monies are available for former industrial and contaminated areas.

Funding and investment appraisal

A balance between value and development cost should be achieved to make a project viable. This may either be carried out by measuring cost and assessing the value, or alternatively by calculating the end value and working out a budget to suit it. In either case the value must exceed the cost to justify the project. The method and source of funding is important in order to assess cost. There are several tools that may be used to assess this on construction projects.

Payback method

This is a measure for assessing project incomes against project capital cost and possibly running cost. The payback period is the point at which future incomes equalise the capital costs expended. So in a simple example a new building cost including fees and fitting out is £1.2m and there are returns from production profit of £300 000 per year, providing a payback to cover the capital cost at the end of the fourth year of income. In tabular form the cumulative income looks alike as given in Table 3.2.

This method shows how long it takes to payback (year 4), but fails to indicate unless incomes are progressively predicted, what final profit is made. It also does not indicate the declining value of a sum of money which is paid later. This means that unless values are discounted then there is an

Table 3.2 Payback method

Capital cost (year)	Income	Cumulative income	Cumulative net income
0			−1 200 000
1	300 000	300 000	−900 000
2	300 000	600 000	−600 000
3	300 000	900 000	−300 000
4	300 000	1 200 000	0 Payback
5	300 000	1 500 000	300 000 Total income = £1 500 000

unrealistic evaluation of the profit received as money in the hand can be invested. This applies just as much to the use of company reserves as it does to borrowed capital.

Accountancy rate of return

This method calculates the sum of all the income flows (+ve) and the capital cost outflow (−ve) and calculates any net return which comes to the business. This is usually expressed on a yearly basis as a percentage. Thus in the earlier example, the net profit would be £1 500 000 − 1 200 000 = 300 000 over 5 years which is £300 000/5 = £60 000 per year. The return is therefore $(300\,000/1\,200\,000) \times 100\% = 25\%$ return on outlay.

Again this method takes no account of the declining value of whether a sum of money received in the future and even worse does not take account of how many years it takes to get a profit. The figure of 25% return again is misleading, but is directly comparable with other projects calculated the same way.

Discount cash flow (DCF) method

Discount cash flow method is similar to accountancy rate of return, but it operates a discount value equal to the cost of capital. So if money can be borrowed on average by the company at 10% then the present value (i.e. the value today of money received in n years time) is the reciprocal of compound interest:

Compound interest can be expressed as $P(1 - r)^n$.
Where n = no. of years invested; r = rate of interest expressed as a fraction; P = principal sum invested.
∴ Present value can be expressed as the inverse of interest rate
i.e. $1/P(1 - r)^n$.

Table 3.3 Net present value method for capital cost and five-year income (£)

Capital cost (year)	Income	Cumulative income	Cumulative net income	NPV@10%
0			−1 200 000	
1	300 000	300 000	−900 000	
2	300 000	600 000	−600 000	
3	300 000	900 000	−300 000	
4	300 000	1 200 000	0	
5	300 000	1 500 000	300 000	−57 058.15

This will be a discounted value taking into account the number of years in the future it is received. The rate of discount of 10% shown in column five of Table 3.3 reflects the return available from investing the cash and also includes a premium to allow for the level of risk of the investment. This rate is set by companies with reference to the cost of borrowing for them, even if the interest rates change.

Column four shows the undiscounted net cash flow. When the discounted income and the costs for all years are added up cumulatively we have what is called net present value (NPV). This is summarised in column five. Because the income is received in the future each year's income has been adjusted downwards using the earlier formula, or tables, which are available. This rate reflects the rates of interest and when inflation rises proportionately with interest rates, it is a reliable indicator for comparison purposes.

Now we can see that the payback over 5 years at a 10% discount rate is negative compared with the undiscounted column four and reflects therefore the cost of not receiving the income until future periods. The NPV method is much more robust than the two previous methods for incomes that are received over an extended period of time, though there are problems in being sure that the predicted discount rate is correctly reflecting the cost of capital, over the period of 5 years projected. If inflation moves with interest rates in the same direction (normal), then there is no problem. If as in some economic scenarios it moves counter to interest rates that is, getting larger as interest rates stay the same then this can not be assumed. It is also possible that cost of capital may rapidly change, moving away from the predicted amount and if a range of conditions is likely then a sensitivity analysis must be carried out to see how sensitive the return is to an upward charge in capital.

Internal rate of return (IRR)

Internal rate of return is the discount rate in the previous method which will return a nil NPV that is, break even. This is often used by investment analysts to compare with the cost of capital which the company incurs. The company will take a view on what return over and above the cost of capital they require, that added to the cost of capital is called the *hurdle rate*. So a cost of capital of 10% and an expected profit level of 10% means a 20% *hurdle rate*. This will pay a contribution to company overheads of say 2% and provide a net contribution to profits of 8%. However, a project with more chance of not receiving the returns will have a higher premium rate. Another risk which might be reflected at the time of making the decision is how volatile the interest rates are. Both these may cause adjustments to the *hurdle rate*.

Taking the given example and allowing another year of income so that we have a positive result for the NPV. It can be seen in Table 3.4 that the effect of adding another 300 000 income at the 10% income rate has added only approximately £154 000 (57 058 + 96 889) to the NPV, because of the substantial discount at the sixth year.

In order to bring the positive NPV to break even that is, 0 we raise the discount rate until that happens. In our example that will be nearly 13% and this is the IRR for the project. This suggests that the project will only make an additional 3% towards overheads and profit, which may not be enough for the company to go ahead until the cost of capital for them reduces or the project income can be enhanced or capital cost reduced.

The IRR can be misleading for the same reasons as the discounted cash flow method. When comparing different projects or options it is important to test how sensitive each option is to changes in any of the factors. If a small change say in interest rate indicates a large change in the NPV then this project becomes more risky, especially if the returns are achieved over the long term.

Case study 3.2 is an example of a comparative appraisal which has been carried out using discounted cash flow and it focuses mainly on the direct financial costs and benefits.

Table 3.4 Example of IRR based on the given example, but with six-year income (£)

Capital cost (year)	Income	Cumulative income	Cumulative net income	NPV@10%	IRR (rate at which NPV = 0)
0			−1 200 000		
1	300 000	300 000	−900 000		
2	300 000	600 000	−600 000	96 889.28	12.98%
3	300 000	900 000	−300 000		
4	300 000	1 200 000	0		
5	300 000	1 500 000	300 000		
6	300 000	1 800 000	600 000		

Looking again at the University example in Chapter 2 (Case study 2.1) for student accommodation it initially applies the DCF method for option appraisal, but also looks at other issues which are relevant to selection.

The requirement here is for providing good quality up to date accommodation for 400 student beds. All options have been reviewed against the 'do nothing' option as a base for comparing their suitability. Five options for providing accommodation on the existing site and one on a new site, have been considered and the costs of these options is over set over 27 years. The existing site has 72 hall spaces and is big enough to redevelop for all 400 residences, but this would mean knocking down existing academic services. Options are reviewed including the cost of reinstating the academic services elsewhere if required. The options are:

Option 1   'Do nothing'.
Option 2   Get rid of all existing halls of residence and then pay leasing costs to use privately provided residential units. This would emerge as a lease charge for the University plus the management costs.
Option 3A   A split site service where 72 hall spaces are refurbished on an existing site and 328 are new built elsewhere.
Option 3B   A split site service where existing halls are demolished and 150 spaces are rebuilt on existing site and 250 new built elsewhere.
Option 4A   A single site service where 72 hall spaces are refurbished and 328 new spaces are built by demolishing rest of site.
Option 4B   A single site service where existing buildings are demolished and 400 spaces are rebuilt completely on the existing site.
Option 4C   A new residential site for all bed spaces with existing bed space accommodation on the existing site converted for academic space.

All the results in Table 3.5 are given as NPV compared with 'do nothing' and they are compared for three different types of procurement. The do nothing option assumes that there will be high ongoing maintenance costs and these must be carried out by the University. In the table/chart they are reduced to nothing to make other costs more easily comparable. The NPV values on all (3) and (4) options are carried out over 27 years including a two-year build period before occupation, a 7.28% discount rate, which represents real cost to the

University and a positive allowance for transferred risk in the case of LB/FM and DBFO options.

Table 3.5 Comparison of options using NPV

		NPV over 27 years compared with the 'do nothing' costs £'000			Procurement options
		UBR	LB/FM	DBFO	
Option 1	Do nothing	0[a]			This option actually is the most expensive, except for conventional funding on the two new build options
Option 2	Head lease	2962			
Option 3A	Split site refurbish and new	827	3537	3337	
Option 3B	Split site new	692	3647	3072	
Option 4A	Single site refurbish and new	1568	4061	4080	The 'non-own' options here create the biggest savings
Option 4B	Single site new	−622	3322	2686	The all new DBFO options are more expensive than the head lease
Option 4C	Single alternative site	−794	3080	2185	

Note

a Positive figures mean costs are less than 'do nothing'. UBR is the conventional loan finance procurement (actually unavailable to University); LB/FM is the private build with facilities management and lease back to the University with a service charge; DBFO is the PFI agreement with private build and facilities management with building reverting back to the University after capital/service charge has been levied for 25 years.

Analysis of option appraisal

Although conventional loan finance is not available it is interesting to note that when all the costs are taken into account over a 20-year period this is not actually a cheap option and the University might as well use a DBFO to allow outside private provision of accommodation for 25 years. The head lease will give the University a newer facility than the 'do nothing' option, but it will not give the University any residual value for an asset and the costs will continue after 25 years at a high level and the University will also retain quite a lot more risks and the responsibility for maintaining the properties. The head lease is more expensive than the other 'non own' options.

The private build with leaseback compares favourably with the DBFO option. However with DBFO, some opportunity savings may be made if the University is able to sell off land, which can be developed profitably by the DBFO provider as an alternative project, for example, luxury housing. Also the provider might be able to use the student accommodation for holiday or conference lets, giving dual use on catering, communal or leisure facilities made available on the site, during the student holidays and reducing the DBFO service charge. These financial analyses should also be subject to a sensitivity analysis on the main variables as in Table 3.6. These may affect the rankings and the table indicates the extent of the changes tested.

The effect of construction costs increasing favoured the head lease option more, but essentially the DBFO and LB/FM options still remained the cheapest and are the least affected by uncertainty.

This type of appraisal gives a strong financial base but other benefits or circumstances may also influence the final decision to go ahead. The sensitivity analysis is of particular importance to test the effect of variations in the assumptions made and in public projects estimates may be put on non-tangible benefits as suggested in the first example. The operational and through life costs are discussed in more detail in Chapter 8, but these will be important to the development decision especially if the client is an owner and user of the building. Discounting already incorporates assumptions for inflation and therefore, maintenance and running costs well into the future feature less strongly in the equation where there is a large discount rate. In private development projects, the net returns are compared with the company *hurdle rates* (break even plus a suitable margin) to ensure that the risks of development are adequately covered by the margins expected.

A project with more than normal risk shown by the volatility of a sensitivity analysis is likely only to be approved in the case of enhanced margins in the appraisal. Lending institutions will also apply risk assessment when considering borrowing, which is purely secured against

Table 3.6 Sensitivity analysis for financial factors on option appraisal

Variable	Change tested (%)
Construction costs	± 30
Operational costs	± 20
Student rental levels	± 15
Occupancy levels	± 3
Land values	± 20
Interest rates	± 50

the project outcomes. Private development is often more short term in its view of returns, looking for payback after a short period.

Developer's budget

The investment decision for a project is strictly in the feasibility stage. The basic costs of a construction project are the land, the design fees, the building costs and the fitting out costs and there are often costs for additional advice. Other things however may influence a project costs such as inflation, the interest rates available for borrowing money and the risk associated with getting a return (this will affect the rate of interest which is offered on lending). Cash flow is also important and if payments are up front they will mean an extended borrowing period. For example, land often has to be bought early, but if a deposit can secure an option on the land with payment later at an agreed rate this will be preferable.

Residual method   A developer's budget sets out the costs and expected income for a project and includes the percentage return which a developer would want from the sale or rent of the building. The value of the land is the residual sum which is the difference between the costs and the income. A typical developer's budget is laid out in Table 3.7 for an office block on city centre land at £1m per acre.

The given figures assume £10m on average is borrowed during the build programme for 2 years assuming a straight line of time versus spend. It also assumes that the building is sold immediately and that the loan is paid off.

If the building is kept for rent then borrowing costs continue and rent income builds up more slowly. Because of the time value of money, both income and borrowings are then discounted to give NPV. Borrowing costs are often judged to be neutral with present value gains. The building

Table 3.7 Developer's budget for office block

	£'000
Building costs	20 000
Design fees @ 15%	3000
Other consultant fees @ 3%	600
Statutory fees @ 2%	400
Infrastructure costs 1% of building cost	200
Building total	24 200
Expected return on costs 20% to cover risks	4840
Borrowing costs £10 000 @ 5% over 2 years	1025
Building Costs and returns	−30 065
Income	31 000
Residual value of land	935

becomes an asset on the balance sheet. The value of the building may increase over and above the discount value, but this could only be recognised in the accounts for profit if it is prudently revalued, or the building is sold.

The cost of a project will double in 7.25 years at an inflation rate of 10% (this can easily be ascertained from the same tables that are used to assess present value).

External factors in feasibility assessment

Construction projects do not exist in a vacuum. They are influenced by external political, economic, technological and social factors which either directly affect the project conditions, like the price of materials or the going rate for labour, or they affect the client's business and impact upon the scope and specifications of the project. For example, the need to get to market earlier may reduce the time scales for the project to take place. Clients are constantly reviewing their investment to increase or reduce their capacity to suit market demand. If interest rates go up it may well affect the volume and speed of house building that takes place. If there is a new regulation which makes building more expensive then, this may reduce a developer's capacity to invest. The government may also increase real estate investment by giving tax incentives for building in certain areas.

The APM Body of Knowledge [5] refers to external factors such as the project environment and defines it as the context within which the project is formulated, assessed and realised. It is important for the project manager to understand the influence these have on the feasibility, strategy, design implementation and the outcomes of the project and they should be evaluated when recommending certain courses of action and when assessing options.

External factors in construction projects arise because they do not normally exist inside a single business. They are likely to have an external client and they are likely to work with other external organisations in order to deliver the outputs which the client needs. This model is shown in Figure 3.5.

Figure 3.5 Projects done for external clients.

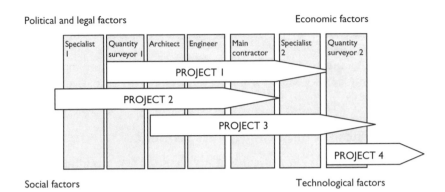

Figure 3.6 Complex relationships.

It is also very likely that a building organisation works on more than one project, for example, an architect is likely to get more than one commission at any one time, and that a larger contractor will succeed in winning more than one contract. Thus we can see that the model should be developed to show this and to show that more than one organisation is involved.

In the model in Figure 3.6 the architect, the engineer and the main contractor are the same on three of the projects, but the quantity surveyor shares only two. The specialists will be assigned to projects in accordance with the need. The same external environment influences all four of these projects. The four external influences are often known as the PEST or STEP factors.

Political and legal factors

Political factors are connected with government policies which might have an influence on the project . These policies cover all sorts of areas, but a few of them are given as follows:

- Fiscal policy covers government tax and spending plans affect the sort of buildings that come on the market and the incentives there are to build.
- The level of skills in the industry can in the long term be affected by the policies for training, although many companies are not facing up to their own responsibilities to sponsor training.
- Regeneration policy means that there are plenty of incentives to invigorate and develop city centres. Many companies have been drawn into dealing with brown field sites and decontamination of land.

- Building regulation changes have forced more concern with energy conservation.
- Land fill tax has encouraged recycling and less waste and aggregates taken to landfill sites.
- The requirement for adapting all buildings for better disabled accessibility has generated a lot of updating and alteration work.

These policies and regulations have a major effect on building design, the types of work available and the methods and resources which are used by contractors.

Economic issues

The economy covers inflation and interest rates and these in turn have an influence on the growth of the economy (GDP), the ability of clients to invest and spend money, the level of house prices and tenders for business and commercial contracts, the value of stocks and shares, the rate of employment and the funding which is most economically available. Other things which might affect the way that companies invest, or do business are the borrowing limits that they have and these are directly related to the profits they can make. The government either controls interest rates, or allows the central bank to set these according to economic need.

Sociological factors

These are related to the fashions that people have and can therefore affect the market demand and the proportion of money that is put into housing and other spending. Communities may also put pressure on developers, contractors and designers to meet societal norms which they feel are acceptable. Environmental concerns show themselves in more energy saving designs, using environmentally sensitive materials and being dependent on less resources. Governments might try and influence this by the use of fiscal policy and incentives.

Technological factors

These are issues which relate to the advances in technology which can affect the methods and materials which are used in construction. They may be the prerogative of the client, the designer or the individual specialists in response to new opportunities. These factors allow for innovative factors and they may or may not be important to the client's future business and are unlikely to be so influential in the post design stages of the project. IT systems affect the communication capacity on the project and integration of individual systems can be important for efficient information flow.

Project scope management

The CIOB Code of Practice recognises project scope as a key requirement of the outline project brief, but this is only the starting point for scope management. During project definition it is useful to state what the brief does not cover, as an aid to setting the boundaries for the project.

The PMI Body of Knowledge identifies five stages of scope management which are initiation, scope planning, scope definition and scope change control. The key issue in scope management is determining the obligations and responsibilities of the project team. Scope refers to two aspects:

- *Product scope* the definition of the product features and functions. This is provided in initial form by the client and is developed by the project manager and the design team. It is closely connected with the design brief.
- *Project scope* which refers to the work which must be done in order to deliver the project with the features and functions specified and is the prime concern of the project team, guided by client and project constraints.

Product scope may be in prescriptive or performance terms. The more the brief is described in performance terms, the more there is room for developing it and adding value. This gives a sliding scale in construction from a client who describes their business requirements such as 'produce 100 cars per week', to a client who hands over the drawings defining the location, the building type and the preferred material specifications. In practice, a client will use their experience to specify key production components and put building design in the hands of a team of specialised designers. Project scope will be covered in Chapter 4.

Building value and risk allocation

Managing risk and value at feasibility stage

Eliminating, reducing or protecting from risk at this early stage is closely allied with the VM process, which seeks to adjust the brief in order to optimise VFM. There is scope here to produce radical solutions that reduce or eliminate risk by more holistic solutions that are properly tied in to client objectives. Later on when the design is firmed up there is less scope to provide alternatives. A problem solving rather than a solution generating culture is now in place. This is done by using VM workshops.

Figure 3.7 shows the relationship between scope to change, cost of change and the stage of the project life cycle. However there is no guarantee that

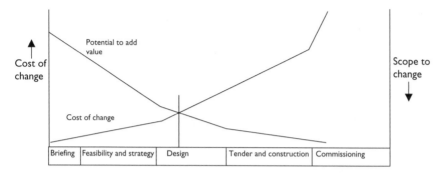

Figure 3.7 The cost of change.

initial assumptions for the scope and use of the building will not change with the demand for the product. Part of RM is to build in flexibility and contingency so that change is possible. Some building projects are more prone to this than others. More is covered on RM in Chapter 8.

Changes need to take place at the feasibility and strategy stage of the project otherwise substantial abortive costs are incurred due to the effort which is wasted in redesigning or worse still in abortive orders or construction work. VM can be used to test the original assumptions made in the brief and early design.

Risks are assessed in the feasibility stage and may be mitigated by applying strategies which reduce them or by choosing different options in order to make feasibility and viability more robust. Risk allocation is particularly important for making sure that residual risks are the responsibility of those who are best able to manage them. This increases the probability of keeping the risk low and also those who have the most experience in managing the risks will price them competitively without a cover contingency. Chapter 4 covers the identification and allocation of risk and Chapter 8 deals with the process of RM and VM in more detail.

Conclusion

Project definition is all about communicating a project brief well, being flexible, carefully assessing scope and delivering a product which suits purpose, but which will also maximise value and be effective. Many projects have failed due to inappropriate design of buildings which have failed to accommodate client expectations. It is clear that there are many strategic issues that should be considered in the project definition stage to inform the design so the early involvement of advice on the construction process and specialist design is good practice.

There is often a lack of communication and modern technology has the capability to produce good quality 3D visualisation so that clients do not have to depend on 2D drawings which they do not understand and are set at scales which do not facilitate spatial planning.

Project appraisal is used to test financial feasibility, and is useful for comparing different options, but have the right options been chosen? Quantitative methods are often presented as infallible, but clients are not warned of the factors that are particularly sensitive to changes in the market such as interest rate changes, technological and statutory updates and obsolescence over quite short periods, that impact heavily on feasibility.

Clients are often unaware of the balance between the operating costs and the savings that can be made in increasing capital costs so that these can be substantially reduced. In short, professionals have often been criticised for their reluctance to give advice, that might seem unpopular, so that clients are unprepared for steep price rises, which exceed budgets and cause expensive late changes.

Taking external factors into account is vital to the success of the project definition stage. Political, economic, social and technological factors can easily change so it is wise to build in flexibility and contingency factors.

References

1 Royal Institute of British Architects (RIBA) (1999) *The RIBA Plan of Work as amended 24/5/01*. RIBA Services, London.
2 CIOB (2002) *Code of Practice for Project Management for Construction and Development*. 3rd Edition. Blackwell Publishing, Oxford.
3 Gray C., Hughes W. and Bennett J. (2004) *The Successful Management of Design*. Centre for Strategic Studies in Construction, University of Reading, UK.
4 Hellard B. (1997) *Total Quality Management in Construction*. Thomas Telford Publications, London.
5 Association for Project Management (2000) *Body of Knowledge*. 3rd Edition. APM, UK.

Chapter 4

Strategic issues

This chapter moves on from looking at the product and its definition and introduces the major decisions which decide *how* the project is carried out and gives an overall view of strategic planning and control *systems* of the project. The overall strategy is sometimes pulled together in a PEP or master plan, which guides the team on the organisation, time, cost, quality, information, risk allocation and identifies the planning and control systems. An approach that recognises the relationship between the strategic subsystems will help to provide a more integrated approach and smooth the path of the project manager who needs to co-ordinate these systems.

The objectives of the chapter are:

- Introducing strategic systems and their relationships.
- Develops scope management and change.
- Understanding the master planning process.
- Defining and establishing control systems.
- Identifying strategic risk and its allocation.
- Understanding cost planning and cash flow systems.
- Choosing a procurement strategy.
- Quality planning.
- Understanding and improving the information flow.

There will be other chapters for design management, RM and VM, project organisation, leadership and teamwork, construction performance, supply chain management and safety and health and the environment (SHE). Project strategy is defined as 'the route to get to where we have determined to go'. The strategy concerns the organisation of people and tasks.

- For people the questions are 'How do we achieve effective communications?' and 'Who is responsible for what?' and 'How can we achieve the quality standards consistently?'.
- For tasks the questions are 'How shall we procure the work and materials?', 'What are the sequencing priorities?' and 'What control systems will suit best?'.

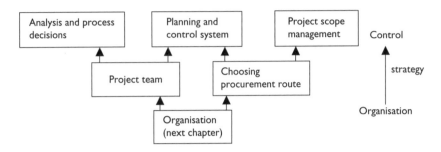

Figure 4.1 Elements of the strategy stage.

Strategy should take an open systems view and heed:

- the impact of the external environment;
- the effect of the project on overall organisational objectives;
- the effect on the community.

Not taking external issues into account has often led to the failure of projects. Although the external environment is not in the control of the project manager, it can be predicted so that a response can be made. For example, the shortage of skilled labour locally could lead to the incorporation of a wider sub contractor tender list to ensure a proper service. The stage runs alongside the feasibility rather than following it. Standards for the time, cost and quality must be formulated. We shall refer to the final developed strategy as the master plan.

The Figure 4.1 shows the main elements to be established in the strategy stage, together with the employment of the project team to suit.

Strategy depends upon organisation and control. Key control processes are managing project scope, procuring the contract, setting up control systems and making decisions. First we need to consider how a project breaks down into systems.

Project systems

The term systems is used as an analytical tool to work out what inputs are needed to achieve the objectives. Each system has an input, a managed transforming process and an output. For example, the construction process has an input of raw materials, a transformation, a process of assembly and an output, which is a finished building. An open systems approach recognises that systems are influenced by what is going on outside, for example, the availability of skilled labour or sufficient plant.

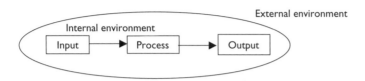

Figure 4.2 The basic system.

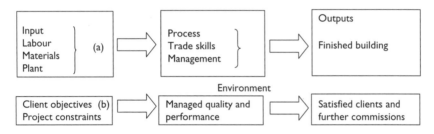

Figure 4.3 Overall construction system: (a) assembly and (b) satisfaction.

The Figure 4.2 shows what is called an open systems approach recognising the environmental influence.

There is an overall system and there are sub systems. In the case of construction projects, Figure 4.3(a), shows how the overall project might look for the assembly system described. The construction system inputs consist of resources to carry out the construction. The process is the assembly of the building using the resources and that includes the project management and the output is the finished building. In Figure 4.3(b) management is still an important part of the satisfaction system, but it is aimed at achieving a good quality building which will satisfy the client and recommendation and further work. This system can be set up for any of the sub systems of a project.

Project subsystems

Each part of the life cycle can be seen to have a subsystem with inputs, process and outputs. The design process consists of designer inputs (e.g. architect) who carries out the design processes. Their output is the design drawings and specification within the client and planning requirements.

The project life cycle divides into distinct stages in which each have an input, process and output. Within each of these stages there is a key client approval which needs to be obtained before proceeding to the next stage. In some cases that approval may be phased. Figure 4.4 shows this scenario.

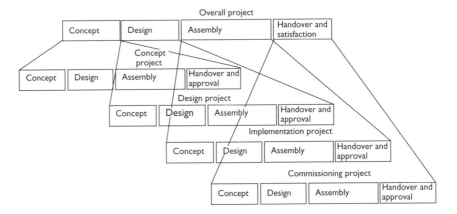

Figure 4.4 The project subsystems.

Source: Adapted from Burke [1] Figure 3.6, with permission.

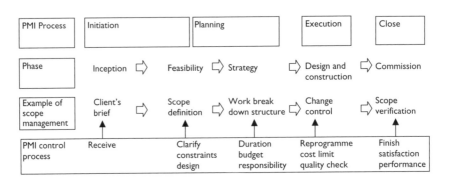

Figure 4.5 Integration of project process subsystems with life cycle.

The breakdown of each life cycle has been simplified. It shows you the interconnection of product sub systems in a simplistic way, as it is quite possible that the different phases of the whole project overlap. These systems help give a good foundation for understanding the project scope.

The PMI [2] overlays the earlier subsystems as also having five strategy or process groups in each of the phases. These are initiating, planning, executing, controlling and closing. They break down in each process into a further subdivision of core and facilitating processes and these identify specifics, although all these apply in a general way to all four phases mentioned earlier. This chapter is particularly concerned with the processes of planning, controlling and execution that *integrate* the design and assembly (construction) phases. These are related across the phases in Figure 4.5.

Scope is a specific, which is given in the inception stages as a client brief, defined in the feasibility stage, broken down into its elements in the strategy stage and executed in the assembly stage. Across all these stages change control takes place.

Figure 4.5 clearly shows how planning and control progressively apply against the various stages of the life cycle, using the example of scope management. The following sections look at the processes in more detail.

Project scope: planning and controlling

Product scope in construction is often referred to as the design brief and was considered under project definition. This may be in prescriptive or performance terms. The more the brief is described in performance terms, the more there is room for developing it.

Project scope covers a much wider remit and includes at least the time management as well as the implementation of the design to the right level of quality and the resource planning for product delivery. The feasibility stage will also be concerned with establishing project scope. There are five stages mentioned in scope planning, which are hierarchical:

- Establishing the project contract or charter.
- Scope definition.
- Planning.
- Scope change control on, for example, business capacity, functional efficiency, appearance, accessibility for customers and staff, user comfort, environmental performance and low maintenance and running costs for success too.
- Scope verification.

The project contract or charter   formally recognises the existence of a project and provides formal reference to the business need that the project addresses and by reference to other documents, the product description or project brief. Constraints are recognised in the contract conditions and will refer for instance to a contract sum, a defined time period, recognised behaviour and protocol and assumptions for managing the contract.

Key documents in construction are the contract which defines the time period and standard conditions of contract, drawings, specification, a contract sum received from the contractor and a contract programme.

In construction it is easy to see the building as only a product, but construction projects are also a service to the client and success depends upon delivering the building on programme, fit for purpose, on budget, with no significant defects, easily maintainable and without a poor safety record.

Scope definition   is the next level of detail in order to facilitate control of the project and together with *scope planning* breaks down the project into different functions or sections. Scope planning identifies the relationship

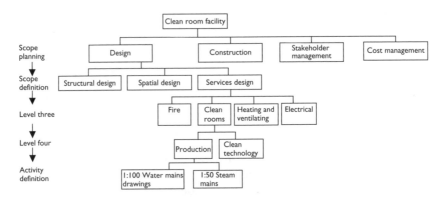

Figure 4.6 WBS for a clean room facility design.

between the activities in the form of a programme of critical path analysis to co-ordinate the numerous activities and facilitate decision making, giving a basis for control. In particular, it should identify the project deliverables and project objectives which can be monitored.

For example, spatial design and environmental design are two elements of a breakdown related to user comfort. As such they are separately delivered, but also need to be co-ordinated. Work breakdown structure (WBS) also creates a hierarchy of subsequent detail as indicated in the clean room project in Figure 4.6.

The primary aim of the lower levels of the WBS is to provide a structure for allocating individual responsibility for some of the service requirements. At the higher levels it helps to identify stakeholder requirements. It may also be used for allocating organisational or senior management responsibility. Breakdowns may not be functional as indicated in the given diagram, but may be related to phasing or to subcontract packages.

Scope verification is checking the achievement of the breakdown structure and should be connected to a checking and monitoring procedure including progress review, audit, testing and inspection. Formal inspection takes place at the end of a project or a phase.

Scope change control manages change in the scope of work that was initially established in the budget. This becomes most critical after the scheme design, but significant changes in scope will have a knock on effect in the scheme design and detail brief. They may occur due to

- an external event such as a new government regulation;
- an error or omission on the part of the client in defining the scope of the product required, such as the need for additional welfare facilities to meet staff agreements;

- a value adding change, such as reducing running costs by taking on a technology which was not available when the scope was first defined;
- a planning permission qualification, such as the requirement for more underground parking or a change in brick type.

These may in turn disrupt general progress and impact budget, sequence, design fees or finish time. A scope change management system needs to track these changes and their budget implications. Such changes need to be fed back through the planning process such as health and safety implications, co-ordination of different aspects of design, drawing adjustments and cash flow.

It does not include those changes which are of a developmental nature during the design period. These changes can be defined as a natural part of a continuous improvement culture to optimise a solution or to make choices between options.

The development of the master plan

The APM Body of Knowledge links the project master plan with the key success criteria and suggests that key performance indicators (KPIs) should be established in order to monitor the plan as a basis for control and evaluate its success. It is a plan for action and establishes the control criteria. In the PMI Body of Knowledge and the OGC Gateway Review™ process the master plan is called the PEP. The plan answers three main questions:

- How? The plan investigates the procurement process, establishes a quality, sustainability and health and safety plan, tracks the flow of information and the use of IT to meet client objectives.
- How much? Develops a cost plan and a cash flow to suit the client's budget and fund availability.
- When? Establishes a programme of key events (milestones) to establish key decision points (gateways) to ensure an agreed finish date is achievable and to determine resource usage.

ISO 10 006 [3] and the OGC guidelines for the gateway process give some guidance as to what should appear in the PEP.

The time schedule is an iterative process. Iterative means that better schedules supersede not so good ones. Programmes are reprocessed as better information becomes available or details are changed. The iterative process will take the form of establishing targets[1] and basic areas of performance responsibility, which are accepted by each party with budgets determined for each area. Freeze dates for any design changes need to be agreed on a procurement programme in order to avoid abortive costs and to establish client commitment dates. The freeze date should be directly proportional to the procurement lead in times and how this relates to ordering and execution date of the process. The iterative process depends upon good teamwork between many participants.

The master plan is a combination of these individual schedules and budgets adjusted to suit the overall budget and programme constraints. A 'hardening' of the plan takes place after consultation and approval by the client or project sponsor. At this stage further significant changes are formalised by a change order and should be at the prerogative of the client. The main stages of approval are project go ahead (after feasibility), planning submission and detail design.

Contingency planning

Contingency means planning a back up, when things go differently from the original plan. Contingency planning is a way of allowing for adverse changes and generally means making monetary allowance for items beyond control or introducing a formal RM system. However flexible programming, insurance, risk transfer or mitigation may also be used individually or together. The approach should be specified and the potential problems should be identified together with the reasonable precautions taken.

The concept of the 'last planner' is an iterative one developed by Ballard and Howell [4] to look realistically at the real completion of activities and not those that were on the wish list of an unrealistic plan. His research showed that work actually completed on a weekly basis achieved between 27% and 68% of the master programme allocation. In practice last planner means the development of realistic weekly plans of work that work towards a flexible replanning to suit achievable progress each week, but provide a planned ability to pull back time lost on the master programme.

This process accords with the feed forward control process, which may also be used for cost and design control.

Risk identification and allocation

It is important to recognise that uncertainty will exist and should be assessed, reduced to acceptable levels and each risk managed by the party who is best able to deal with it. Risk transfer can be expensive if it means that risks are managed by those who are inexperienced, or who do not have the influence to reduce its impact or probability.

Typical risk areas are client changes due to business predictions or user requirements, subcontractor default, technical failure, strikes or other labour shortages, bad weather, critical task sequences, tight deadlines, resource limitations, complex co-ordination requirements and new complex or unfamiliar tasks. External political, economic, social and technological issues can also affect a current contract. Many of these are not covered under insurance.

It is important to assess and manage risk in the context of the feasibility of the construction project. Risks which appear in the business case can be put into four categories according to Smith [5] – these are financial, revenue, implementation risks; and risks connected with the operation of the finished facility:

- Financial – interest rates; payback periods; borrowing facilities and internal financial structures; currency and stock markets; dividend expectations.
- Revenue – market demand, tolls or development risk.
- Implementation – physical conditions, obstructions and accessibility; construction resource availability, time delay and quality; design appropriateness, changes and standards; technology and provision for change and update.
- Operational – operational and running costs, maintenance requirements, training and aftercare, legal responsibilities and updates.

It is very difficult to get a universal categorisation and experienced clients such as the NHS have developed standard risk registers that categorise risk to allow comparison, but allow flexibility for the project team to add and subtract as appropriate. Edwards [6] has added others and some of these are included later in allocation.

Allocation of risks in the construction context

Allocation of risks is very dependent on the procurement system and the chosen contract conditions.

Political and legal   liabilities could be allocated to any of the parties and insurance is used to cover events such as professional indemnity, third party, fire and damage which are compulsory. Political events often effect the project financially in its operating or financial structure. For example, health and safety planning is important to avert accidents and health incidents, prosecution and compensation costs.

Financial risks   are allocated to the financial sponsor of the project, who is usually the client in the traditional construction procurement. In PFI, DBFO or turnkey developments this transfers to the provider of the facility. Other risks include interest level predictions changing significantly. Equity funding may not be taken up as expected and stock market conditions can destabilise company borrowing levels and cause lenders to be more wary and more expensive.

Revenue risks are allocated to the client or the facilities manager. Traditionally revenue is a risk for the project client, but this is important for DBFO projects which have toll collecting agreements, although there are often let out clauses, the main issue is cash flow. For example, the Second Severn Crossing is run by a private consortium that charge tolls, but are restricted to toll escalation within inflation index linked limits. However, in order to strike the deal with the Skye Bridge crossing the time span for collection of tolls before handing back to the Scottish Office was extended to cover the extra costs of certain environmental design changes. Market changes or poor market conditions can severely affect revenues.

Implementation risks are those most closely associated with the construction process of design, tender, construction and commission. They are traditionally shared between the client and the project team according to standard conditions of contract. Broader forms of procurement move all but scope changes back into the realm of the designers and contractors. Design and build and design and manage lock the design and construction responsibility together and provide much better protection for the client from design development changes that often create abortive work with no increase in value to the client. This area will be followed up in the chapter on RM and VM.

Technology risks are to do with teething problems, changes and updates during the concession and prime operation period. These are usually at the risk of the client, but are related to the design role also, where the project team has a performance brief and is dealing with the development of new products, or using them in innovative ways.

Operational risks are shared between the designer and the client and relate to weak and inappropriate design, poor briefing, lack of co-ordination between design and fitting out requirements and poor use or misuse of the building. Plant performance has a major effect and again is related to design. However, the fitting out and quality standards specified by the client may have led to operational unreliability, expensive breakdowns and downtime, excessive running costs or even a de-motivated workforce. Case study 4.1 illustrates this type of risk.

— *Case study 4.1* Design risks

An operational problem occurred in a brand new office block for a government agency where an open plan, sealed and air-conditioned building had been designed and built. The commissioned building was handed over to the client without fitting out. When the computers were installed there were far more than designed for and the extra heat generated meant that, in hot weather, the air conditioning was unable to adequately cool the building, so staff were faced with difficult working conditions during the summer.

A major electrical breakdown was also caused by flooding into the basement electrical switch gear, putting systems out over a frenetic weekend. Insufficient back up, lack of co-ordination and operational changes gave the whole project a bad name and wrecked initial productivity in the new office. Managing risk for the operational conditions is one of the hardest jobs for the project manager as they often do not have access to the maintenance team at the design stage and dependence is placed on standard building loads.

Table 4.1 Table to show breakdown of allocated risk

	Client	Design	Contractor	Shared
Planning delay	X			0
Design change	X	0		
Client scope change	X 0			
Planning qualification	X		0	
Higher cost of construction	X		0	
Industrial action			X 0	
Latent defect			X 0	
Heating failure		X 0		
Level of use of rooms	X 0			

Note
Traditional contract (X) build, lease and operate (0).

In Table 4.1 the risks have been allocated between three parties in a contract to provide student accommodation on the basis of traditional and on the basis of a build, lease and operate. It shows how risk is allocated quite differently for different procurement strategies.

The allocation of risks to the party that can best manage them is an important principle. It has the potential to move from a risk aversion culture, where risk is passed on to the first level of incompetence, to a value inducing culture where the risk allocation has been properly agreed and negotiated.

It is reasonable for an inexperienced client to pass on all the construction related risk, but it is expensive for a client to pass on the risk of, say, the level of occupation of the rooms to the contractor. The contractor would cover generously for a risk in the area of room letting for which they have no knowledge or prior experience. The project manager is in a strong position to guide and advise the client in this position for improving the VFM.

As the project progresses towards the scheme design, so the scope for change reduces and after the brief is frozen the cost of any change rises rapidly. At the inception stages there is the greatest level of uncertainty and risk evaluation becomes firmer as more is known about the project. Risk evaluation and RM is covered in Chapter 8.

Project control

The control system is critical to the health of the project and its choice should influence the planning processes rather than the other way round. Initially, the client will specify project progress reports which are critical to maintaining their confidence in meeting project objectives. They will have an interest in the design, the cost, the performance and the programme. Frequently, experienced clients also see the project team make up as critical to effective control, wishing to vet the personnel at interview stage. It is also likely that they will insist on supplier proposals at this stage for planning and proper control of the project. As well as the earlier point stated, clients may also be interested in a number of other controls such as environmental management, health and safety and relationships with neighbours.

There are two approaches to control discussed here – feedback and feed forward systems. In construction, the former method is used much more often and is shown in Figure 4.7. Unless instantaneous feedback is obtained, the system may not produce information quickly enough to take action in a unique project situation. This is because a project's activities take place over a relatively short period of time and if control is based on finding out when things have gone wrong then corrective action might arrive too late.

The feedback control system briefly consists of setting up targets, monitoring the performance, measuring the gap between planned and achieved, identifying the problem and taking corrective action.

There are a number of issues with feedback systems which inhibit the efficiency of the feedback control system:

- It can take time before an output is reliably measurable.
- Monitoring is not normally continuous with immediate correction. For example, a thermostat would give immediate feedback of a change of temperature and simultaneously switch the heating back on.
- Control is reactive and time is spent on investigation. For example, labour productivity, plant suitability and availability and materials delivery are all very different problems.
- Projects have limited repetition and delayed correction means that action will be applicable for a limited period only.

Feed forward control indicated in Figure 4.8 shows how monitoring mainly anticipates the problems prior to taking avoiding action.

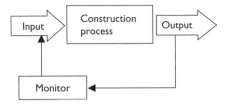

Figure 4.7 Feedback control system.

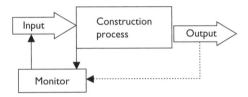

Figure 4.8 Feed forward system.

In the feed forward system the key components are to

- set up targets;
- anticipate problems with meeting targets by discussing with team, who will know their concerns, at regular intervals before the event;
- take prior management action in order to avoid or mitigate the problem;
- monitor performance to make sure action has worked (feedback);
- make further adjustments, if necessary.

Programme control

A typical feed forward system is operated through the establishment of advance meetings to co-ordinate the activities of different specialists and to consider resources and information availability and the resolution of current problems. It tests the robustness of individual plans and concentrates on interfaces where the performance of one contractor will have an effect on another. It operates on the latest information available and allows for contingency planning if there are any doubts over performance or issues outside the control of the project staff.

Successful operation depends on openness in discussing problems, a 'no blame' team culture to overcome problems and a commitment to the project as a whole (Case study 4.2).

Case study 4.2 Slippage in design programme

Figure 4.9 illustrates a typical situation for a project falling behind programme and not meeting the critical construction start date on site. In a feedback system, the programme will be measured at time now (square box) and the programme will be discovered to have dropped behind. If this is a critical delay then the activity is part of a series of activities where none of those activities have any or very little slack. Reprogramming the project to finish the stage on time depends upon feeding in extra resources (plant or labour), re-sequencing the logical progression or overlapping activities to a greater extent. This is a traditional response.

Where logic is known through the use of a critical path technique (see Appendix, Chapter 5) a prediction can be made that it will complete late if it proceeds at the current rate of progress (i.e. using the same resources), to allow feed forward to take place.

In feed forward, communications will be better and design problems will be picked up before irretrievable lateness becomes a problem. It is wise to investigate the reasons in the first place so the original problem does not persist.

Design must be largely complete to go out to tender, which means that additional designers must be drafted in to solve this problem under both systems. Both systems can fall down if there is a personality clash or a lack of communication. Here the targets will not be monitored tightly or over optimistic or misleading information will be given.

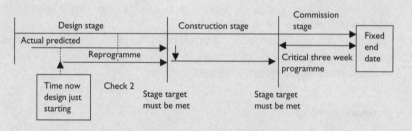

Figure 4.9 Reprogramming within the critical stage dates.

Reiss [7] regards a plan as having three main uses:

- a thinking mechanism 50%
- a communication tool 25%
- a yardstick 25%.

He puts the emphasis of planning on thinking through the work ahead in a systematic way and enhancing the manager's communication ability. In feed forward, the emphasis is on thinking through the implications with others so that communications are implicit. The feed forward method (prediction) is more efficient as long as there is an openness to discuss problems and to deal with them realistically with a commitment to flexible planning. There are other methods of control that are helpful in considering effective management.

Management by exception is a simple principle for releasing management time monitoring the formal control system. It only requires inputs outside a ± band of conformity to target as shown in Figure 4.10. Detailed control is delegated to the first line manager.

This system allows for a band of control within which the project manager does not interfere. It recognises that 'actual' will vary from 'planned' at times, but that it is the responsibility of others to let the project manager know where performance is outside the limits. This is still a feedback system, but its strength is that the project manager can have more time for strategic issues and forward thinking. Delegation of this nature has a motivating effect in the right climate, allowing responsibility and development of more junior staff. The weakness especially for construction projects is the variable quality of supervision in the context of specialist contractors and the critical effect of lots of parallel activities if one goes out of control. It is also difficult with one off projects to accurately monitor and give timely feedback on time, cost and quality conformity to plan. Therefore, it may not be realised that an activity has gone outside its control band.

Management by objectives (MBO) has a similar goal in transferring responsibility to a lower level and cutting down the level of supervision required. Here, a junior manager will be required to identify performance targets for their section of work for which they are accountable and to report back at longer intervals on their achievement. This can be motivating for the section manager, as it empowers them to make important decisions. It also requires perception and trust on the part of the senior manager in agreeing achievable targets, which stretch and do not strain the manager. It works best where there are clear objectives.

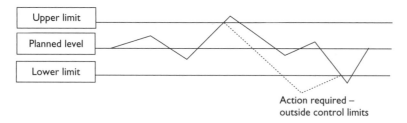

Figure 4.10 Control bands.

Cost planning and control

Cost management aims to control the project budget which is set after the VM process and to predict the timing of the payments (client cash flow). Budgets set for the design and construction stages should be monitored and kept within the cost and cash flow limits set. The key areas of change are design changes and client scope changes. Again, there are two systems of control – feedback and feed forward.

Many systems are based on creating a project cost plan with elements or cost centres, collecting cost data by monitoring actual costs and providing a monthly feedback report to the client and then looking for ways to make savings at a later stage, where there is an overrun on a cost centre. This is called a feedback system and creates problems in projects where there are relatively short activities with limited opportunities to make savings within the latter part of the element. Feedback control is often applied to a specialist package as a whole. The problems are:

- Contingencies are built in as a sort of cushion with resultant uncontrolled extras.
- It provides clear information about problems, but often too late.
- Clients are made to pay for lack of planning by project team.

Once the budget breakdown has been allocated to specialist elements, early expectations of cost breakdowns are used to limit expenditure in these proportions. If radical VM takes place, contortion can occur and a second stage of cost planning is needed, which allows new information about actual prices to adjust elemental budget allocations, as a dynamic control system applied to the design development.

In a feed forward system there is forward planning and warnings can be made. It predicts areas of critical pressure by assessing the risk and focusing cost management efforts.

A process of control stages is indicated in Table 4.2, which encourages the early prediction of 'cost critical areas'.

From a clients point of view no cost increases are welcome, which do not produce extra value for the client. Danger areas are:

- Misreading the market tender price.
- Buildability problems inducing contractor delays.
- Poor planning creating an unpredictable cash flow.
- Poor tender documentation inducing additional claims.
- Abortive work due to late changes or instructions.
- Claims due to disruption or late information or client risks.

The best way is to anticipate problems before they occur, so that avoiding rather than corrective action can be taken. For example, if cladding chosen is

Table 4.2 Cost management as a feed forward system

Phase	Process	Life cycle stage
Outline cost plan	Elemental cost breakdown based on outline specification. Cost checks on design as developed	Feasibility
Scheme cost plan	Elemental cost plan based on approved design, following VM process. Cost checks on detailed design. Value engineering	Strategy and design
Cash flow forecast	Monthly commitment based on procurement programme	Detailed design
	Update client on contractor progress. Earned value analysis	Construction
Firm cost plan	Based on contractors tender. Tender reductions if required	Construction
	Interim valuations and change management	
Final account	Contractors tender plus extra work Final valuation of all approved work	End of defects liability

found to cost 30% more than the scheme cost plan, the first reaction is to find a cheaper one to contain price escalation. If it is cost checked at the point of choice then it may be possible to look at a reduction in another element of the building which has less effect on the value, reduce the amount of cladding, or re-orientate the building to make a cheaper cladding possible.

Feed forward allows a value decision based on all client objectives and not just cost. For example, alternative cladding may have more costs in use, such as energy loss, cost more to fit, wear out sooner, delay the contract or be less environmentally acceptable. Cuts in other areas may cause equal problems in which case the budget may need to be exceeded or some leeway on completion time may help to reduce costs also.

Cost control measures

There are a number of cost control measures which may be used by the contractor and they break down into systems based on budget reconciliation and systems based on cumulative cost compared with cumulative earned value.

1   Measures to control budget levels in cost planning. Typical feedback systems are:

  - Cost value reconciliation based on the monthly or stage valuations process, which measures the level of spend with the value of the same priced activities, for example, bill of quantities items inclusive of variations.
  - Cost centre control which is a more sophisticated system and allocates all expenditure to a cost breakdown by activity, for example,

subcontract package or trade and may also break down into elements such as labour, material and overheads.

2 Cash flow which measures the scheduled value for all costs showing this as:

- A monthly cost plan of scheduled value against time as in Figure 4.11.
- A curve for the actual value earned according to the project valuations. This is often called *earned value*. In an ideal world this would be the same as the scheduled value, but assumes exact compliance with programme on time and in sequence.
- A reconciled actual expenditure curve which discounts expenditure not yet valued (prepayments and retention), or value gained not yet paid for (accruals). This then makes it fully comparable with the valuation figures and should differ by the margin expected on the project.

Figure 4.11 shows a graphical representation of a client that pays the contractor monthly shown by the black lines under the dotted curve. These have been cumulated to give the dashed cumulative value curve and compared with the cumulative costs of the contractor in the solid line. The contractor is currently making a small margin.

There are two forms of feedback which can also lead to a feed forward prediction based on the current rate of spend.

1 *Cost slippage* which measures the difference between the actual expenditure and the actual value. This indicates whether the margin is currently being achieved if gross figures (including retention and any other valuation deductibles) are used.

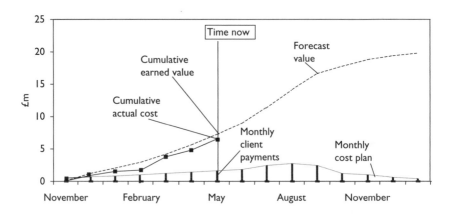

Figure 4.11 Earned value cash flow chart (from contractor's viewpoint).

The cost performance index (CPI)

$$= \frac{\text{actual value}}{\text{actual cost}} \times 100\%.$$

For example, if the actual value is £2.5m and the actual cost is £2.2m and the margin to be achieved is 10%, then we have

$$\text{CPI} = \frac{2.5}{2.2} \times 100\% = 113.6\%.$$

Therefore the margin required is being exceeded at the current rate of remuneration. All variation needs to be properly reconciled with the scheduled amounts.

2 *Programme slippage* measures the difference between the scheduled value and the actual value, indicating whether the project is running behind or in front according to the expenditure. This is quite a coarse form of control as cost is not directly connected to unit production. So material prices differences and delivery times may distort the true picture. However, it is a leading indicator and a tool for prediction of completion at the current rate of spend. The schedule performance indicator (SPI) = scheduled value/ actual value. The formula for calculating time taken at the rate indicated by the spending rate is

$$\text{Predicted time period} = \frac{\text{scheduled value}}{\text{actual value}} \times \text{scheduled time.}$$

For example, if the schedule value on week six is noted as £3m and the actual value is £2.5m and scheduled time is 25 weeks then we have:

$$\text{Predicted time} = \frac{3}{2.5} \times 25 = 30 \text{ weeks.}$$

Therefore the project is predicted to be five weeks behind schedule if the spend rate continues as it is. This is related to the contractor's account, but may provide good information for the client.

Reporting systems to control costs for the client

There are three stages for cost reporting which are:

- design development cost checking in the project definition stage;
- confirmation of commercial price and budget when tenders arrive;
- cash flow control measures during the construction phase.

The client's budget is not finite and any movements need to be planned and managed. At least 80% of the costs are decided in the design development stage so this becomes a critical stage to control.

In the case of the client, cash flow management is critical. They need to know dates for financial commitments so funds are made available to pay. Sudden changes in work sequence which speed up payments needs to be agreed and signalled early to allow arrangements to be made. It is likely that the inexperienced client in particular will latch on to early estimates of cost and will find it much harder to adjust to the pattern of payment to meet early procurement of equipment and materials. Clients may also wish to make changes in scope to manage their budget and to respond to changes in external factors which will change the early cost estimates.

According to the CIOB Code of Practice, *cost reporting systems* need to provide the established project cost to date, the anticipated final cost of the project which should be fixed within reason, the future cash flow and any risks of expanding costs should be reported and any potential savings. The last three fit in with the idea of feed forward and provide room for proper budget reordering, so that the final cost is contained.

Reporting needs to deal with two stages to assist the client's financial planning. They will need to know:

- the last date they can change their minds (commitment) and they will need to know approximately how much they have to have each month to pay the bills;
- they need to be informed of what the financial implications of any scope changes are.

Figure 4.12 indicates the typical cumulative cash flow and the progressive nature of the client's commitment. The costs will change but it can be managed to make savings to cover escalation, so that the final budget is unaffected.

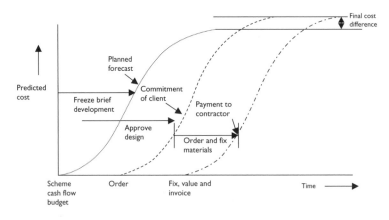

Figure 4.12 Connection between client cash flow forecast, commitment and payment.

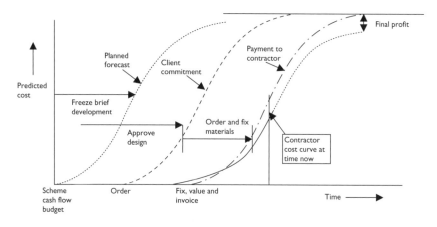

Figure 4.13 Relationship of contractor cash flow and earned value with client.

In Figure 4.13 the contractor cost curve has been added to show the relationship of earned value, which is what is paid by the client and the development of a profit as indicated.

Inflation and the possession of fresh information may have a significant effect on expanding the budget and a proper contingency should be allowed for inflation and the client advised of the danger of delaying outside the original time frame.

Summary of cost control

The principles for budget and cash flow control are:

*   Time spent at design stage on identifying risks to costs, so specific rather than general contingency allowed, for example, ground investigations, option appraisals to reduce risk.
*   Contain design inflation (80% significance) in a VFM framework.
*   Integrate design and construction risk to one co-ordinating party if possible.
*   Work together to give accurate advance costing information for any changes to the budget, so decisions are made on basis of knowledge of expenditure, share savings.
*   Transfer risk to contractor to cover reasonable unforeseen circumstances, but risk they are unused to managing will be budgeted expensively.
*   Give reliable information on the distribution of the payment cash flow schedule so that funding is easily available.
*   Consider life cycle costing, if possible.

Feedback cost control systems need to react quickly enough to give the client information, so that they make decisions on the basis of cost. Where accounts are made transparent to the client as in partnering, then information about costs and profit are easily available. If they are not available, as in traditional procurement, then feed forward systems, where cost of a possible change are calculated ahead of the time when go ahead is required, have the same effect. Close collaboration is needed to be aware of the commitment dates. Procurement and contract types should take account of the degree of flexibility that is required by the client.

Procurement principles

Procurement of the design and construction work is about reconciling the client's objectives for the project with the particular characteristics of the procurement system chosen. Each of the procurement options has implications for risk allocation, project organisation and sequence. There are many tables which have been prepared, which match the attributes of a procurement option to the client's wishes and there is one in the CIOB Code of Practice, which has 17 attributes measured against four main categories of procurement. These categories of traditional, design and build, management contracting and construction management are by no means comprehensive as they poorly cover clients who wish to shed more (e.g. PFI or prime contract) or less risk (e.g. measured term or cost plus contracts). They do however give a range of risk options. The purpose in this chapter is to look at the underlying principles of the procurement of design and construction services rather than the detail of a particular system which may in any case be adapted for the project.

Table 4.3 Client priorities linked with procurement features

Client priorities	Example of procurement features which suit
Project time and speed of mobilisation	Overlap design and construction, e.g. management fee
Single point responsibility	Design and build, prime contract, or PFI
Degree of client control over design	Independent designer, e.g. traditional
Cost certainty	Single responsibility, e.g. design and build or PFI
Ability to make late changes	Transparent pricing, e.g. New Engineering Contract (NEC)
Early involvement of contractor	Management fee, e.g. design and construct or construction management
Reduce confrontation	Partner agreements, e.g. use of NEC or prime contract
Facilities contracted out	DBFO, e.g. PFI

The principles which govern procurement type are:

- Degree of client involvement, priorities and client objectives.
- Risk allocation client side versus supply side.
- Balance of competition and negotiation.
- Tendering process (e.g. single or two stage, open versus selective, closed or negotiated).
- Whether framework agreements or repeat projects required.
- Degree of integration required.
- Degree of collaborative practice and partnering.
- Controlling life cycle costs facilities in or out.

The main procurement types are matched with the priorities of the client. These features may be met in more than one of the standardised procurement types or contract forms as shown in Table 4.3. Alternatively, a tailor made procurement could be developed amending contract conditions to suit. This is not usually recommended, because of the difficulty of working with conditions which are untested.

The choice of procurement system then is a major issue in meeting the client's requirements and it is important to make sure that they understand the implications of the different approaches.

A table to help proper selection of procurement systems can be found in HM Treasury Procurement Guidance No. 5, Procurement Strategies, which provides a matrix for scoring evaluation criteria for a given project against different types of procurement. This theme is developed in the Achieving Excellence[2] guides. A similar table can also be seen in the CIRIA guide on managing project change.

Chapter 8 will consider procurement from a RM and VM perspective.

Contracts

Standard contract forms can be linked with certain types of procurement, but there are further choices here in the contract conditions. The New Engineering Contract (NEC) contract has been designed to give an alternative approach to the Joint Contract Tribunal (JCT) forms with an emphasis on more collaboration. The design and build contract links best with this form and it makes particular allowance for the contractor and the client to develop close relationships (Case study 4.3), by amending the contract programme (not finish date) regularly to suit the latest information available. It also makes provision for open accounting of changes, where change events are signalled early and by keeping the accounting transparent with agreed prices before changes in scope are agreed. The language of the contract is much more open and is designed to encourage more trust between the parties. This is a contract routinely used by many public clients and has advantages for developing collaboration, providing a single point of contact for the client, although the client may also choose to appoint an executive project manager.

The FIDIC International contract gives a suite of contracts to suit different procurements, but provides greater protection for contractors working internationally. PPC 2000 partnering contract gives a broad based contract to support partnering teams.

Case study 4.3 Contract flexibility

> The client was a Hospital Trust requiring a medical training centre which was to be funded by a commercial mortgage with strict budget targets. This was a medium sized contract and was built on several floors in a tight city centre site within reach of other hospital buildings. It was planned to fund the mortgage repayments by letting out areas of the building to related commercial activities that would pay a market rent. The Trust decided on the use of the NEC contract conditions and let the contract on a design and build basis, novating the outline design architect to the successful design and build contractor. An ideal commercial tenant needed to move in early (who was to provide specialist medical training facilities) and, together with the pressure for space in the hospital, the contractor was asked if they could finish the top floor of the building early to enable this to happen, but not to increase the project budget.
>
> Solution: Because of the transparency of the budget it was possible to work out an affordable solution, which cut down the fitting out of one floor to create early access to the top floor, by commissioning the lift early and fencing it off from the rest of the site. Fire escape access was also approved by doing the same to one of the two fire escape stairs. Some contractor usage of the lift to lower floors was agreed.

Quality planning and satisfaction

Maylor [8] distinguishes between conformance and performance planning. This may be described as the difference between quality assurance (QA) and quality improvement. In practice, this means defining an effective system which is transparently attained by sufficient checks and balances (formal systems are audited) without suffocating the service or the delivery. An effective system is much harder and requires a consideration of the expectations and the perceptions of the customer and other stakeholders. In bespoke construction projects, this should lead to a dedicated system adaptation using the project constraints and client requirements, making the system dynamic.

A non-dynamic project based QA system renders it, at least frustrating and at worst harmful, to the efforts of project staff operating within them. Accountability should be assigned at operational levels for conformity and at project director level for effectiveness. This will be achieved better on larger projects by using a responsibility matrix which identifies all members of the delivery team as well as specialist co-ordinators and makes them responsible for specific tasks, which are co-ordinated so that no job is overlapped or left uncovered.

Quality also needs to be associated with specific customer requirements so that the commissioning and testing of the building is fully connected with this strategy planning stage of ensuring that the finished building is fit for purpose and all systems work. This is not a static compliance with customer specifications, but a proactive process in checking that the specification which is being quality assured fulfils requirements and maybe improves the status quo by increasing the durability or suitability. Too often, buildings are failing because requirements are not interpreted fully. The strategy for defining and controlling the quality is to maintain close relationships with the client and to manage the stakeholders in accordance with their influence and power. This confirms the need to define the brief properly and to develop value.

Figure 4.14 makes clear the relationship of the quality systems with meeting the customers' requirements. It is clear that there are a range of quality outcomes and if perceptions are equal to expectations there will be satisfaction. Maister's [9] law states:

Satisfaction = perception − expectation

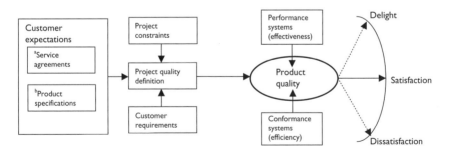

Figure 4.14 Quality planning and delivery.

Source: Adapted from Maylor (2003:167).

Notes
a Service is defined as based on stakeholders expectations and perceptions.
b Product is in conformance with measurable specifications.

If there is a negative difference between perception and expectation then there will be dissatisfaction. If the difference is positive then there will be customer delight. The promises which are made in terms of how and when (service) are also strongly related to satisfaction. A customer may well be happy with a longer contract time or a later start on site if what is delivered is not later than advised. One of the problems with competitive tendering is that it often promises more than what is delivered resulting not only in claims, but also in dissatisfaction. This is a lose–lose or win–lose situation for contractor and client.

Benchmarking is used as a way of maintaining and improving quality. KPIs have been developed within organisations in order to measure, objectively if possible, the level of quality achieved. They are more easily measured where long term partnerships exist and set uniform parameters. More will be developed on this strategy in Chapter 5. Clients like the use of external measures such as European Foundation Quality Model (EFQM), third party accreditation as in ISO 9002 and the National Housing Building Council (NHBC) buildmark. They relate to these more easily and the industry needs to consider the use of these (see Chapter 6).

Information and communications management

Information management supports transactions and decision making. Decisions may be programmed or non-routine and may be required at strategic, tactical and operational levels of management. Information is specific, but starts with processing raw data into a management information database which supports common decisions. Information systems break down into transaction processing, decision support systems (DSS), executive information systems (EIS) or ad hoc enquiries. Figure 4.15 shows the hierarchy of these systems in relation to the database and the level of

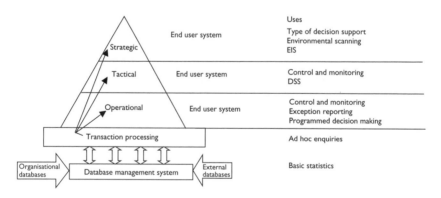

Figure 4.15 Hierarchy of MIS in relationship to the database management system.

Source: Adapted from Lucey (1991) Management information Systems. DP Publications. p. 264.

management use and the context of the decision in proportion to the level of management use.

A database is a collection of structured data. The structure of the data is independent of any particular application [10]. A database management system may draw on organisation and/or external databases. It integrates the different sources and structures data in a format for the outputs which are required in the organisation. Information needs to be timely and appropriate for use in either transaction processing or decision support. A transaction process is routine such as the financial accounting process to record payments and receipts and to present final accounts. Management accounting on the other hand needs budget information to make strategic decisions about investments and project selection.

Decision support systems are in place to support rather than make decisions and therefore need to be more specific to the end user requirements so that they remain helpful and used. An example of a DSS relevant to projects would be an update showing the progress critical activities. This would pinpoint where resources are needed most critically to pull the project back onto target.

Certain questions are relevant in the design of DSS such as: What type of decision? Is it for long term or short term? How long have you got to make the decision? How is it integrated with other decision makers? These questions can be applied for the process of setting up project management information systems (PMIS) which are unique to the project type.

PMIS need transaction analysis like progress reports on programme and budget updates, and decision support such as those supporting RM and VM. Knowledge based systems (KBS) may be used to support estimating and integrated systems may be used to connect computer aided drawing (CAD), costing and planning systems, though these are very much in their infancy. Integrated information in its most advanced form is the subject of much research and object orientated models which define generic elements which appear in design, planning, estimating and financial systems are still a long way off.

In the case of projects, DSS is a key requirement and because of the temporary nature of projects needs to feedback quickly or preferably feed forward to be effective for decision making. In the PMIS the databases will include cost, time and performance data as well as many other databases to serve the transactional processing such as accounting, inventory, buying, estimating, health and safety etc. Table 4.4 shows a simplistic structure for a PMIS.

Information may be hard or soft. Hard properties are those that can be defined, measured or assessed in a tangible way, for example, the cost and production outputs of various expansion options against the market forecast. This may include the attachment of various levels of uncertainty.

Table 4.4 PMIS Matrix

Project databases	Functional data base			
	Cost	Time	Quality	Health and safety
Project 1	price/bed			accidents/week
Project 2		m²/h		
Project 3			±3 mm	

Soft properties are more imprecise and may depend on individual taste and value. For example, for a company moving to a new HQ out of the central business district, it may be necessary to consider the incentives that will be required to move staff away from city centre facilities. The PMIS should adjust outputs to match the client's business case parameters. For example, the client might want the price per bed rather than the price per square metre.

Socio-technical systems are a way of combining the technical hard data of production outputs with the soft data, which include the expectations, aspirations and value systems of the organisation's workforce. A knowledge of both constitutes a more holistic decision making process.

Expert systems sometimes known as KBS are computer based systems designed for making more complex decisions more accessible to non-experts. They are built up by rationalising the steps made by an expert to take the non-expert through the decision structure based on the answers given. They can build in what if scenarios as well to provide instant sensitivity analysis. Solutions are derived by its link with a database, for example, estimating price book. A price book on disc is not a KBS, nor is a computational system such as software that generates a bill of quantities. The user is not surrendering their judgement, but is using the system to enhance judgement and improve decision making. Case study 4.4 shows the use of a KBS.

EIS provide selected summary information for director level. In the project context, these need to be standardised across projects for comparison. Key features should be easy to use, give rapid access to the data that feeds into the report, provide data analysis – trend and ratio calculations and a quality presentation – interesting and understandable, for example, use of graphs and colour.

The project manager also needs to provide data to the client in a format suitable for them to monitor their business plan. For example, if a change has been requested in the scope of works then the effect of this change could be worked out by the client in terms relevant to their own EIS, prior to approval. Issues such as client commitment, cash flow market considerations and production targets may or may not override extra cost.

Case study 4.4 Elsie

The Elsie expert system interacts with the user to determine project cost. It finds out the geographical and relational location of the project, the floor area, the quality of the finished building, the degree of external works, the nature of the building use, the type of design and the structural form and the programme time available if critical.

It then assesses an updated database of costs such as the Building Cost Information Service (BCIS) in order to give a range of current construction prices. It automatically adds statutory and professional fees and allows a land price to be added to give total project estimate for that location, building type and size etc. A report is generated outlining constraints, elemental cost breakdown and uncertainties in particular circumstances.

It is not an econometric model so depends on the significance of the current and updated projects used in the database accessed for its accuracy. A further sensitivity analysis might be necessary for the effect of interest rate changes, inflation, local factors and non-standard specifications.

It does not take direct account of market supply and demand which might affect the competitiveness of tender prices or the procurement system used. The tenders may also reflect the soft factors such as maintaining client bases, partnership agreements, client preferences and prestige.

Grant regimes and tax breaks may be fed into the Elsie database if they are locational, but not if they are negotiated.

Conclusion

Project strategy is critically considered at the feasibility stage of the life cycle as well as its ongoing development and implementation throughout the implementation stages. Typically there is a lot of uncertainty even after feasibility studies are complete and the overlapping strategy stage identifies, prioritises and allocates risk where it can be best managed. Further discussion on evaluating and responding to risk to optimise value systems is dealt with in Chapter 8.

The master plan typically includes a cost plan, a quality plan and a risk assessment which provide frameworks to optimise the client objectives and recognises the project constraints.

Cost planning needs to take into account an on-going value exercise to suit the iterative nature of the design and the cycle of developing it to its final approved post planning application stage. The initial outline cost plan budget becomes an important marker for the client and market changes, the design changes and the client changes are often not seen as reasons for an inflated budget which means that value has to be generated and design changes made to suit.

A strategy for managing the programme time at this strategic level of detail is more dependent on reaching critical milestones than it is in calculating durations and the planning needs to be iterative and flexible to allow for this. A strategic view at this stage involves being aware of the linkages between the design, the construction, the procurement, the commissioning and the client fit out stages. In a climate of 'time is money':

- A fast track strategy would allow a certain amount of overlap between these stages dependent on the nature of the work with impacts on the level of risk. The design management needs to be very tight to reach a series of critical package milestones. Commitment to early packages orders will mean early involvement of a construction manager.
- A fast build strategy will focus on the need to complete each of these stages sequentially and squeezing the timing of each of the stages so that there is a proper handover between them. This has more risk in resource management as the critical path of the project becomes more critical.
- A procurement strategy sets the constraints for the delivery sequence of the work.
- A commissioning programme formalises the sequence for the end stages of testing and client fitting out and occupation. The beginning of commissioning is the end of construction for each holistic phase and so commissioning needs to be considered first. Quality depends on proper testing and a holistic approach to the whole building in order to hand over with zero serious defects.

The quality planning process is a strategy for meeting client's requirements and expectations and should address service quality as well as product quality if perceptions are not to fall short of expectations. This is best achieved by close relationships with the client through the continued VM process. The quality improvement process has a place both in the formal post project evaluation (final chapter), but also in the development of a project organisational culture, where lessons are continuously being learnt and simultaneously being applied. The basis for this is covered in Chapters 6 and 12. Benchmarking is covered in Chapter 5.

A project information system needs to be integrated across the supply team and keep the client informed. In a format which integrates with their own reporting and monitoring systems for asset management and the business. One of the developing issues is the compatibility of CAD and the document distribution systems so that information can be received electronically and simultaneously in order to integrate say design and construction approaches. This helps to change the culture of partial information or information systems that are separate for the project team and client reporting. The sharing of project accounts transparently with the client is becoming an important part of the partnering culture.

The setting up of structures which promote teamwork and integration of the project packages is critical for good communications, responsible health and safety and for creating if possible, a blame free culture. An enlightened strategy here would be the bringing together of the design and construction at an earlier stage. Chapter 6 on organisation and culture and Chapter 7 on engineering the psycho-productive environment take this further.

A project handbook may be useful in order to integrate the various responsibilities and duties. A good example of a handbook framework is given in CIOB Code of Practice.

Notes

1 The APM Body of Knowledge links the project master plan with the key success criteria and suggests that KPIs should be established in order to monitor the plan and evaluate its success.
2 OGC Achieving Excellence in Procurement. www.ogc.gov.uk. These guides provide comprehensive information on a variety of associated topics also.

References

1 Burke R. (1999) *Project Management: Planning and Control Techniques*. 3rd Edition. John Wiley & Sons, New York. p. 32.
2 Project Management Institute (2000) *A Guide to the Project Management Body of Knowledge*. Project Management Standards Committee, Newtown, PA.
3 International Standards Organisation (2003) 10006. *Quality Management*. Guidelines to Quality in Project Management. The ISO Standards Bookshop.
4 Ballard G. and Howell G.A. (1997) *Improving the Reliability of Planning: Understanding the Last Planner Technique*. www.ce.berkeley.edu/~tommelein/LastPlanner.html (accessed 12/11/2004).
5 Smith N.J. (ed.) (2002) *Engineering Project Management*. 2nd Edition. Blackwell Science, Oxford.
6 Edwards L. (1995) *Practical Risk Management for the Construction Industry*. Thomas Telford Publications, London.
7 Reiss G. (1997) *Project Management Demystified*. 2nd Edition. E & FN Spon, London.
8 Maylor H. (2003) *Project Management*. 4th Edition. Pearson Education, Edinburgh.
9 Maister D.H. (1993) *Managing the Professional Service Firm*. Free Press, New York.
10 British Computer Society, Kingston, UK.

Chapter 5

Construction performance and planning

The construction of the project is the most interesting phase when work actually starts on site and some highly visible progress is made in one of the last major stages of the project life cycle. Construction however often starts too soon because there is an over-eagerness to proceed with the demonstration of progress on the project and to make reassuring promises about the programme completion. It is vital though to have properly completed the planning stages so that materials are ordered in time to meet the date when they are needed on site, sufficient labour is available to meet the resource levels planned for and plant and equipment is available at the right price and specification when we want it. How many of us have been kept waiting for that new house when we reserved the plot at foundations level and the brickwork and the roof seemed to steam ahead, but it was an awful long time between seeing the roof and getting that call to arrange to move in? Construction companies, to remain competitive, need to benchmark their performance and to introduce a continuous improvement programme that keeps them abreast of the expectations of their clients following recent reports in the industry. The chapter concentrates on the performance of construction and the potential for improvements. Case studies will be used for best practice.

The chapter objectives are:

- Reviewing the performance of the UK construction in general and its strengths and weaknesses.
- Understanding the principles and the potential of benchmarking measurement techniques.
- Assessing construction performance, using case studies.
- Applying the principles of lean construction.
- Appendix – the critical path method (CPM) and its use.

The performance of individual contractors

Construction is under pressure today to 'improve its act', but there is a great variation in performance on construction projects. Many clients have had good experiences, but equally many have been let down on promises that they believed to be binding and have seen extended programmes, suffered cost overruns and have experienced quality problems. Projects such as the Channel Tunnel or the Millennium Dome and the Scottish Parliament have become headline news, seldom for their feats of ground breaking engineering and architectural innovation, but for going over budget and programme. Hong Kong International Airport and Petronas Towers in Kuala Lumpur are other international projects, which have failed to get more than a passing mention in spite of their great success. Looking at the credentials of one of these (Case study 5.1).

Case study 5.1 Great structures

The 'Chunnel' is a joint Anglo-French venture first muted in Napoleonic times and is, at 31 miles (50 km), one of the longest tunnels in the world joining Calais in France and Folkstone in England under the sea. It was built in three years at a cost of $21b, which is 700 times more expensive than the Golden Gate Bridge in San Francisco. It uses tunnel boring machines two football pitches in length and could bore 250 ft × 7.6 m diameter in one day, hundreds of feet below the seabed. It is in fact three tunnels with an East and West bound tunnel for a high speed rail link taking trains at 100 miles per hour; a central tunnel acts as an escape route and service tunnel and there are cross over tunnels in between. It opened in 1994 and in the first five years, in spite of some operating problems, it took 28m passengers and £12m tonnes of freight. Thirty-one passengers caught in a train fire were all safely evacuated through the central tunnel.

In the UK, the Latham Report (1994) succeeded in providing a snapshot of industry performance that fell short of the productivity achieved in other industries and set up working parties under the Construction Industry Board (CIB) to see through specific process reforms that would help the industry to improve its act. These improvements were fundamental like outlawing unfair contractual practices and encouraging a more open partnering approach. It also put in place a regime for better skills and management training and encouraging the registration of consultants and contractors so that clients had better information for selection. The industry was also challenged to meet tough improvement targets in a five-year period and to improve its image.

The Egan Report (1998) follows on from Latham and suggests that good construction is very good, but many clients have a poor view of construction performance because of failures in the system. This shows up in the fact that over half of all contracts either finish late, or over budget or are of inferior quality. The Egan challenge was to reduce construction prices by 10% year on year by improvements in the process and to raise the profit margins of contractors at the same time, as failures in competitive process were diagnosed as driving waste into the system and lowering the motivation of contractors who were in sectors outside housing and development and working on very low margins.

The structure of the industry is fragmented between design and production, which has often meant that the client has been confronted with separate design and production teams and sometimes a plethora of planning advisers. It has inhibited an integrated approach to delivering value to the client and prevented an openness and honesty, because of a lack of trust between the separate teams, a limited accessibility to the client and an extended communication channels. The structure has often led to a confrontational and a contractual approach.

Case study 5.2 is a construction best practice profile covering the period over which these two reports have been current and indicates the take up of issues that have become important to competitiveness.

Case study 5.2 Benchmarking

The company has a £500m turnover and 40 offices throughout the UK and employs 3000 people which puts it into the large category for contractors. They have adopted the maxim 'if you look within the sector you are always going to follow. You need to look outside to lead'. This has caused them to make leading edge improvements by looking to solutions which have been adopted by firms in other industries who have made substantial improvements.

Over a period of 12 years they have steadily looked at a series of initiatives that have built upon each other and led to an integrated process, such as prompt project completion commitments, process definition and re-engineering, building an understanding of client requirements and the development of customer service units, standardisation and lean thinking.

With the publication of the Latham and Egan Reports they revisited processes and introduced a programme of lean construction to further remove wasteful activities from the whole value chain. This led to an integrated management system combining quality, health and safety and environmental management together and individual rewards in meeting Egan targets and they use KPIs to demonstrate this.

They believe the key to continuous improvement towards integrated construction is:

- Top down commitment – the direct involvement of a director in driving each of the improvement agendas;
- Bottom up workshops and team building – the involvement of all in the simple on the job improvements;
- Measurement – use of KPIs to demonstrate where improvement is taking place. They measure:
  - likelihood of repeat business
  - user friendliness of the business
  - environmental awareness
  - level of defects.

Critically, the company shows a financial improvement in its results in the period 1995–2000 having fast growing profit margins, a four fold increase in turnover with equal earnings per share increase and 85% repeat business. They also have a 5 gold medal health and safety awards and a registration for integrated assessment by the BSI for quality, safety, health and the environment management systems.

(Adapted from the Construction Best Practice Profile [1])

Research by the Construction Industry Training Board (CITB) [2] indicates that construction managers do not have good skills for managing stakeholders which they define as the process of managing customers, employees and suppliers. The importance of realising maximum value to clients, end users and other stakeholders and to exceed client expectations is also a theme for the Egan Report (2002), appearing in their vision statement.

Performance benchmarking

Benchmarking is a method of improving performance in a systematic and logical way by measuring and comparing your performance against others, and then using lessons learnt from the best to make targeted improvements. It involves answering the questions:

'Who performs better?'
'Why are they better?'
'What actions do we need to take in order to improve our performance?'

The idea of benchmarking is to compare the performance with another of a known standard. A benchmark (or a standard) in business is like Olympic qualifying standards – unless you meet a certain minimum, you are unlikely to be a competitor. The standard is also likely to move upwards in a competitive market.

These benchmarks may be based on industry, national or global standards. They may also be compared between organisations in the same sector, the same industry or against different industries. The choice depends on the competition which you have. Measuring performance in construction may be based on:

- internally set targets which are often based on continuous improvement between projects or on a time line or issued as a target for best practice in the business;
- against another in the construction industry who are carrying out similar activities, who are the best in class;
- financial performance, for example, by comparing the profitability or turnover growth ratios against others;
- generic targets across industries which represent excellent management achievements, such as the levels of training, business results and customer satisfaction.

The best in class term refers to a real performance in a company's own defined market place. Measured KPIs may be set against average industry performances, but to be competitive more than average is required. Being in the top 10% of companies is a requirement for the best in class.

Continuous improvement is the concern to keep up with the best and to offer a better service to the client in the belief that it produces returns. It involves developing a culture of innovation, a target setting, a continuous monitoring of standards and a reviewed performance. As expected, it also recognises the role of the client and the construction/design team working more closely together to produce improvements and adding value. To be successful, benchmarking should be focused and a senior manager needs to select critical areas for improvement which will impact on the project as a whole. It also needs challenging goals that are achie able, a willingness to change and a persistence in getting there. It is not always going to be possible to improve over one project.

Benchmarks can be applied to various business parameters such as levels of profitability, respect for people and the environmental impact of the company, or to project parameters such as or customer satisfaction predictability of time and cost plan forecasts, number of defects, accidents recorded. It is continuous improvement that we are interested in here and systems have been set in many national construction industries.

Benchmarking standards for the UK construction industry

Egan issued a challenge to the UK construction industry by setting up some improvement targets that are effectively benchmarks. These are:

- Decrease defects by 20% per year.
- Reduce accidents by 10% per year.
- Reduce time periods by 10% per year.
- Increase VFM (client gets more for their money) by 10% per year.
- Increase contractor productivity (turnover per head) by 20% per year.
- Increase contractor turnover and profits by 10% per year.

Larger construction companies have tried to give a lead, but a *measurement* system is required to validate the claims of those who say they are giving better value for money. To be successful VFM must be sustainable and this means a win–win situation in which contractors profits are increased and a client gets more for their money. Innovation and process improvement then is an important part of the objective.

In the UK, a benchmarking facility is available based on nationally prepared profiles of performance for the construction industry as a whole and for particular sectors, such as housing and refurbishment. These are published as graphs annually. Constructing Excellence co-ordinates a self help service for the measurement of KPIs relevant to the Egan targets and the KPIzone [3] provides a calculator to compare average construction performance in key areas with your own. The M4i runs a complementary service for benchmarking by collecting KPIs from demonstration projects and publishing results. In conjunction with present selection criteria KPIs may be useful predictors for the client and critical marketing tools for the contractor.

Effectiveness is related to meeting set objectives. It is important that there is a wide and measurable range of data for industry comparisons. There are three areas of concern with the use of KPIs.

- There may not be a direct relationship of indicators with performance.
- The past performance may not be repeated for all sorts of reasons on future projects. For this reason, they must always be interpreted in the present context and not as a magic formula. A standard comparable format is needed to make them useful and give incentives to use them, but where suppliers wish to use their own, this is much harder to certify.
- There may be disagreement about which indicators show excellence. Many league tables produce distortion in comparison where companies have different strengths or weaknesses.

The initial 10 construction performance KPIs averaged for all sectors of the industry are measured each year. They are also available for different sectors of the industry such as housing, new build, refurbishment and building services. More generic sets are available for the environment, health and safety and people. They are a different set which apply to construction consultants.

Table 5.1 shows the median (exactly half achieved this result or better) results of the scores over the four years between the two Egan Reports for a survey of construction companies. These KPIs have also been measured for the so-called demonstration projects and show a major improvement on the standard average for construction.

Crane [4] outlined a comparison report on the demonstration projects that were set up after the Egan Report. These are year on year 2000–3:

- Client satisfaction with product: 81–90%
- Client satisfaction with service: 76–86%
- Accidents on site dropped: 716–428/10 000
- Construction time improvements: 10% saving
- Profitability ahead: 5.6–7.0% (25% better).

Table 5.1 The construction all industry KPIs comparison between Egan Reports

Parameter	Median		Demonstration 2003
	1998	2003	
1 Client satisfaction – product (questionnaire 1–10)	8	7.9	9
2 Client satisfaction – service (questionnaire 1–10)	8	7.5	8.6
3 No. of defects significant faults at completion[a]	3	7.2[a]	
4 % Change in predicted budget			
– design (start to finish)	0	0	
– construction (start to finish)	1	0	
5 % Change in predicted time			
– design period	+18	0	10
– construction period	+9	+2	
6 % Profitability (pre-tax margin)	+3.2	5	7
7 Productivity (£ turnover/person)	30 000	37 000	
8 Accidents (No./100 000 employees)	997	1500	
9 Construction cost (%) (cf. similar last year)	−3	+2.5	
10 Construction time (%) (year on year)	−1	+1.2	

Note
a Roughly equivalent as previous 1–4, now 1–10.

To calculate construction cost change year on year

Table 5.2 Case study particulars (case study of two offices)

	Office A (3rd quarter last year)	Office B (3rd quarter this year)
Raw cost office	£10m	£8m
Size of building	10 000 m²	7500 m²
Inflation index [BCIS 1983 = 100]	447	478
Quality	100	105 (5% greater)
Location	Outer London 1.03	SW England 0.99

The following example shows a hypothetical case study for the headline benchmark for the year on year underlying cost saving. It uses data from two similar types of project built 12 months apart and making adjustments to the variable factors of cost inflation, location, size of facility (in square metres) of floor space and quality level. Agencies such as BCIS are able to publish cost inflation, locational variations in cost and ranges of quality prices to help in arriving at the adjustments. In this case, an office in different locations of different sizes and underlying quality has been used. Adjusted cost is the comparator. Data is shown in Table 5.2 to calculate construction cost change year on year. The objective is to measure the change in real construction cost, over 12 months, for a particular contractor.

Step 1 Calculate the normalised cost of the two offices.
Note, you should normalise both buildings for size (per m²) and location (equal to 1.0). You only normalise office B for inflation and quality difference as shown in Table 5.3.

Table 5.3 Case study calculation for adjusted price

Adjust both for size	£10m/10 000 m² = £1000/m²	£8m/7500 m² = £1066.67/m²
Adjust both for *location* of building	Outer London	Bristol (South West)
Outer London 1.03 SW England 0.99	1000/1.03 = £970/m²	£1066.67/0.99 = £1075.98/m²
Adjust for *quality* of building Building B 5% more quality	High = £970/m²	High+5% [normalise by × 100/105] £1076 × (100/105) = £1024.76/m²
Adjust for *inflation* Building A index = 447 Building B index = 478	£970/m²	£1024.76 × (447/478) = £958.30/m²
Normalised cost	£970/m²	= £958/m²

Step 2 Calculate the construction cost indicator by comparing the normalised cost of the two buildings.

$$\text{The construction cost indicator} = \frac{B - A}{A}$$

$$\therefore \text{Construction cost indicator} = \frac{958 - 970}{970} \times 100\%$$

$$= -1.2\% \quad \text{i.e. better performance}$$

This means that this contractor is becoming more efficient, but by only 1.2% which is less than the Egan target of 10%.

The above example indicates the calculation of just one of the KPIs – the efficiency of cost and a similar calculation can be done with time. Care in interpretation is required, as the normalisation exercise does not cope with all the physical, or market differences, between two projects. To overcome different ground conditions the costs for comparison are often compared above ground. However, the role of project management is to reduce extraneous costs by perhaps reducing access problems, orientating the building to minimise shading/heating costs and negotiating efficient planning authority – developer agreements. Market conditions may also be different, creating competitive tender prices. Comparisons

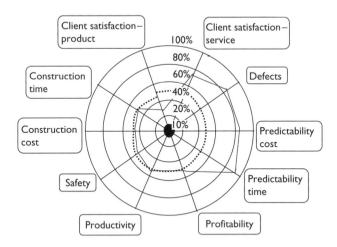

Figure 5.1 A spider diagram of the 10 all industry KPIs for a contractor.

Source: Adapted from the Constructing Excellence, KPIzone.

Note
Each radial arm represents a benchmark. The radial zones represent the benchmark score. Those that achieve <50% (below dotted line) are below average and those >50% are above.

are therefore more reliable where two buildings for a similar client are compared.

The profile given in Figure 5.1 indicates good scores in defects and predictability and poor scores in client satisfaction for the product and improvements in cost and quality. Since the defects are low, it is more likely that the briefing process has not clearly provided an effective building. If this had occurred then it might also have dragged down the client satisfaction with service of the design party to the contract. Thus client satisfaction surveys may be influenced by a halo effect or the opposite.

A scorecard might have more of a weighting on client satisfaction and defects and therefore provide a much better score for this particular contractor.

Construction performance

This section uses Case studies 5.3, 5.4, 5.5, 5.6 and 5.7 to give illustrative examples of performance in action. They are real examples of good practice in the construction stages of the project life cycle. They illustrate areas of logistics, prefabrication, quality control, programming and package management.

Case study 5.3 Logistical planning on construction management fit out

The site is a seven floor section of a 12 storey office building with double basement and plant room adjustments, for a prestige client. At the height of the work 600 people were working on the site on 30 separate packages including a logistics package. The packages were let and managed by a construction manager but each package contractor had a direct contract with the client. A separate fee was payable to the construction manager. Cluster co-ordination was done by a lead specialist contractor for each zone of work such as the ceiling, floor zones and the kitchen and canteen area.

The construction manager develops programmes for each floor broken down into quadrants as a way of co-ordinating the zones and monitoring progress of the work. These programmes become the basis for the trade contractor co-ordination. For communication a trades dated finishing programme is established in order to monitor completion of individual parts of the programme. This provides a latest finishing date for each of the trades in a particular sector. The programme is tight with most of the trades running on two floors and overlapping the previous trade on each floor. A 25 week turn around for each of the follow on trades in a 38 week programme was planned.

In order to control quality, a mock up of the office finishes was carried out as a package at the start of the contract. One problem area for the quality has been the drywall finishes. In some cases, these have been redone three times to meet a demanding specification and this has caused problems to delay following trades.

This project is an example of a highly intensive work programme making any hold ups affect the critical path. The hands off nature of the construction management function has only been possible with the setting up of well prepared packages and a well administered change control system. As most decisions for significant design change have to be made in New York where the client representative is, this has the potential to delay the site team if the problems are not anticipated in good time or client changes are frequent. However, the collocation of an integrated design and construction management team on site has helped in getting design decisions implemented immediately. Giving ownership to the trade contractors by making day to day co-ordination their responsibility has provided space for more advanced planning and contingency management. It is probable that the pure construction management created by letting out all work including logistics has allowed more objective strategic decisions to be made by the construction/project manager, but this may have cost the client more up front.

The case study on logistics indicates the close association that there is between the design and construction stages in deciding the most advantageous methodology in running the project. The construction stage tests the effectiveness of the strategy and pre-construction design and planning stages. The project manager is faced with their most complex organisational challenges as the supply chain opens out to a wide range of specialist contractors and interacts with different parts of the design team in the detailed design. The speed and complexity of construction means that production targets need to be clearly stated and closely monitored. Feed forward control is more effective than feedback given by benchmarking measurements.

Case study 5.4 Prefabricated construction

The superior product team has been set up as a consortium between contractors, designers and client to provide completed corridor units which fit inside a structural steel frame. Each prefabricated unit weighs 6.8 tonnes is 4 m long and 7 m wide (Figure 5.2).

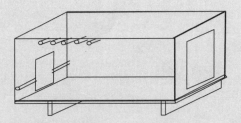

Figure 5.2 Diagrammatic view of a corridor unit.

The unit can be double stacked or divided into two narrower corridors of 3.5 m wide. It is externally clad and glazed to provide a finished waterproof product bolted within the on site in situ steel frame. The units are complete with modular plug and fix heating, electrical and IT services. The steel frame is stand alone and is configured to the route of the corridor which generally links up between two buildings. In order to install the units within the frame, rails are fixed to the frame at the height of the floor of the corridor and units are lifted into the frame by leaving out a section of the top frame structure. Units are lowered onto the rails and pushed along them until they meet up with the last corridor unit and are bolted together, sealed and services joined. This action is repeated for each corridor unit in both directions up to the access holes in the steel frame. Tolerance is critical allowing the steel to be slightly out of true longitudinally, but requiring high tolerances on the rail level and the plumb of the structural columns, to which the units are bolted top and bottom. Once a corridor is completed between the two buildings, the services can be hooked up live and the services tested and commissioned.

The production line has been set up in a factory close to site and units are stored for at least the capacity of a night's work. Night time is usually when free access is guaranteed to the building works and is least disruptive to the client's business. A single unit size has been designed to fit on the back of a rigid bed lorry and is similar in size to a large container that is open sided, so needs to be braced for safe transport. A unit is lifted straight off the lorries and into the in situ structural steel frame (Figure 5.3). A multi-skilled production gang receives the unit, manhandles the unit along the rails and installs it. Specialist 'Hiab' equipment is used to load and unload the units, saving the constant hiring of cranage. The fixing rate varied from 2 to 8 units per night shift as a learning curve took place. Approximately two modules were produced per day in the factory.

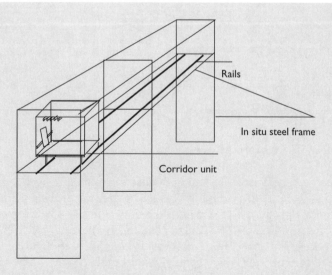

Rails

In situ steel frame

Corridor unit

Figure 5.3 Steel framework with unit within.

The installation programme was quite flexible, because there was a need to cope with delays for installation to suit on site progress and client operations. The critical areas of management are caused by the limited storage capability and the need to deliver components to the factory in job lots to suit a flexible delivery schedule. This meant that forward planning was required. This was often achieved by negotiation of extra site visits if stored units were mounting up. To maintain quality and motivation, the factory teams and installation teams were interchanged.

Design changes resulted from a better integration of manufacture and installation. Quality improvement led to the re-engineering of brackets in order to make fixing easier and cope with lesser tolerances on site. The factory management stop work on an ad hoc basis when feedback is given in order to give the whole workforce 'one point lessons'. This group learning process ensured equal delivery to all on a check, prevent, repair basis and saved the need for quality inspectors. Another problem was the deflection of the structural steel with the fixing of units in the middle of spans, creating movement after two corridor units was complete and stressing the sealing between them. Both the design of the sealing and the rigidity of the structure were reviewed. Adjustments to the painting methodology were also made to ensure consistency of finish. The manufacturing unit is now in the process of extending its operations to other sites needing the same units and is developing its transport fleet to cope with a longer turn around.

This case study illustrates the way in which a flexible installation programme has been achieved by the coming together of the client, the contractors and the designers to design a particular standardised component (corridor units) to give a fast build scenario to meet a continuing client requirement for limited site access. This could also be applied productively to any type of repeat unit needed in a project, or preferably a series of projects. Another example of this is the new Severn Bridge crossing where heavy deck units were standardised and manufactured on the bank in order to crane directly onto barges for fixing. In addition, there is a culture of continuous improvement for the factory production and this has included motivational as well as technical considerations.

Detailed pre-planning drawing together designer, main contractor and manufacturer as is evidenced in the case study on prefabrication provides a measure of confidence which works well with the very limited access that is afforded on site and the speedy installation that takes place once the units arrive. Case study 5.5 looks at quality.

Case study 5.5 Quality control of piling

This site consisted of a piling mat of 1523 12 m and 15 m piles to be placed in a series of concentric circles across the foot plate of the building. Continuous flight auger piles were used. The tolerance for the piles was critical, as the scientific use of the building required high floor level tolerance and foundation movements were critical to this. High loadings were expected. An inspection and test plan (ITP) was devised and supervised by the managing contractor. First line responsibility for quality and setting out was the responsibility of the specialist contractor. The managing contractor acted autonomously on small discrepancies, but any discrepancies discovered in the piling outside a band of tolerance required structural engineer involvement and possible redesign of the piling pattern to compensate. To ensure the quality of the piles several checks were made:

- A testing programme for the integrity and movement of the piles.
- A testing programme for the strength and slump of the concrete.
- A record of the piling operations indicating the pile concrete profile and wet concrete pressures.
- An interpretation of complex clay and chalk ground conditions to predict different piling methodologies and concrete slumps.

Each check gave clues as to the success of the pile and it was the last two measures which might be partly preventative and reduce the cost of remedial action by the more immediate action that was possible. All data and test results identified the pile.

Concrete could be rejected if it was not within certain slump limits and/or cube tests under strength. The slump needed to be higher if cast into chalk which sucked out the water from the concrete making it stiffen quicker and cause problems for the lowering of the reinforcement cage into the concrete. The seven day results were tied to other data about the pile from the computerised piling log and the 28 day strength predicted. A bad result or prediction would lead to re-boring the pile.

Proof testing was carried out on 5% of piles in situ, by the use of 23 h load tests to check the friction resistance. Cheaper integrity testing was carried out on all cured piles using non-destructive testing and measuring toe seat deflection. Any sub-standard piles found would be supplemented with further adjacent piles.

This regime needed a full time quality checker on site on behalf of the main contractor. The cost of this was offset by also giving them a planning and monitoring role to predict the rate of piling progress and to co-ordinate safe, uninterrupted access for three piling rigs, their associated equipment and to ensure access for concrete lorries. The checker also carried out supervisory roles for ensuring the progress of other works and progress with the master programme. All work was completed in 11 weeks at an average rate of 30 piles/day and a maximum of 48 piles/day.

It could be argued that in an enlightened role, the accountability for non-compliance could have been put upon the piling contractor and there would be no need for inspection. However, the huge delay and financial consequences of the foundation works being found faulty at the user stage justified the extra cost of a single well trained checker with adequate management backup.

This case study indicates the intricate nature of quality control and the need to supervise it arises out of the tight tolerances required by the client, but value has been built into the process by using the supervisor to also co-ordinate other work. This is an example of feedback control that is used effectively.

The first three reports indicate the interaction of quality and time constraints in particular. Best practice has been around reducing cost as well at a sustainable profit level for the contractor, which means considering the reduction of project waste, not just in materials, but in the process. Rework and the duplication of management tasks is a waste of resources. This points to better pre-planning and better communications. Case studies 5.6 and 5.7 look at the systems which have been used to control the programming and procurement to give planning flexibility and to add value.

The project is a specialist facility for physical research. The contract is worth £80m. The building is circular with many specialist packages and high tolerance requirements. The programme time is 55 weeks for the main building though there are other phases including an office block and enabling works. It involves the use of many specialist package contractors.

In the main project a master programme gives overall coverage of the 55 week construction. It is based on the production needs of the seven key packages that provide 80% of the work. This amounts to 100+ items. There are also phased handover dates for each of the sectors. The contractor monitors this programme every week and provides a monthly report to the client on each item and the overall progress. The master programme is linked in with the procurement programme in order to determine when packages need to be let. Table 5.4 illustrates the hierarchy of the programming.

In order to break down the detail on relevant packages in progress a 12 week rolling programme is produced. This is adjusted to the current progress and also reviewed every six weeks. Tactically, it aims to retrieve any programme slippage from the previous 'stage' programme. These are also monitored and presented to the project manager for strategic assessment of the reasons for any problems.

A two weekly programme is also produced and agreed with specialist package contractors to show the immediate work schedules and their co-ordination with other contractors. This provides an awareness of the interfaces and critical path of other contractors as well as an opportunity for *feed forward* with immediate risk and contingency management. The meeting provides an opportunity to discuss any problems and slippage to programme and to work out what impact they have and how the programme may be retrieved.

Table 5.4 Hierarchy of programme control

Programme type	Programme output
Master programme	Overall report to client. Contract start and finish
12 weekly programme	Specialist contractor control and tactical adjustments
6-week stage review	Control and management action
2 weekly specialist co-ordination programme	Co-ordination and work schedules

This hierarchy of control in the table recognises the dynamic nature of programming in order to retain confidence in the programme as a control mechanism for bringing work back onto programme where there is a slippage. The baseline programme is retained for end and not intermediary control. The short-term programmes remain realistic statements of agreed methodologies and target dates using the latest state of progress and can, therefore, benefit short-term control and feed forward systems. Case study 5.7 looks at package management.

Case study 5.7 Package management

On the same contract, the contractor is responsible for the procurement of specialist subcontractors and their co-ordination and there is a separate contract with lead designers:

- enabling works which include site preparation, roads, external works and piling;
- the building works which have been divided out into the office block and the physical research facility.

The main contractor tendered competitively on the basis of drawings and specification for both the enabling works consisting of site works and piling which they immediately started. The building works were tendered later and they were also selected. They are responsible for managing the procurement of specialist subcontractors and also have an input into the buildability of the design. They have direct contact with the designers and the client representative. The client is responsible for equipping and testing the facility.

The main specialist packages are piling, ground works, steelwork, concrete superstructure, roof, curtain walling, mechanical and electrical, raised floors and ceilings/partitions and a procurement programme is used to determine and control the information flow for letting out packages.

The main documentation is sent out on a competitive lump sum basis to approved suppliers on a competitive basis. They receive drawings and a specification prepared by the design team, a scope of works and a programme of works with about 15–20 items, prepared by the contractor procurement team. The contractor team raised 170 queries from the tender drawings before they were finalised. The contractor comments on the design drawings before sending them out from the point of view of buildability. The specialist contractors provide a schedule of rates and may wish to offer alternative specifications or brands. In the event of tender queries, decisions are distributed to all tenderers.

In the case of the curtain walling, the procurement manager went out to five contractors on their approved list and short listed two from the tenders received on the basis of selection criteria including price, but not exclusively based on price. In the event, these two were not the lowest price because of the specialist nature of the work. This led to further negotiations with the main contractor design manager that allowed a best value final price to be proposed and presented for scrutiny by the project management and approval by the client. The final signing off of the contract was the responsibility of the principal contractor at project level.

In this case study, the process employed for key subcontractors, was more proactive and provided

- a basis for transparency for the client;
- specialist contractor involvement in the detail design to use their expertise;
- more ownership to the supplier with ensuing commitment to quality and time constraints.

Where savings were identified they were passed on to the client on the basis that there was no design liability for the main contractor. Client extras might also be identified at an earlier more cost effective stage. As a general principle handling, access and storage were included in the specialist package, to minimise general site overheads and preliminaries.

Construction programming and control

It is often perceived that a programme carried out at the beginning of the project, with limited information, will still be able to predict accurately each detailed activity timing. It is also expected that there is only one logic for the assembly of a construction project. Neither of these is true and a progressive system like that described here allows more respect for the planning process and allows response to current data.

The *overall* programme should be used as a framework with the critical start and finish times of the projects and its major sub sections. It sets out subcontractor start and finish dates for the procurement of major suppliers and long lead in material deliveries and a basis for reporting to the client. This is the baseline programme and is often prepared using the CPM which is in the appendix to this chapter.

The *procurement* programme is a co-ordinated programme with the designer based on getting information in time to meet the lead in times of the overall programme. It provides a backward linkage from the date that materials and components and specialists are required on site. Thus, if steel was required on site in the eleventh month and there was a lead in time for design and manufacture and delivery of six months, then information is required in the third month to give time to tender and get competitive quotes and place an order.

The *stage* programme provides a quarterly sub frame responding to new information and predicting the impact on activities and reprogramming to contain the correct progress within that stage as per the overall programme. This programme takes into account major changes in scope and looks at programme acceleration or alternative logic and critical path, giving time to agree major alternative methodology and its implications on health and safety and resources.

The *monthly* programme fine tunes the stage programme and gains agreement and sign off for subcontractor start and finish dates. It shows the detailed integration of activities between them on a detailed basis, by discussion with them. This can be used with the system of interface planning indicated later.

The *weekly* meeting provides an opportunity for the site team to meet all the supervisors involved in the current work and to make final adjustments to avoid last minute problems which are anticipated. On a weekly basis, there is an opportunity to discuss whether any of the common problems of lack of staff, late deliveries, availability of lifting facilities, inefficient management, lack of information and poor integration of work are likely to occur and looking at the impact of unexpected events on planned activities. This also gives ownership to the planning process and not imposition. This, to be successful, will show the true strength of the teamwork to help each other.

Feed forward control is most important with time because of the programme constraints. It is used in cost control where the contract permits and the client requires to formalise the cost of design and construction changes before proceeding. The importance of tackling slippage on a regular basis to bring the programme back into line cannot be over-emphasised. Feed forward has the same framework, but tries to stop slippage by early intervention.

A method of critical path programming is introduced in the appendix to this chapter.

Interface planning

Project management should include systems for the proper co-ordination of subcontractors within the work place; this also can be exacerbated by the need to co-ordinate different design drawings:

- Co-ordinating follow on work between different contractors and establishing agreed sequences and co-operative work in areas where many trades have to go on side to side.

Figure 5.4 Workplace flow chart for production control.

- Allocation of work to separate specialist packages must not leave gaps or overlaps.
- Space co-ordination of designers, for example, different services and structural features in the ceiling spaces.
- Health and safety co-ordination in role by the principal contractor.

It is not uncommon for conflicts to arise and a proper system to define the boundaries for responsibility and the hierarchy of authority should be in place. A typical work place planning system concerns a workplace which has sequential procedures with two or more different trades doing work integrated with another trade. This could be with follow trades working in the same space. For example, partition walls followed by first fix carpentry electric and plumbing, followed by plaster, followed by screeds, followed by second fix carpentry electric and plumbing systems, followed by painting and followed by carpets. Each trade needs to take responsibility for timely checking of own and previous work and to be accountable for standard of work and timely delivery of the work space to the next contractor. For example, in the case of painting there is a need to check early that an acceptable foundation that takes the finished product is being provided. The work flow from one trade to another will need controls (Figure 5.4).

The main contractor is responsible for dividing the work areas up into suitable areas of work to provide maximum resource smoothing. They will also be responsible for central access scaffolds, conflict management and ensuring that health and safety method statements are available to all affected contractors.

Conclusions

This chapter has looked at the production process with special reference to the need to benchmark the process and make it better. Benchmarking is measured by the outputs, but also by the responses that customers give. This also provides a basis for evaluating the best practice cases which are presented earlier in the chapter. The old way of looking at production which required that a system of planning and monitoring was in place to satisfy the basic needs of the clients for progress updates on the programme and the budget targets is no longer enough because now clients perceive that service improvements are possible and that more fundamental changes are required to the construction process to give them VFM. This has led to a more holistic look at systems such as prefabrication, quality management, defects reduction and indeed more committed leadership and team building to reduce waste in the process. The project manager has a significant responsibility in their unique influence over the design and construction functions by bringing savings that better integrate design into construction and generate possibilities for a more open culture in contracting organisations.

Contractual and cultural changes in the normal confrontational approach between designer and contractor and client and contractor in the industry have been called for regularly since the Banwell Report, that have not materialised. However, there are indications that major change is emerging for some.

Forces for change that may break this mould and catalyse widespread change in contracting are the threat of greater international competition, more client willingness to develop integrated teams, negotiated contracts, the need to conserve natural resources and the greater awareness of clients (public and private) for the need for controlled competition to meet public expectations for sustainability and investment for the future.

Forces which act against change are business uncertainty and short-termism and the culture of a throw away society with its avid thirst for new facilities and instinct for bargain hunting. Buildings are wanted cheaper and there is not yet a widespread commitment, trust or belief that this can happen to the financial benefit of client and contractor. Where framework and partnering agreements exist there is often mistrust by the supply chain that continuous improvements are brought about by regressive contract conditions between themselves and the managing contractor making long-term partnerships less sustainable. More about this will be discussed in the supply chain chapter.

Greater respect and long-term commitments between client and their contractors are a key to cultural change and the development of trust in the relationships. There are many good examples of the development of trust for large and especially small contractors who depend critically on repeat business. However as large numbers of clients are one off clients, ways of ensuring better short-term relationships also seem to be critical and it is this area of collaboration which needs to be developed to get away from the 'cowboy' image, the 'Dutch' auctions and claims culture generated by cut throat competition and one off opportunism. A realistic first cost with an opportunity for reduction needs to be distinguishable from a low fake price. This is an ethical and not a market issue so that savings can be generated on a pain or gain sharing basis, using earlier contractor involvement, value engineering, supply chain innovation and open book accounting.

For example, a small contractor carrying out a small office extension who has won the contract in competition would have priced for a maximum sum on an output specification – the tender price – and signed up to a commitment for open book accounting and a 'pain or gain' clause for target savings to be made on the basis of innovative savings. Savings would initially arise from specialist contractor inputs for component or detail design changes and a VM exercise which took place in the second negotiated stage of tender. A contract could be signed on the basis of an 80%:20% client/contractor share of the gains. Later events which delayed the programme or escalated the budget could be defined by the use of the NEC contract for a no blame compensation event to either party. The cost of this would be made clear within 10 days so that the client would be free to compensate any agreed escalation in budget with savings in the specification in other areas. Further savings could be suggested on the 'pain or gain' basis. No retentions would be held on the contractor as a basis of trust with the open book accounting. This of course would also be on the basis of no retentions being held on the supply chain. Programme escalation would be at the risk of the contractor except in the case of client changes. Liquidated damages, however, would not be applied except in the case of proven loss of business income.

Appendix

Critical path method (CPM)

The appendix looks at the networks which are used to determine the logic of the time scheduling. It assumes that you are familiar with a Gantt or bar charts which show activities as bars against a horizontal time scale. Figure 5.A1 indicates a 5 day activity starting at the beginning of day 1 and finishing at the end of day 5. Figure 5.A2 shows a second two day activity that starts at the beginning of day 6 and finishes at the end of day 7.

An activity indicated in a Gantt chart does not indicate a logical relationship between activities – it is only assumed to be linked to other

Figure 5.A1 Gantt chart activity.

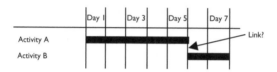

Figure 5.A2 Two Gantt chart activities.

activities, when they have a finish point which coincides to the starting point of a successor. This is imprecise and they may be connected logically to more than one activity.

On complex projects where this is the case a proper indication of logic links may be drawn using the CPM. There are two main formats – the arrow network and the precedence diagram. The former is sometimes called activity on arrow, because the arrow is the activity and the node is a point in time. The latter is called activity on node because the activity is the node and activities are shown linked by the arrow. Neither has a time scale, but may be converted into a bar chart. The main types of links are:

- An activity starts immediately after the predecessor finishes, for example, formwork to a beam followed by reinforcement.
- An activity starts after a time lag, for example, a long trench excavation takes a few days, but drains are laid from the following day.
- An activity can start at the same time as another starts, for example, when foundation is finished the floor slab formwork and the brick work can start.
- An activity can finish at the same time or at a time lag following the finish of another, for example, the pipe above can only finish and be tested after the trench has been excavated.

Critical path analysis (CPM) precedence diagrams

This is a network for linking activities together logically in sequence and analysing the combined durations of the activities and identifying activities with critical time or resource constraints. There are several definitions which are useful.

1   *Activity* is an operation or task carried out using a resource and having duration. In the precedent method they are represented by a box (Figure 5.A3).

Activity

Figure 5.A3 Precedence activity – activity on node.

2   An *event* is a point in time (no duration) and will mark the beginning or the end of an activity. Marked by the beginning or end of a box.

3   A *link* is a line which links the activities together to make a logical sequence from left to right, which means Activity A must be finished before Activity B starts (Figure 5.A4).

Figure 5.A4 Links between activities.

4   A *lead or a lag* links activities together which overlap; it means Activity B can start 5 days after A starts and Activity B cannot finish until 5 days after A has finished (Figure 5.A5).

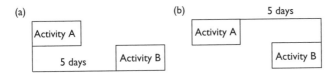

Figure 5.A5 A precedence: (a) lead link and (b) lag link.

This technique can be used to build up a complete map for the logical sequence of various activities. More than one activity can be dependent on another, for example, a concrete beam can depend on the completion of reinforcement and formwork. One activity can also allow two or more to start.

5   *Calculate the critical path* This is the longest path through the network and joins all the longest activities together. The longest route through the net is $9 + 6 + 7 = 22$ days which means that A–B–D is the critical path and that C can start one day later without affecting the end date (Figure 5.A6).

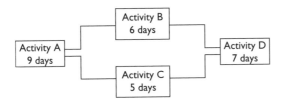

Figure 5.A6 Critical path.

- The *resources* used in CPM are labour, plant, materials and cash, for example, Activity A could be the ground floor walls and have a resource of five bricklayers and if the total cost of the labour plant and materials is £10 000 then this is also a cash (lump) resource.

If there are other activities with the same resource, which have to be carried out at the same time then the resources are added together to give a total resource. It is much better to float activities with the same resource so that they do not overlap.

- *Total float* is applied to an activity or a milestone date and represents the flexibility to start an activity later without affecting the longest critical route through the project, that is, C has a one day float from the diagram above because A–C–D is a shorter route than the critical path by one day.
- A *critical activity* is one which has no total float, that is, Activity A, B and D.
- *Free float* represents the flexibility to start an activity later without affecting any other activity. This is much rarer as it only occurs where an activity is dependent on more than one activity. In this case, C also has one day of free float.
- *Negative float* is the period which a critical activity (or an activity which has been made critical) has been delayed. The effect is to lengthen the longest critical route through the project, unless corrective action is taken, for example, if the Activity C was delayed by two days it would become supercritical by one day.
- A *supercritical activity* is a way of describing an activity with negative float. The only way to avoid delaying the whole project is to reduce the time taken by a supercritical activity, by using more resources. It is no good using more resources on another route.

References

1 Constructing Excellence (2001) *Mansell Best Practice Profile*, http://www.constructingexcellence.org.uk (accessed 15 October 2004).
2 CITB (2000) *Managing Profitable Construction: The Skills Profile*. CITB Research, London.
3 Constructing Excellence (2004) http://www.kpizone.com
4 Crane (2003) Crane in Australia – Part Two. CIOB International News. Article 4180. September. CIOB, UK.

Chapter 6

Organisation, culture and leadership

with Martyn Jones

This chapter looks at organisational and leadership theory and the development of structures and a culture that can further integrate the project organisation and ensure its effectiveness. This leads to the concept of excellence or world class, which requires effective leadership. Project teams are usually made up from different functional departments within the same organisation or from other organisations who are brought together to tackle a specific project over a set period of time. The challenge in construction projects is to develop a culture that *integrates* the various organisations involved and ensures that there is a synergy built up in the team that also removes adversarial attitudes. This is even more important with ongoing changes in construction following Latham and Egan and the greater emphasis on integrated teams and collaborative relationships.

This culture needs to promote good relationships with the client and to provide the opportunity for the client to be involved in the VM and risk allocation of the project. The structure of a project will vary through the life cycle, but during design and construction there are a large number of different organisations working with each other who need to structure their interdependency and adjust to external and market influences. Chapter 7 is complementary and considers the softer skills of people management.

The objectives of this chapter are to:

- Identify different structures under which projects may be run and to discuss their impact on the running of construction projects – traditional and current approaches in construction.
- Consider the culture and values of various organisations and to consider the factors which are important for creating an effective project culture that meets the needs of the client and motivates the project team.
- Define external factors and their effect on organisational structure.
- Briefly identify partnering principles.
- Create a model for continuous improvement and benchmarking world class.

- Examine the theory and practice of effective leadership.
- Case study and implications of integrated project teams and consider the principles of integration in making organisation change in projects.

Organisational structure

Organisational structures differentiate specialist skills and seek to delegate tasks in a way that ensures coverage and prevents overlap of responsibility. Classically, a chain of command is created with the unity of command and a regulated span of control of 5 to 6, so that everyone restricts their relationships. Later research by Woodward [1] indicated that an average span of control for unit production was 23 and this leads to less layers of authority and better communications. It is also clear that many structures are not so formal in the chain of command and networking between different departments and levels is critical to modern communications. Project organisation in particular has cut through the classical principles and set up alternative structures because of the need to work with many different organisations at different levels of authority and to closely co-ordinate interlocking work.

The organisational structure for a project essentially breaks down into three types: functional; matrix; or project based as shown in Figure 6.1. The matrix organisation arises because team members are accountable to functional and project authority. The matrix organisation may be weak or strong in projects depending on whether the emphasis of authority is with the functional manager (weak matrix as in panel (a)) or with the project manager (strong matrix as in panel (b)). Some organisations operate as project organisations, where all activities are project based as in panel (c).

Construction and engineering project teams have traditionally employed a strong seconded matrix with specialist functional roles drawn from professional specialists with pronounced differences in values, attitudes and behaviour.

Of particular interest in the context of this chapter is the degree of authority that the project manager exercises as shown in panel (d) of Figure 6.1 and therefore the scope of his or her leadership. From panel (d) this can be seen to increase the more there is a direct link between the line management and the project manager. If they have to use staff from other functional departments or organisations they approximate more to co-ordinators. If those staff are seconded to the project then the authority role is strengthened. Table 6.1 has been built up to indicate the differences and focus for the project manager.

There are advantages and disadvantages to each of these organisational structures. From a pure project point of view there is instability for staff as

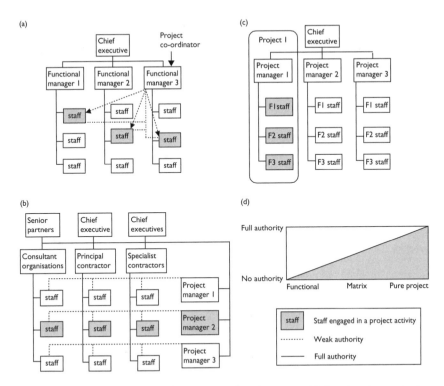

Figure 6.1 Functional, project and matrix organisation types: (a) functional organisation; (b) seconded matrix organisation; (c) project organisation; (d) comparative authority.

projects come to an end, sometimes causing inefficiency as organisations look for other work and staff get attached to their next job. For matrix organisation the project personnel have different calls on the use of their time and project managers need to fight for priority use of personnel, making progress slower than expected. In function based projects the project co-ordinator has to be diplomatic in calling on the use of other functionally based personnel who are primarily doing non-project activities and needs to firmly agree on resource usage.

Construction

Generally it is assumed that construction projects have pure project organisations, but due to the specialisation of organisations in the industry and the division between design and assembly, the authority structure is likely to approximate more to matrix unless staff are dedicated to a single

Table 6.1 Influence of organisational structure type on projects

	Project type			
	Functional	*Matrix*	*Seconded matrix*	*Pure project*
PM authority	Little or none	Low to moderate	Contractual	High to total
Project personnel involvement	0–25%	15–60%	50–95%	85–100%
Position of PM	Subservient to functional manager	Subservient to equal	Leadership and direction	Executive decisions
Title for PM role	Project co-ordinator/ Project leader	Project manager/ Project officer	Project manager	Project manager/ Project director
Type of project	Minor change to existing product e.g. maintenance upgrading and improvement	Implementing change to work organisation e.g. space planning and implementation	Most types of new build and refurbishment project. Design and construction inputs. Facilities management	Very large construction projects created as business units or singular specialised projects such as IT, roll outs, redecoration and refits
Project manager focus	Diplomatic approach, integration of functions	Negotiating adequate time allocation. Good communications	Creating an integrated team. Planning and leadership	Productivity and teamwork

project. Functional staff are employed by separate organisations. This set up is sometimes called a specialist project organisation as the project manager buys in specialist skills for the project. As they deal with organisations, they are essentially buying a service which is subtly different from employing a person as the organisation is more autonomous in deciding how that service might be supplied, conforming to a performance contract. In practice, project managers prefer to have a particular person as team building and motivation is an important part of the productivity of the project.

Although some staff will be totally committed to the project, for example, the project manager, the main supervisors and the personal staff, others will spread their time between different projects which their own organisations are committed to. For example, a Quantity Surveyor (QS) might be committed to valuing work on two projects and an architect might have two design commissions. It is also possible for involvement in the project to be limited and not for the whole life cycle except for key players like the job architect.

The most common form of organisation in larger construction projects can be described as the seconded matrix organisation (Case study 6.1), as the most direct project personnel spend all their time on one or more projects and integration of the team is recognised as an important element. Projects in the spend range of £20–40m require significant project organisation.

The chapter on supply chain management explores the challenges of the network of relationships and hierarchies.

Case study 6.1 Typical seconded matrix organisation

This seconded matrix is a new build £26m project in the South West for a national company HQ and has the following seconded matrix organisation (Figure 6.2) run from the site. They need to cover most of the roles with project staff who may be a mixture of dedicated or staff shared with another project. Some shorter-term staff such as the engineer may be employed through an agency.

The staff were directly employed by the design and build company to manage the site. In order to ensure that all job responsibilities are covered unambiguously then job responsibilities should be allocated to each. In addition, some services will be supplied by separate design personnel who are not based on the site. The organisation is an example of a decentralised site with all executive control operating through the project manager who directly reports to the Director and has formal accountability for the actions of the company.

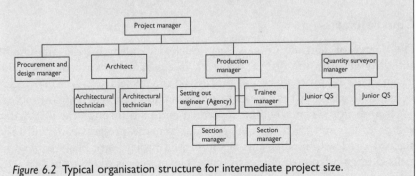

Figure 6.2 Typical organisation structure for intermediate project size.

The following roles emerge:

Project manager and health and safety officer	Executive control of design, construction, planning and finance, specific accountability for site health and safety
Quantity surveyor manager	Financial control of valuations, change control and estimating

Junior surveyors	Re-measure and progress on site, procure and check materials
Design and procurement manager	Initiation of design, management of design process, procure and tender supply chain, select subcontractors, method statements
Architect, service engineer and structural engineer	Scheme and detail design, quality inspection and compliance
Production manager	Overall control of production, compliance with BREEAM rating required, health and safety oversite and method statements implementation, quality improvement
Section manager (and trainee)	Day-to-day management of relevant subcontractors or section of work, manage dimensional checks and order bulk materials, safety inspections and records. Technical queries. Access requirements
Setting out engineer	Setting out grid, co-ordinate stations and level benchmarks, checking critical dimensions and levels after setting out by steelwork, bricklayers, cladding and ground workers. Check base, floor and roof levels, plumbing in lift shaft and main elevations and columns. Setting out of drains, roads and car parks and level checks

Centralisation

The degree of centralisation is very much dependent on the amount of authority given to the project manager. In *centralisation* a contractor will try and impress procedures and process on the project set up. Major decisions will be made by the contract or construction manager and contractual issues and supply chain will be appointed and paid through them. Day-to-day co-ordination with the supply chain and the project team will take place at site level. Smaller and medium-sized projects are likely to be centralised to gain from economy of scale. The culture is likely to be less well defined and akin to the culture of the largest organisation in the project. The client will critically communicate through the project manager, which is likely to be the design leader or the design and build contractor.

Decentralisation is a delegation of project control and supply chain procurement to project level and allows a more autonomous culture to be developed and a closer understanding between the project team members who although dealing with different levels of project management will benefit from a closer knit relationship with all decisions being made at project level. In this structure, a strong leader is required who can handle decisions

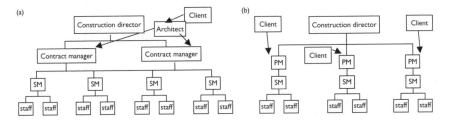

Figure 6.3 (a) Centralisation and (b) decentralisation.

on the construction and design streams and provide reporting of all direct to the client. The client's main point of contact is the project manager who will distribute all instructions and be responsible for the flow of information between the project team, although the client will hold contracts with individual members of the project team. This elevates the authority and seniority of the project manager. This a common situation for large projects in construction. Decentralisation is compared with centralisation in Figure 6.3. Case study 6.2 shows a typical organisation for a large project that altered its policy to take account of the environment of other projects it had.

Case study 6.2 Major development projects organisation structure

A large developer in the West Country has developed its portfolio of building land to take account of what was becoming available and what would be granted planning permission to take on mainly 'brown field' development. On a marketing perspective they noted the rising demand (and prices) for 'smart residential' property in city centres and the planning authority's desire to see past industrial sites in their city centres decontaminated and reused for residential and leisure. Over a period of time, they have moved from smaller 'green field' sites to large longer term mixed use schemes. This required the use of standardised footprint designs with a housing mix to suit the local housing market, to more comprehensive schemes which needed a master plan to satisfy the planning authority. This depended on building and attracting tenants or buyers for commercial and leisure and even industrial use and sequencing the whole to suit the general constraints of the market, but one of the major planning qualifications was to assure social provision, employment and balance mixed use development to enhance city life.

One such site is a £300m budget over a build period of four years with a cash breakeven eight years after the initial start of decontamination and regeneration works on a messy industrial site around obsolete docks. The preparation works included clearing away a disused power station and electrical substations, demolishing chemical factories and remodelling the dock land area to maximise the amenity to the new residential and leisure facilities being provided. In its place 7–9 storey flats were to be constructed adjacent to the dock area, a new marina and basin, the creation of park and wild life areas, luxury house plots and higher density town houses, together with a new transport exchange, shopping and factory facilities and the possible reinstatement of a railway link. The traditional well defined housing scheme had become a regeneration plan in its own right and mainly at the developer's cost.

This has led them to move away from the typical centralised structure required for the small to medium-sized housing scheme, where sites are serviced with site managers calling off bulk purchase and using standard house designs. These were replaced by a suite of operational and executive site offices on the 'muddy side' and a smart sales centre to sell the houses segregated fully from the site traffic. Two directors were housed alongside the site staff to provide strategic direction for the future on going planning, the production and design strategy and the sales and marketing. The project operated as a business in its own right with direct accountability to the MD.

Subsequent to this case study research, the company decided to centralise the directors away from the project to give them a multiple project portfolio due to the go ahead of a similar sized project in the same area. This indicates the dynamic nature of project organisation to suit the varying business circumstances, which in this case was taking the opportunity to capitalise on personnel experience by broadening a portfolio. The sales office and site offices remained, but the strategic control of the project was made more remote. It is important in the design of project organisations to allow for changes during the life cycle of the project.

Although we have looked at some case studies to illustrate project arrangements, many of the influences for the organisational structure and culture of projects come from the environment, which is now considered.

Mapping the project environment

Project teams do not exist in a vacuum. They exist in a wider system which includes organisations, governments, competitors, suppliers and customers. They are affected by legal, economic, social, political, cultural and technological forces. As pointed out at the beginning of the chapter, this wider

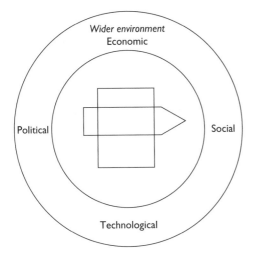

Figure 6.4 Elements of the wider environment.

environment within which projects are undertaken has become more significant and more demanding.

The main elements of the wider environment are shown in Figure 6.4. Awareness of the wider environment is now seen as being more important than ever to organisations and project teams and vital to their long-term success. This has made the management of projects more challenging. In turn, today's project leaders need a much greater strategic awareness of the environment within which their projects take place. The four elements of the environment have been discussed in Chapter 3.

The organisational context

The project environment includes everything external to it. Maps or models are a good means of developing awareness and understanding of the wider environment and identifying what forces are at play. Developing an understanding of the external environment is a tall order, but this can be done by constructing a simple map. The map starts with the project as shown in Figure 6.5.

Figure 6.5 The project.

The next step is to decide where to map the project in relation to the organisation. This is often problematic. Do the projects take place within the organisation, outside it or do they straddle the boundaries of the organisation in some way? The main situations are shown in Figure 6.6.

Take the case of the project within the organisation. Here the project sponsor will be from within the organisation. The project manager will normally select his or her team from within the sponsoring organisation with the aim of making them a cohesive whole and ensuring that the interests of all the stakeholders are balanced. Thus teambuilding is a fundamental role of most project managers.

Where the project is within a single firm or organisation, the challenge here is to organise people from different specialisms or functional departments (e.g. marketing, finance or engineering) into an effective project team. There are a number of organisational structures that have been compared in Table 6.1.

However, in many contexts, such as in construction, for example, even this complex view of organisations and projects is too simplistic. It is common for a number of independent firms or organisations to be engaged in a project as shown in Figure 6.7.

The challenge for the manager of projects involving a number of organisations is to bring together individuals from different organisations or firms

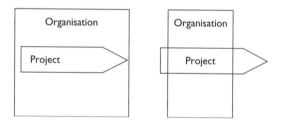

Figure 6.6 Relationships between organisations and projects.

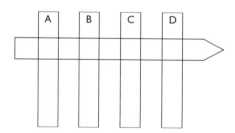

Figure 6.7 A project involving a number of independent organisations.

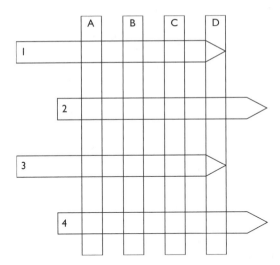

Figure 6.8 Complex matrix of firms and projects.

(rather than different functions or specialisms within the same firm) and form them into an effective project team. The firms involved are autonomous firms, which are organisationally independent in terms of the project. It is likely that each firm or organisation will have conflicting aims and objectives and varying degrees of allegiance or commitment to the project. Effective project management offers a means of unifying overall processes and presenting opportunities for participants to place the demands of the project above those of their own enterprise.

This complex kind of matrix structure raises two significant management issues: managing in the context of the culture and context of the firm or organisation and that of managing projects. In addition, such projects are often undertaken by an amalgam of firms which may well change from project to project – as is often the case, for example, in construction. This can give rise to all sorts of problems as the situation creates a potential for conflict between the needs of each firm and the project.

Figure 6.7 has also greatly simplified the situation because it relates to a single project. In practice, a portfolio of projects may be involved as shown in Figure 6.8.

Again this is a simplified view, as it implies that the projects shown are all being undertaken by the same firms. In reality, this may not be the case as there will often be different mixes of firms involved for each project.

Case study 6.3 provides an inside view of how one developer organised the multiple projects on an ongoing development.

The project takes place over five years on a prime 16 acre derelict brown field site worth £300m in the centre of a provincial city. The site is very sensitive and planning permission has only been agreed with very stringent conditions to maintain site lines and to properly connect the city centre to the waterside and to ensure full public access. The proposals are for leisure, retail, commercial and residential use. The latter is core to the private developer in terms of cash flow and profit. The development links directly to a highly popular leisure sector and a significant commercial venture already in place on the site. The site is in a desirable location, which faces onto the old Harbour waterside. A master plan is in place.

The Developer does not employ an extensive project team, but seeks instead to outsource the design and tactical project management. The first phase is procured by three design and build contracts. On each of these contracts there is an out-sourced team for the client, consisting of a cost consultant, an architect and a project manager. These will report to the programme director who co-ordinates the whole. Due to the sensitivity of the site to the conservation of the city character and heightened public interest, a development director has concentrated their efforts in key marketing of the subsequent phases of the development and bringing them through the planning process and maintaining good relationships with the public and community. The programme director has taken on the implementation and later marketing stages of the development and is maintaining the master programme for parallel and serial projects. Parallel contracts have impacted on each other and needed to comply with the overall constraints of the site. This has led to a loose structural organisation, which also needed to have a flexible response to the market forces which will determine the speed of completion of the development and the logistical access and environmental constraints which affect the methodology and sequence of separately let contracts.

The design has been spread out between architectural/engineering practices to reflect specialities and inject variety. Leisure, commercial and residential blocks have been allocated and the master planners have been retained for more detailed involvement in one area. The process is managed by phasing and allocation of different buildings to different contracts, so that design management is linked to the procurement route chosen and is not developed centrally. The programme director is the one main player who co-ordinates the production design for the Developer. Figure 6.9 indicates the emerging organisation structure.

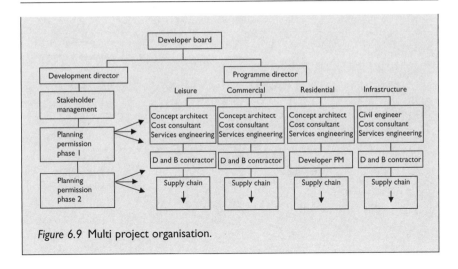

Figure 6.9 Multi project organisation.

In this more complex amalgam of firms, developing a greater understanding of organisational culture and behaviour and the relationships that exist between firms is seen as being increasingly important in raising the performance of the whole project process.

Corporate culture and behaviour

Corporate culture consists of an organisation's norms, values, rules of conduct, management style, priorities, beliefs and behaviours. The current fascination of business with organisational culture began in the 1980s with the work of writers such as Peters and Waterman [2], although earlier writers such as Blake and Mouton [3] had argued the link between culture and excellence.

Silverman [4] contended that organisations are societies in miniature and can therefore be expected to show evidence of their own cultural characteristics. However, culture does not spring up automatically and fully formed in response to management strategies. Allaire and Firsirotu [5] argued that it is the product of a number of different influences: the ambient society's values and characteristics; the organisation's history and past leadership; and factors such as industry and technology. The main determinants of business culture are shown in Figure 6.10.

Culture is a term that is increasingly used but poorly understood. This is because an organisation may have a number of cultures within it and that even a detailed deconstruction would provide an inaccurate picture of the totality of the organisation.

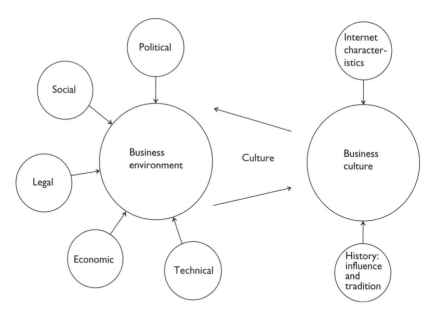

Figure 6.10 The relationship between business environment and business culture.

Harris and Moran [6] have identified a number of determinants of cultural differences:

- Associations – the various groups with which an individual may be associated.
- Economy – the type of economy affects the way individuals conduct themselves at work, how they feel about achievements and their loyalty to their employer.
- Education – the types and amounts of educational opportunities.
- Health – recognises the impact of the health of workers on productivity and effectiveness of the organisation.
- Kinship – the family and its importance in the life of the employee.
- Politics – the political system impacts on the organisation and the individual worker.
- Recreation – recognises the role of leisure time in the life of a worker and his/her family.
- Religion – in certain countries this can be the most important cultural variable related not only to the workplace but also to the daily lives of people.

These factors must be taken into account by leaders and managers in their style of management and leadership. Fordist and Taylorist approaches favoured coercion where functional devices and authority are used to force the individual to carry out a particular task and in a non-thinking way. With the shift to post-Fordism there is more emphasis on co-operation and the education of the individual as to the reason why it is in their own interest to perform at a high level. This is the focus of the humanistic movement.

Handy [7] has usefully grouped organisations into four cultural types which help to identify the different characteristics, but these are too simplistic for organisations to have an exact fit and indeed it is possible that several cultures will emerge and this is a particular issue between the 'office' and the project.

A *power* culture is frequently found in small entrepreneurial organisations such as some property, trading and finance companies. Such a culture is associated with a web structure with one or more powerful figures at the centre, wielding power.

A *person* culture is rare. The individual and his or her wishes are the central focus of this form of culture. It is associated with a minimalistic structure, the purpose of which is to assist individuals who choose to work together. Therefore, a person culture can be characterised as a cluster or galaxy of individual stars.

A *role* culture is appropriate to bureaucracies, and organisations with mechanistic, rigid structure and narrow jobs. Such cultures stress the importance of procedures and rules, hierarchical position and authority, security and predictability. In essence, role cultures create situations in which those in the organisation stick rigidly to their job description (role), and any unforeseen events are referred to the next layer up the hierarchy.

Task cultures, on the other hand, are job- or project-orientated; the onus is on getting the job in hand (the task) done rather than prescribing how it should be done. Such types of culture are appropriate to organically structured organisations where flexibility and teamworking are encouraged. Task cultures create situations in which speed of reaction, integration and creativity are more important than adherence to particular rules or procedures, and where position and authority are less important than the individual contribution to the task in hand.

It is the last two forms of culture, role and task, which are most frequently found in organisations.

(Extract from Handy [7])

Many small businesses are run as power cultures. In many projects and for much of the project cycle, the role culture prevails as the tasks are routine. Projects run in this way may be effective in undertaking the task but often at the expense of the project team and the stakeholders. The role approach has an emphasis on structure, which although it allocates tasks and responsibilities clearly, they can be bureaucratic and obstructive to change. The task cultures place more emphasis on the team.

Project managers need to be aware of culture in managing project, particularly where different cultures are involved. Weak matrix management often fails to produce the anticipated results because the role culture of the organisation is overlaid with the task-centred culture of the project. In relation to quality, the procedures associated with systems are often applied less effectively in task, person and power cultures, but are most effective in a role culture with its formalised and rule-based approach. The advantages and disadvantages of Handy's four types are shown in Table 6.2.

Construction project cultures

Table 6.2 suggests that a construction project bests fits into the task culture because a project is task-orientated. This is not unreasonable, but different project situations can achieve characteristics that bend them towards roles rather than teams. A contractor's organisation can be quite large with a strong matrix culture between different functions such as quantity surveyors, accountants, project management, plant supply and planning. A strong functional influence can result in the role culture of meetings, reporting procedures and approval forms.

The construction project has the involvement of several organisations in the supply chain and these may conflict culturally, or be actively organised by the project manager into a project culture, sometimes called a virtual organisation. Many small organisations contribute to projects and they may exhibit a strong person culture and this may lead to conflict. Fragmentation of project tasks may take place due to the very different culture of learning there is between the different professions of engineering, architecture and construction management, the legal profession and accountants. This can be attributed to the clash of different organisational culture types. A small architectural or an engineering organisation is likely to be a partnership enshrining a culture of entrepreneurship or a loosely coupled organisation of separate stars. Team building is a key issue here.

Dubois and Gadde [8] point out that there are both tight and loose couplings in supply chains. Tight because of the direct effect of late delivery or poor performance on the other parties and loose because of long lead times and intermediaries. Loosely coupled supply chains are co-ordinated with a subsequent reduced control of key aspects such as programme and budget. Within firms they are likely to operate on more than one project

Table 6.2 Benefits and concerns for Handy's cultural categories

Culture type	Comments and example	Advantages	Disadvantages
Culture of personal power			
Power	Frequently centred around a single personality in the spirit of entrepreneurship e.g. property development	Very focused objectives with the potential for creativity and fast growth	Control is operated from the centre and some may feel strait jacketed if they don't catch the vision
Culture of inter-dependence			
Person	A loose coupled organisation of equals. Gain from shared common facilities. Some shared aims and interests e.g. partnership	Management and responsibilities are decentralised allows flexibility and a forum for networking and recognition. All individuals are stars	Synergy is difficult to achieve. One tries to take over. Lack of loyalty, breakups or low-retention rate
Culture of position authority			
Role	Bureaucratic depending on procedures and rules and formalised structures e.g. civil service	Stable, predictable and everyone knows where they are	Lack of flexibility and innovation
Culture of team building			
Task	Aligns to the matrix organisation. Focus on getting the job done e.g. project based work	Satisfying relationships and efficient working organisations formed and reformed to suit needs	Short lived, hard to produce economies of scale and to develop expertise in depth. Competition for available resources

Source: Adapted from Handy [7].

and these may have different cultures for the reasons given earlier. A culture may be built up, but is difficult without a commitment to supply chain integration.

Loyalties are given to the client in terms of repeating work with the client and the project culture comes second unless the client becomes part of the project team. Individual agreements may create competition in the team and a lack of transparency and sharing of information. If it is correctly focused and controlled, a strong partnering culture could be built up with repeat work or client partnerships over several projects.

As we have already seen, projects and organisations do not exist in a vacuum. They exist in a much wider system, which includes governments, competitors, suppliers and customers. They are also affected and shaped by legal, economic, social and technological forces. It is this wider external environment which facilitates, constrains and threatens activities – and, of course, provides opportunities.

The external environment, strategic issues and forces for change

Clearly, organisations and project teams have less influence over the wider environment than their market environment. Day-to-day actions for running the organisation and projects are, therefore, more likely to be governed by the market environment than by the wider environment. Elements of the market environment such as competitors and suppliers will, however, be influenced by the wider environmental forces.

The market environment

Figure 6.11 shows that projects and firms exist within a market environment. As can be seen, the map of the market environment comprises a number of elements including customers and clients, the labour market, wholesalers and retailers, suppliers, Trade Unions, Professional Bodies and other groups and associations. What must be borne in mind is that a firm's ability to make profits is largely determined by the structure of the market in which it operates. Clearly, some markets and industries are more profitable than others.

A major feature of most market and sectors of the economy is increasing competition. This means it is important that organisations are aware of their competitors. They need to know who their organisation's rivals are and are their key clients and suppliers! This is necessary because their actions can have many consequences for an organisation. Leading analyst Michael Porter [9] identifies five forces that determine the level of competition and underlying profit opportunities within an industry. These forces apply within the industry and affect organisations and projects.

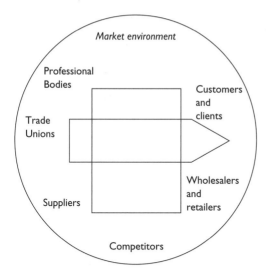

Figure 6.11 Elements of the market environment.

Industries are highly competitive when there are many competitors or a small number of equal size, with little to differentiate between products and services, fixed costs are high, economies of scale are important and industry growth is static or low. Porters' five forces are:

- Highly competitive industry – A highly competitive industry will reduce profitability as firms keep prices low to be competitive.
- The threat of substitutes – Most industries compete directly or indirectly with other industries, which offer substitute products or services.
- Ease of new entry to the industry – Where entry barriers are low an industry's profit potential will be limited because new entrants can enter the market and compete.
- The power of buyers – Powerful buyers can increase their profits by negotiating lower prices, improved quality, better service and delivery from their suppliers.
- The power of suppliers – In some circumstances where there are a small number of suppliers, the suppliers may be able to make excessive profits by forcing customers to pay high prices.

Awareness of these factors within an industry enables firms to forecast, recognise and respond to changes in the market environment. A firm can position itself to respond to threats and constraints created by the structure and nature of the industry in which it operates. However most sections of the economy are characterised by significant market changes.

The role of partnering

From the early 1990s, a significantly different form of organisational structure began to emerge which came to be known as partnering. Partnering was identified by the Latham Report [10] as a way forward in dealing with the conflict in many construction projects. Partnering is defined by Bennett and Jayes[1] as

> a management approach used by two or more organisations to achieve specific business objectives by maximising the effectiveness of each participant's resources. It requires that the parties work together in an open and trusting relationship based on mutual objectives, an agreed method of problem resolution and an active search for continuous measurable improvements.

Partnering is based on building up trust and is normally brokered by the signing of a voluntary pact. Essentially it is a risk sharing approach and is not directly connected with any particular organisational structure, but rather with a more open culture and effective leadership to ensure that the trust is reciprocal and respected. Partnering potentially gives savings arising from cultural change, which improves collaborative activity based on co-operation and reduced conflict that cuts out costly contractual procedures. Strategic repeat work with the same client helps this. Johnson-George and Swap [11] notes that trust is a central concept to alliances and partnerships in mainstream organisational studies literature. He points out that partnering comes at a cost and that this may well be associated with the risk of trust breaking down between the parties, because trust requires them to be vulnerable to the actions of another party. This is on the expectation that the 'trustee will perform actions important to the trustor, irrespective of the ability to monitor or control that party' [11]. This cost needs to be factored against the savings. Certain measures have been developed to seal the trust commitment and reduce the risk of opportunism including, perhaps surprisingly the use of a contract! It is important though to recognise there are different levels of trust and this is discussed in greater depth in Chapter 12.

In many cases partnering uses open book accounting, so that costs and profits are differentiated transparently. It also makes it possible to determine the basis for any profit sharing through either:

- a joint venture partnership between key suppliers; or
- a risk sharing 'pain:gain', 'loss:profit' incentive scheme between supplier and client.

The former is a sharing of an equal margin of profit in proportion to the work share of each joint venture partner. This incurs single accounting for work done. Progress payments are divided in accordance with the value of work done and so are easier on a pre-agreed stage payment basis, for example, completing foundations triggers payment. Monthly payments require an interim account from each of the partners which needs to be agreed. 'Pain gain' agreements means that the client is prepared to share any savings made on an agreed target budget, on the basis that the supplier partner is prepared to share any overrun. The proportion of gain : loss is agreed in the contract.

These methods represent the 'hard nose' of partnering and are not essential to the gains that can be made in normal collaborative working. Gaining repeat work with 'valued clients' never used to be called partnering and yet it is a basic tenet of good business practice to increase certainty of success when making an investment in extensive tendering costs, and most prefer a negotiated approach. This has been apparent by the behaviour of many large contractors in the industry turning away work gained from single stage open or selective competitive tendering.

Partnering at its best can induce real value into a project and share the benefits of savings (up to 10% is claimed for single projects), but it can be one sided where a dominant client induces lower prices *out of the profits* of the suppliers for the promise of keeping them busy. Continuous improvement and value savings are often achieved over several projects in what is called a strategic partnering or a framework agreement, where a small number of partners share a repeat client's workload and keep each other competitive. Real savings of up to 30% are possible over a gestation period and are made by identifying and cutting out waste in the process, reducing the learning curve of subsequent projects, gaining a better knowledge of the customers real value system, encouraging supply chain synergy and possibly by bulk buying power.

A common 'experienced' client approach is to look at reducing the equivalent project tender for successive jobs by 5% based on identified production or process waste cutting. This can be done by using lessons learnt and applying these to the next project. If both sides are to gain then 10% savings should be identified. One of the keys to this is by harnessing savings through the supply chain, but it is hard to keep complete supply chains constant from job to job and to expect the whole hearted involvement of large networks of suppliers in a very volatile construction environment, where easier 'pickings' might be made elsewhere. Indeed some criticism of profitable partnering is that value is lost due to the lack of competition. This criticism can only be countered by the use of reliable benchmarking (Chapter 5) to measure improvements, that is independent enough to be trusted by all sides.

Customer focus

Peters and Waterman [2] argue that excellent companies really do get close to their customers, while others merely talk about it. The customer dictates product, quantity, quality and service. The best organisations are alleged to go to extreme lengths to achieve quality, service and reliability. There is no part of the business that is closed to customers. In fact, many excellent companies claim to get their best ideas for new products and services from listening intently and regularly to their customers. Such companies are more driven by their direct orientation to the customers rather than by technology or by a desire to be the lowest-cost producer/provider. This is how Peters [12] describes the lack of customer focus in many types of organisation.

> Look through clear eyes and you'll find that almost all enterprises – hospitals, manufacturers, banks – are organised around, and for the convenience of the 'production function'. The hospital is chiefly concocted to support doctors, surgery and lab work. Manufacturers are fashioned to maximise factory efficiency. The bank's scheme is largely the by-product of 'best backroom (operations) practice'. I'm not arguing there's no benefit to customers from these practices. The patient generally gets well, the car or zipper usually works, the bank account is serviced. And enterprises do reach out, sporadically, to customers – holding focus groups, providing toll free numbers to enhance customer dialogue, offering 'customer care' training to staff. But how many build the entire logic of the firm around the flow of the customer through the A to Z process of experiencing the organisation? Answer: darn few!
>
> (Extract from Peters [12])

Over the past decade or so, more and more companies have sought to change their external relationships by developing closer and more harmonious links with their customers and suppliers. These changes are taking place due to a growing climate of opinion that customers and suppliers working co-operatively must be more beneficial than more traditional adversarial purchasing relationships [13, 14].

Projects which need a business approach to organisation

Case study 6.4 indicates a large project which has responded to their customers' major requirements by creating their own strategic business unit. This is a specific way of giving customer focus and financially protecting the parent company in the event of uncontrolled loss on mega projects. It is also a way of responding to the risks in the wider external environment.

Mega Projects where there is a spend of many millions per month are often run as special projects and are allocated as strategic business units with their own profit centre with full accountability. These will be determined where they represent a significant proportion of the turnover of the company. One such £300m project for a government organisation was built in two years by a PFI consortium led by one of the large contractors in the UK. The project involved the moving into new buildings, of an almost complete government department on a phased basis, decanting staff from existing buildings on the same site. The construction works included moving major power lines, putting in new roads and diverting traffic flows, releasing land for development and the design, construction and fitting out of high technology facilities. During construction the cash flow went up to £22m/month at the peak of the contract when 1200 people were working on the site. At the start of the project and to suit phased handover, the speed was much reduced to deal with site and client constraints and this required careful management of contract personnel.

There was a 150 strong project management team, which was part of a joint venture company set up to deal with the design, construction and management of a large supply chain. This company will also continue to manage the facilities for the next 25–30 years of its operation. Major car parks, recruitment teams, training facilities and health and safety systems were created to cope with the workforce and visitors to the site. The project organisation is divided up into procurement management, design management, financial management, sectional management of the construction and facilities management. A senior project director provided business and strategic leadership supported by a small team of project managers and sectional construction managers. In addition, a general manger of the joint venture company was responsible for the strategic level contact with the client throughout the life cycle of its operation.

A unique logistics network was designed into the new building to receive supplies on the non-secure side of the building, where they are vetted and to then to deliver them by train to various parts of the basement where they were taken by lift to the relevant section requiring them. Security is often a major issue on projects and a security reception was set up to provide identification, escort and give health and safety induction for all visitors and new workers on to the site. Fire escape routes with fire points and segregated pedestrian and plant routes were designated to back up the health and safety and security requirements on a large site.

Although this project is one of the only very few in its size category it indicates the importance of creating a fully competent integrated team for a large business undertaking in a very short period of time to meet. The client required evidence of a world class organisation who were able to meet the exacting requirements of a very tight programme, budget and phased changeover programme to move staff, whilst maintaining security.

Experienced customers are becoming more discerning in their requirements for getting the quality they want and have looked for methods for choosing between contractors. The following model is one that has been used widely to identify those who are able to give the best and most sustainable service.

World-class performance

What makes a top performing or world-class company. Studies show that a typical profile of world-class performers within the post Fordist era includes the following characteristics, strong leadership, motivated employees; extremely high customer satisfaction ratings; a strong and/or rapidly growing market share; highly admired by peer group companies and society at large and business results that place it in the upper quartile of shareholder value.

Achieving business excellence demands the following:

- organisational learning;
- farsighted, committed and involved leaders;
- a clear understanding of the company's critical success factors;
- unambiguous direction setting;
- flexible and responsive process management;
- people with relevant knowledge and skills;
- constant search for improving the ways things are done.

Self assessment has become a popular approach in assessing performance. It is a way of looking at how well a company, an organisation, a branch, a unit or a project is performing. It enables an organisation to look right across all its activities, set a stake in the ground to represent its performance to date and determine what it now needs to make improvements in that performance.

There is growing use of the European Foundation for Quality Management (EFQM) model for business excellence as a basis for self assessment and the award of the European Quality Award (EQA). It is based on the principle that in order for an organisation or team to succeed, there are a number of key enablers on which it should concentrate to achieve improvement goals. It also needs to measure its success through a number of key results areas.

The key enablers by which people judge themselves are:

- how well the organisation is led;
- how well its people are managed;

- how far its policy and strategy are developed and implemented by its leaders and people;
- how well it manages its resources and develops and manages its processes.

The key results areas by which the EFQM model then measures successes are:

- how far it satisfies its customers;
- how well-motivated and committed is its workforce;
- how the local and national community outside the organisation views its activities in terms of its contribution to society;
- its key business results including profit, return on capital employed, shareholder earnings, achieving budgets.

An overall view of the model is shown in Figure 6.12.

Using the EFQM model, organisations can score up to 1000 points on their performance. Customer satisfaction (up to 200 points). This reflects the importance of providing customers with a product or service which delights them. The best companies in Europe are scoring around 750 points. A score of 500 points is extremely good, and would equate with one of the best in the UK.

There are several ways in which self assessment may be carried out. The most usual are for the organisation to hold a self assessment workshop or complete a questionnaire. For a more detailed assessment, and to gain recognition for their achievements, the organisation can produce a written award style application report, against EFQM guidelines, which is then

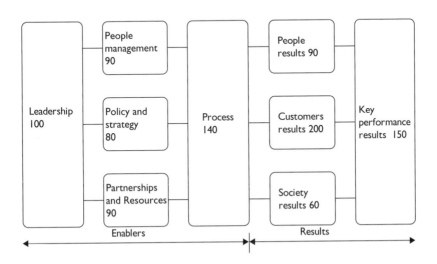

Figure 6.12 EFQM model for business excellence.

Source: '© EFQM. The EFQM Excellence Model is a registered trademark'.

assessed by trained assessors. A low-cost method of assessment now available is the use of computer based questionnaires. In deciding whether self assessment would be useful for an organisation or team it is important to look at the team characteristics. As a general rule, self assessment has been shown to work well when an organisation:

- has a clear set of customers;
- knows who its key suppliers are;
- has a clear set of products/services;
- is able to measure its own results;
- is run by a management team.

What are the benefits of self assessment?

Organisations which have used the EFQM model do show a commitment and enthusiasm to continue to give a better and better service to their customers, and meet the needs of all stakeholders. Case study 6.5 is a summary of the qualities of a small business EQA winner:

Case study 6.5 EFQM RBD-50

RBD-50 is a Ukrainian road building Small and Medium Enterprises (SME) that has won the EFQM quality award for the qualities of enthusiasm and commitment mentioned earlier as well as its excellent score against the EFQM model. The firm employs 182 employees who own 72% of the company, with a further 27% ownership by the families of employees. It has survived in a difficult economic climate and still managed to show continuous improvement in all of the elements of the EFQM model. It has maintained its position as the most productive player in a competitive road building industry in the last four years and it improved its workload by 250% whilst improving its profits by 15 times. Its main focus on people and change reflects this and it has consistently scored 90% satisfaction in each of 25 areas under the headings of production and overall image. It was also awarded a 100% score in the categories of willingness to recommend to others, innovation, communication and accessibility and it is well linked to its customers. The company has shown a consistent improvement in performance and is well appreciated by its customers.

This focus on leading improvements to reach out to world-class performance may be seen as being right at the heart of the future role of the project manager or project leader. Briner *et al.* [15] argue that:

> The new ways in which organisations are using projects as mechanisms for managing innovation and change, or making things happen fast, have resulted in the rapid rise in importance of a new role – that of project leader.

A study in the early 1990s, by the CBI and DTI, examined 120 of the UKs most successful companies selected on the basis of their profitability, growth and recognised standards in their sector. One of the characteristics distinguishing the leading edge or world class from the run-of-the-mill was their strong leaders who champion change, set targets and are open with customers and suppliers. A further characteristic is that they constantly learn from others, who are committed to innovation and the continuous search to introduce new products and services by exploiting new technologies and ways of securing a competitive edge. It is therefore important, to consider the role of the leader in more detail.

Leadership

Issues of leadership and associated power lie at the core of group life in a variety of contexts. Even the most informal groups typically have some form of leadership within their organisation. The understanding of leadership and power from a social psychological standpoint can lead to a greater understanding of group processes both inside and outside the workplace. A measure of the significance of leadership in the modern world is the vast number of publications on the subject.

Recent definitions demonstrate how our view of leadership has been broadened to include the introduction of major change, giving meaning and purpose to work and to organisations, empowering followers, infusing organisations with values and ideology and contributing to social order. It can be argued that there are three main forces shaping the process of leadership – flatter matrix forms of organisation, the growing number of alliances and informal networks and people's changing expectations and values.

The earliest approach to leadership was the 'traits' approach. It was aligned to the notion that leaders are born and not made. It emerged mainly in the context of the military. This view is now contentious and highly dubious. Maylor [16] argues that great leaders in all spheres of human endeavour have developed skills and attributes to the point needed for the task at hand. Both of these are teachable and although intelligence is one of the few characteristics that cannot be taught, this has rarely been a constraint on success.

The more recent approaches have emphasised the functional or group approach, leadership as a behavioural category and styles of leadership.

The emergence of 'contingency' theory reinforced the view that there is no single recipe for successful leadership. Transformational leadership is the most recent and the most changed and people-focused of the approaches which has permeated many sectors of the economy including construction.

This changing view of leadership has emphasised the need to differentiate between management and leadership. Management is defined as the rational-analytical behaviour of a person in a position of authority directed towards the organisation, co-ordination and implementation of organisational policies and strategies. Their behaviour includes planning, organising and setting up and sustaining administrative systems. On the other hand, leader behaviours need to appeal to follower motives and be interpersonally oriented. Leader behaviours include the articulation of a collective vision, infusing the organisation with values and motivating followers to exceptional performance by appealing to their values, emotions and self concepts. From this it can be seen that managers provide the intellectual inputs necessary for organisations to perform effectively, whereas leaders set the direction for organisations and appeal to ideological values, motives and self-perceptions of followers.

Power and leadership

In an extensive study of leadership the power or influence exerted by leaders over others emerges as a consistent pattern (Stodgill 1974, cited in Ref. [17]). However, leadership, power and position are not always equated with one another in any organisation. Sometimes power is to be found in the most unexpected place.

> Power is the ability of one social unit to influence the behaviour of another social unit and to achieve their preferred situation or outcome.
> (Rosenfeld and Wilson [17])

French and Raven [18] provide a framework for understanding the bases of power:

- Reward – the capacity of one individual to reward another.
- Coercive – the capacity of one individual to operate sanctions or punishments against another individual.
- Legitimate – the extent to which an individual feels it is right to make a request of another.
- Expert – key knowledge or specialised skill.
- Referent – the extent to which others identify with a single individual.

Burke [19] talks about the responsibility-authority gap because of the need of the project manager to cut across formal organisation communication and authority lines, to get things done directly. For example, the project manager has different ways of working with people to get things done.

Formal authority or position power can require things to be done, but is often less productive in the long term than those listed here, which Burke [19] adds to the earlier mentioned:

- Information power, by creating a central mine of information and communications through the project manager, making other's progress dependent on the project manager.
- Cognitive power is the ability to use reasoned argument, evidence and logical consistency to persuade outcomes which commit resources, gain preferential (some would say fair) treatment and best performance.
- Emotional persuasion is done with a knowledge of people's soft points, asking nicely and requesting and calling in favours.
- Personal power is the charisma and strength of personality of the manager, leading by example and having inspired leadership qualities.

What makes an effective leader?

O'Neil [20] sets out what leadership is about as:

- providing meaning and purpose;
- focusing on the right things to do;
- structuring the environment to achieve the organisation's goals;
- getting others to do what you want;
- motivating people towards getting things done willingly;
- enabling others to take responsibility;
- empowering others to do what they think is right;
- helping people to feel less fearful and more confident;
- developing, sustaining and changing the culture;
- having a bigger market and more profitable market share than competitors;
- having the most prestigious products and services on the market.

Van Knippenberg and Hogg [21] argue that leadership has three essential components. First, the would-be leader must convince the other members of the group to regard that person as a credible and legitimate source of influence. This recognises the significance of the status of the leader. Hollander [22] and Hollander and Julian [23] maintain that a leader's legitimacy flows from the perception that the leader is competent enough to help the group attain its goals and trustworthy enough to remain loyal to collective interests and objectives. Once a person has gained the legitimacy of leadership status, he or she needs to develop relationships with followers that motivate and enable them to act to attain collective goals. This recognises that leaders do not accomplish tasks on their own and that in understanding effective leadership, it is necessary to recognise the characteristics of effective followership. This includes ability and requisite competencies that must be recognised by the leader as well as the followers but also the motivation to expend effort to perform at high levels. Finally, the leader needs to mobilise and direct the efforts of the group to make the most effective use of the combined resources of the group in task accomplishment.

It must be borne in mind that there are cultural constraints on what leader behaviours can be used and their effectiveness. Most prevailing theories of leadership have a definite North American cultural orientation: an individualistic rather than collectivist approach; self rather than duty and loyalty; rules and procedures rather than norms; and rationality rather than aesthetics, religion or superstition. There are, however, many cultures that do not share the assumptions on which much of the theory is based and has given rise to an increasing amount of research geared towards understanding leading across different cultures.

Cultural norms are standards of conduct or acceptable behaviour in any given culture [24]. Different cultures have somewhat different priorities, and their leadership reflects their priorities. This reflects patterns of socialisation but it also likely reflects innate biological differences and pre-dispositions. At a physiological level, males and females differ in ways that have broad behavioural implications. Females seem to be attuned to more diffuse internal sensory and external social stimuli while males seem more attuned to and more focused on external objective stimuli. This suggests that men and women may reason and make decisions in somewhat different ways. Men seem to be concerned more with abstract 'truth' and appear to make decisions using objective reasoning. Women, on the other hand, seem to be more concerned with interpersonal 'fairness' and appear to process information and make decisions using subjective reasoning in a more empathic, personal and feeling way. Denfeld Wood [25] argues that leadership is made up of the task-oriented 'masculine' side and the relationship-based 'feminine' side. Increasingly, these two aspects – task and relationships – are seen to be fundamental dimensions of leadership. He goes on to identify the contrasting features of the two approaches including:

- autocratic versus democratic leadership styles;
- production oriented versus employee oriented behaviour;
- directive versus participative leader behaviour;
- reflections of Theory X (negative and directive) versus Theory Y (positive and facilitative);
- decisiveness versus harmony;
- abstract truth versus interpersonal fairness.

In Western society, the masculine principle is more prestigious and helps explain why most managers, male or female, think of political and military figures when asked to identify leaders. With the shift to post-Fordism we have seen some movement from the male to the female aspects of leadership. This is reflected in the components and associated traits of emotional competence. The first three focus on self-management:

- Self-awareness: accurate self-assessment, emotional awareness and self-confidence.
- Self-regulation: innovation, adaptability, conscientiousness, trustworthiness and self-control.
- Motivation: optimism, commitment, initiative and achievement drive.

The remaining two focus on relationships:

- Empathy: developing others, service orientation, political awareness, diversity, active listening and understanding others.
- Social skills: communication, influence, conflict management, leadership, bond building, collaboration, co-operation and team building.

Motivation and project leadership

Maylor [16] argues that the project leader has a responsibility to the organisation and the team members to ensure that they are provided with high levels of motivation in the new challenging environment of projects. There are a number of theories of work motivation. In Fordism and Taylorism, incentives were purely financial. There are considerable advantages but also major disadvantages such as worker alienation and demotivation. The Hawthorn studies identified the benefits of paying attention to groups and attending to their needs. Maslow's 'hierarchy of needs' published in 1943 set out different levels of needs. Process theories focus on the process of motivation. One of the key theories is 'expectancy'. The theory considers that people have a choice regarding the amount of effort they expend, known as the 'motivational force' in a given situation. This will depend on their perception of what they will get and how particular behaviour is initiated.

Reinforcement theories relate to how good or appropriate behaviour can be positively reinforced. Skinner [26] sets out five rules for reinforcing behaviour:

- Praise should be related to specific achievements and be based on current information.
- Praise should follow immediately good behaviour so that the recipient of praise can make the link between action and praising.
- Set targets that are achievable within realistic milestones.

- Intangible praise may be more of a motivator than pay or status.
- Unpredictable praise can be more rewarding than an expected 'pat on the back'.

Buttrick [27] argues that leading in a project is different from leading within an organisation. In line management, the manager or supervisor has the power and authority to instruct a person in his/her duties. Most likely they should have it but often they do not. Project managers have to deliver the project using a more subtle power base more rooted in the commitment of the team than in the authority of the project manager. Teamwork and team spirit is important in line management. It could be argued that it is even more important in projects. Reasons for this include the short time available for 'forming and norming' behaviours and to optimise performance and the fact that many team members are not dedicated to the project as they have other duties to attend to.

The project manager must be the leading player in creating and fostering a team spirit and enrolling the commitment of the project participants. Factors contributing to this are clear communication, realistic work plans and targets and well-defined roles and responsibilities.

> Projects need strong, experienced people to drive them forward and lead those involved. Not only must the project be efficiently administered, there should be a high standard of leadership so that people, both within and outside, accept its goals and work enthusiastically towards its realisation. Management drive of an extraordinary order may be necessary to get the project moving and to produce results of outstanding quality on time and within budget. To assure the necessary resources and support, the project may need championing both within the sponsoring organisations and externally, within the community.
>
> (Morris [28])

Morris [28] points out the increasing emphasis on entrepreneurship, leadership and championing in projects from the 1970s onwards. He predicted that in the twenty-first century, frequent change will become even more pervasive. Social, economic, demographic and environmental pressures will grow. More democracies will mean more political change. Technology and communications will become even more global.

As argued by Briner, Hastings and Geddes [29], there has never been a more comprehensive case for a new style of leadership within projects. Their approach is based on a number of concepts. The first are explicit and relate to managing organisational networking:

- The visible team – the group of people working directly on the project who come together from time to time to make the project happen. Often, as in the case of construction, many of the team members come from outside the leader's own organisation.

- The invisible team – the group of people who contribute indirectly to the work of the visible team and whose contributions are vital to the success of the project. Project leaders neglect the management and motivation of this group at their peril.
- The multiple stakeholders – the people who have an interest in the outcome of the project.

There are also two implicit concepts they identify:

- The organisational context which recognises the impact of organisational factors beyond the immediate project. There are often numerous vested interests, often at odds with each other. It often involves politics, change (where there are winners and losers), commercial and financial implications. Projects often fail because the project manager has lacked sensitivity in relation to these contextual factors.
- The people factor which recognises the importance of people in projects.

They go on to suggest that the project leader needs 360° vision. They suggest he or she needs to look in six directions – upwards, outwards, forwards, backwards, downwards and inwards.

Looking upwards involves managing the sponsor in order to achieve organisational commitment. Looking outwards involves managing the client, end user and external stakeholders (including suppliers and contractors) to ensure that the project meets their expectations. Looking downwards involves managing the team in order to maximise their performance both as individuals and collectively. Looking inwards to review performance and ensure appropriate leadership of the project. Looking backwards involves monitoring progress with the appropriate control systems to ensure that the project meets its targets and that the team learns from its experience.

Shusas

This evolving role in Western project management resonates with the 'shusas' in Japan. The differences in roles between conventional Western project managers and their Japanese counterparts, the shusas, have been compared by Womack *et al.* [30]. Originally conceived as part of the Toyota production system, the shusas are responsible for cross-functional activity, particularly in relation to product development. The key differences from the Western project manager are:

- the level of authority the shusas have;
- the opportunity they have to make a real impact on the project with less interference form senior corporate management;
- the way the role is seen as being an effective career route to senior management;

- the way shusas have power over the career path of individuals engaged in projects including their assignment to future projects;
- the guidance shusas provide in focusing the project team's attention onto the difficult trade-off issues – particularly in design.

What is an integrated project team (IPT)?

The integrated project team is a term used by the Egan Report, but is also referred to in the PMI Body of Knowledge as integration management. The Strategic Forum for Construction (SFC) [31] define a fully integrated collaborative team as

- a single team focused on a common set of goals and objectives delivering benefit for all concerned;
- a team so seamless, that it appears to operate as if it were a company in its own right;
- a team, with no apparent boundaries, in which all the members have the same opportunity to contribute and all the skills and capabilities on offer can be utilised to maximum effect.

This IPT removes the barriers between design and implementation, enabling all parties to be involved in discussion on the principles which will affect them, enabling high-quality decisions to be made on the basis of all the implications throughout the lifetime of the facility in question. It has four principles of

- committed leadership;
- culture and values;
- process tools and commercial arrangements;
- projects and organisational structures.

IPTs break down functional/organisational boundaries by using cross-functional/organisational teams. Running projects in functional parts with co-ordination between them slows down progress, produces less satisfactory results and increases the likelihood of errors. Lateral co-operation is better than hierarchical communications, which can become bureaucratic and slow down the contract. Increasingly, this involves taking people out of their functional or organisation locations and grouping them in project work team spaces (known as co-location). The recent trend has been to tip the balance of power towards the project and away from the organisation.

When selecting the members of the project team, SFC argues that special attention should be paid to enthusiasm and commitment, team attitude and communication skills as well as the technical attributes such as experience and technical qualifications. They recognise that team members often have differing and conflicting objectives and see the project manager as playing a key role in overcoming conflict arising from these different objectives. The project manager should aim to create an environment in which the team member can achieve personal as well as project goals outlines. This means using a problem-solving and no-blame culture where issues are identified, communicated and tackled early in the process.

Some work has been done to evaluate the leadership demanded by embracing the principles of an integrated project team that has been formed on the basis of competitive tendering [32]. The construction industry does not have the culture or track record for integrated team working and tough committed decisions have to be made to gain greater value. Leadership is having the courage to keep the belief when

- fighting off the cynics that believe the lowest cost is achieved by competitive tendering;
- you know that the collaborative relationship that you have with the contractor is not reciprocated through the supply chain;
- the agreed time has to be extended to ensure that value is achieved;
- the supply chain's price comes in well above the agreed budget;
- the supply chain puts more effort in defending why the price has gone up rather than meeting the agreed budget.

Leadership in this context needs to be generous and involved and means:

- building and inspiring a team that understands and is able to deliver the project ideals;
- being happy for each member of the supply chain to make a profit on your project;
- being prepared to work harder on your project to ensure that innovation can flourish and greater value can be achieved by more collaborative working.

A case study in the letting of one of the first project based prime contracts illustrating the concept of an integrated project team in action with this type of leadership is used in the final chapter of this book.

Conclusion

An increasing number of organisations in a number of sectors of the economy throughout the world are now recognising the growing importance of projects and effective project management. As project management becomes more established as a management approach, the development of project leadership as a career, with its own promotional path and professional recognition, will be an emerging feature of progressive organisations. Perhaps, in the future, only those who can develop the wide range of skills and knowledge to meet these challenges will be eligible to call themselves project leader. Traditionally construction has moved from full control of the workforce to a network of different contractor, specialist managerial and design organisations, who will have personnel who work for more than one project.

Construction project organisations have diversified their leadership with the spread in the use of different procurement types and the challenge to find new organisational forms that provide a framework for an integrated approach. Leadership is needed throughout the supply chain that discourages the traditional adversarial approach.

The influence of external factors such as market place competition, customers and the political, economic, social and technological factors has particular impact on how the project is run, as it has a unique timing, location and contract conditions which pass different risks to the project. The culture depends on the preferences of the project team and the client and the culture will be an adjunct of past work experiences and inherent characteristics such as the education, personality and aspirations. Different organisation cultures as discussed by Handy will have influenced their employees. The competition will influence the behaviour of organisations contributing and may restrain a full co-operative environment. Management style and the authority, the position and the power of the project manager will influence the way the communications occur within the structure and the degree of integration that can take place. Porter's model indicates how major suppliers influence prices, but other stakeholders such as the community, may also exert their influence on the project and gain a greater say or force additional accountability. Experienced clients may superimpose certain structures which they believe are critical such as VM, user group meetings and conflict panels. The introduction of customer account servicing, framework agreements, environmental constraints, incentives, supply chain agreements and partnering are driven by these customers, who will also favour the use of specific contract conditions that transfer different risks to contractors. Contractors may respond defensively to pass risk on, use additional insurance and withhold payment to suppliers, or positively to create strategic business units, develop trust, abolish retention, create single bank accounts and offer innovation.

World-class or 'best in class' performance is a stated aim of companies who wish to continue improving performance and will use measurements to benchmark their performance against leading best practice across industries. This requires strong leadership to find out how customers can be satisfied better, to champion change, to set tough objectives that have mechanisms for implementation as well as measurement, to train and develop personnel and to connect all this to improving the 'bottom line' by the elimination of waste in the system. Project leadership needs to be empowered and rewarded for this process of change to be sustainable.

Leadership definitions have been changing to take account of the need to include the role of champions for change and this better describes the more facilitative approach of the project leadership to inspire as well as to provide many management functions. This is especially so, as leadership in projects applies to many different contributing organisations in the supply chain as well as the project manager. Information control is likely to be a position of power, but only works with the use of persuasive and cognitive power (relationships) and this is also connected with motivating the team, which is covered more in the next chapter. A leader will develop their own style, but need to adapt it to different circumstances for effect. Political awareness with a small 'p' may also help this approach and help awareness of harnessing indirect stakeholder's support, as well as managing the visible team. It is, however, critical that the control of resources and tasks is not neglected in the 'new' paradigm.

This leads to the concept of an integrated project team that is working together and has time to develop a more tightly defined culture, in a loosely coupled context where organisations build up their trust in each other and seek to work as a single entity to squeeze out more value from the project. The next chapter looks at the way forward in engineering such an environment and reviews the tricky areas of conflict, negotiation and communications. Is an integrated team realistic and can it provide the solid basis for improvements in an adversarial industry, that has lost productivity?

Note

1 Bennett J. and Jayes S. (1995) *Trusting the Team*. The Reading Construction Forum, UK. (They have developed this principle to what they call second Generation partnering in a later book *The Seven Pillars of Partnering* (1998) which discusses the new demands of clients and the requirement for a higher level of trust and multi-level partnering. More of the latter is discussed in Chapter 12.)

References

1 Woodward J. (1980) *Industrial Organisation: Theory and Practice*. 2nd Edition. Oxford University Press, Oxford.
2 Peters T.J. and Waterman R.H. (1982) *In Search of Excellence: Lessons from America's Best-Run Companies*. Harper and Row, London.
3 Blake R.R. and Mouton J.S. (1969) 'Organisational change by design'. *Scientific Methods*. Austin, Texas.
4 Silverman D. (1970) *The Theory of Organisations*. Heinemann, London.
5 Allaire Y. and Firsirotu M.E. (1984) 'Theories of organisational culture'. *Organisation Studies*, 5(3): 193–226.
6 Harris P.R. and Moran R.T. (1987) *Managing Cultural Differences*. 2nd Edition. Houston, Gulf.
7 Handy C. (1993) *Understanding Organisations*. 4th Edition. Penguin, Harmondsworth.
8 Dubois A. and Gadde L. (2002) 'The construction industry as a loosely coupled system: implications for productivity and innovation'. *Construction Economics and Management*, 20: 621–31. E & FN Spon, London.
9 Porter M. (1980) *Competitive Strategy*. Free Press, New York.

10 Latham Sir M. (1994) *Constructing the Team*. Final Report of the Government/ Industry Review of Procurement and Contractual Arrangements in the UK Construction Industry. Department of the Environment. HMSO, London.

11 Johnson-George C. and Swap W. (1982) 'Measurement of specific interpersonal trust: construction and validation of a scale to assess trust in a specific other'. *Journal of Personality and Social Psychology*, 43: 1306–17. In Walker A. (2002) *Management of Construction Projects*. 4th Edition. Blackwell Publishing, Oxford.

12 Tom Peters (1992) *Liberation Management*. Pan Books in association with Macmillan, London. pp. 740–1.

13 Hines P. (1994) *Creating World-Class Suppliers: Unlocking Mutual Competitive Advantage*. Pitman Publishing, London.

14 Lamming R. (1993) *Beyond Partnership: Strategies for Innovation and Lean Supply*. Prentice-Hall, Hemel Hampstead, UK.

15 Briner W., Hastings C. and Geddes M. (1990) *Project Leadership*. 2nd Edition. Gower, Aldershot, UK.

16 Maylor H. (2003) *Project Management*. 4th Edition. Pearson Education, Edinburgh.

17 Rosenfeld R.H. and Wilson D.C. (2003) *Managing Organisations: Texts Readings and Cases*. 2nd Edition. McGraw Hill, London.

18 French J.R.P. and Raven B. (1960) 'The bases of social power', in D. Cartwright and A. Zander (eds), *Group Dynamics: Research and Theory*. 3rd Edition. Harper and Row, London. pp. 607–23.

19 Burke R. (2003) *Project Management Planning and Control Techniques*. John Wiley & Sons, UK. pp. 297–8.

20 O'Neil B. (2000) *Test Your Leadership Skills: Institute of Management*. Hodder and Stoughton, London.

21 Van Knippenberg D. and Hogg M. (eds) (2003) *Leadership and Power*. Sage Publications, London.

22 Hollander E.P. (1958) 'Conformity, status and idiosyncrasy credit'. *Psychological Review*, 65: 117–27.

23 Hollander E.P. and Julian J.W. (1969) 'Contemporary trends in the analysis of leadership perceptions'. *Psychological Bulletin*, 71: 387–97.

24 Cooper C.L. and Argyris C. (1998) *The Concise Blackwell Encyclopaedia of Management*. Blackwell Publishing, Oxford.

25 Denfeld Wood (1997) *Mastering Management*. Financial Times/Pitman Publishing, London.

26 Skinner B.F. (1974) *About Behaviorism*. Jonathan Cape, London.

27 Buttrick R. (1997) The Project Workout. Financial Times/Pitman Publishing, London.

28 Morris P.W. (1994) *The Management of Projects*. Thomas Telford Publishing, London. p. 255.

29 Briner W., Hastings C. and Geddes M. (1990) *Project Leadership*. 2nd Edition. Gower, Aldershot, UK.

30 Womack J.P., Jones D.T. and Roos D. (1990) *The Machine that Changed the World*. Harper Collins, New York.

31 Strategic Forum for Construction (2003) *Integrated Toolkit*. http://www. strategicforum.org.uk

32 Constructing Excellence (2004) *Focus group on Leadership in the Supply Chain*. Workshop conducted by the Bristol Century Excellence Club.

Chapter 7

Engineering the psycho-productive environment

At least 50% of project management is dealing with people and building relationships and by definition we are interested in managing these relationships to the benefit of the project. In the life of the project the aim should be to achieve better productivity, but also to look at long-term objectives such as retention of staff, building up customer relationships which provide mutual benefits and attracting the best talent for future projects in order to gain a competitive edge through an attractive and effective working atmosphere.

The aim of this chapter is to look at various strategies that would promote project effectiveness and efficiency by managing people well. Traditional leadership will be evaluated and some new approaches will also be discussed in the light of best practice. We shall consider communication, leadership, conflict and aggression, selling and persuasion, team building, problem solving, innovation, negotiation and team selection. With a lot of interest and generic research in this area it is important to make specific applications for the management of construction projects. As it is a complex subject, the chapter will give a broad overview in order to provide a holistic approach to possible changes that could take place.

The main objectives of the chapter are:

- Developing the role of leadership and team building.
- An understanding of the communication process and some techniques, which have been developed to make communication more effective in project relationships.
- Identifying and managing conflict and stress situations that occur in construction that specifically hit productivity.
- Definition of the concept of psycho-engineering and identifying ways that teams can critically improve productivity with the use of Belbin and Myers Briggs and other indicators in determining behaviour.
- To review the use of negotiation and other forms of bargaining in resolving conflict.
- The role of innovation and creativity in construction project management problem solving.

Teamwork

It is important to differentiate between group work and teamwork. Groups are defined as a number of people working alongside each other, but perhaps having little control over their combined outputs. Most work is a group-based activity, but not necessarily team-based. Groups can be informal as well as formal, cutting across the accepted authority structures. Several groups such as trade unions, professional clubs and work teams may work in parallel with some, but not all the same members. A lot has been written on the characteristics of groups starting with Mayo's Hawthorne experiments (1924–32) and by writers such as Likert [1], Tuckman [2] and Argyle [3] and Argyris [4] on informal groups. These cover such things as group types, cohesiveness, communications, roles, role conflict and group behaviour. Allcorn [5] makes a difference between the intentional group and the other groups. He describes the intentional group as non-defensive, where members accept roles and status which is relevant to the objectives of the group and all members are responsible. He contrasts this against the defensive nature of other groups.

A team goes a stage further, where the members are committed to common aims and are mutually responsible for their outcome [6]. They tend to be more autonomous and set up their own terms of reference, often to suit unique clients. Members of the group regard themselves as belonging and having a generic or technical role. It is the latter, which is more aligned to effective project working. However, there is a need to measure the amount of teamwork and the conditions in which it may flourish. In construction, teams are often formed from members in different organisations.

The level of team working could be measured by the impact measured on the overall project by a member dropping out – teamwork might well be improved as well as adversely affected. A *group* of people will not notice the positive or negative effect in the same way, as they will have weak relationships. What concerns the project manager is to optimise working relationships and this comes from experience and intuition, but will be helped by an understanding of the individual behaviour and interactions among members. Different people react in different ways to stimuli and different mixes of people have a different dynamic that may also be affected by the context and the working environment of the project. Pre-occupation with the fine tuning of *ideal* procedures and procurement systems is a setting and not a cure for the optimum delivery of projects and the people environment can be productively engineered to achieve synergistic behaviour and attitudes.

Tuckman's well-known development of team maturity by passing through the first four stages in Figure 7.1, suggests the need for a team to move *past* a normalising or neutral stage to use the synergy of the team to perform, is still a difficult lesson to learn as many teams have stuck at less than potential because they have not got to know or trust each other.

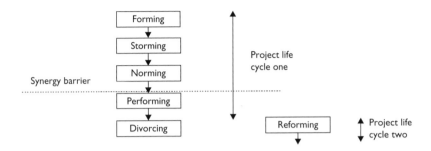

Figure 7.1 Adaptation of Tuckman's four stages of team development.

Critical things that effect this are the quality of the communications, the leadership, the conflict resolution systems and the contractual environment.

Construction has a particular challenge as project teams are often only brought together for short periods and the membership, speed and nature of the task are constantly changing at the different project stages. At the completion of the project the divorcing stage is often added to Tuckman's four, when members of the team usually proceed to new non-related project teams. This is characterised by a demotivating effect as members become concerned with other pressing calls on their time or their search for employment.

Project teams are often fragmented because of the break up of the project stages into client briefing and concept teams, design teams, construction teams and maintenance teams. Because of their common status, qualifications and professional standing, they become strongly cohesive and competitive to other sub groups, creating exclusion or rivalry and developing different sub objectives, which become removed from the primary client requirements. This is a disadvantage if a project manager does not take a life cycle view of project integration.

Walker [7] distinguishes between 'group think' where individuals spend an inordinate amount of time agreeing with each other and 'team think' where there is a willingness to talk through the issues, to be creative and encourage divergent views. Gunning and Harking [8] also emphasise the additional role of the project manager in building the team by 'managing their interactions and satisfying their ego needs' that will also require training in breaking down gate keeping or defensive behaviour and reducing uncertainty and mediating between conflicting demands. They remind us that there is still productive benefit for meeting face-to-face and that the need for these skills is not past history.

Leadership and motivation

Adair's well-known model indicates three aspects of leadership, which are achieving the task, managing the team and the managing individuals. In this

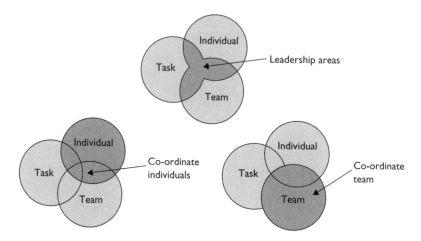

Figure 7.2 Project leadership.

Source: Adair J. (1979) *Action Centred Leadership*. Gower Publishing, London, p. 10. Used with permission of John Adair Foundation.

chapter, the main discussion is around the team and the individual issues which lead to effective project leadership, although it is acknowledged that leadership also concentrates on the co-ordination of the different tasks or disciplines that members of the team are able to contribute. Figure 7.2 shows these emphases using the John Adair model. The aim is to redress the balance and to consider the potential for managing the different human and cultural issues that are brought less consciously into the construction project team. The starting point is to attain client and project objectives.

Cornick and Mather [9] talk about the concept of profitable teamwork and the existence of teams does not automatically indicate better productivity. There is a chance that project team working may contribute to a less profitable outcome and worsening relationships, if clear insight and leadership is not imposed. The project manager is a team member as well as a team leader and the client may also have a similar dual role. Various sub teams may exist in project funding and business management, design, construction and commissioning also. Each of these teams will need to be led, the interfaces managed and the overall direction established. This leadership role is linked by the CIC (2000: 17) [10] to the ability to:

- Know how the construction industry operates and the roles and responsibilities of the parties. This is likely to require technical knowledge from both the client and the supply side of the business.
- Build or recruit a team inclusive of the relevant skills and make the relevant appointments confident that responsibilities are covered and working relationships are smooth.

- Establish a communication system, which is open and honest and engenders trust in the team in order to maintain and develop effective working relationships.
- Use motivation and team building theory to enhance the effectiveness of key project individuals and stakeholders.
- Manage team knowledge and project interfaces to maximise the proper planning and execution of the project to meet project and client objectives.

These abilities represent mainly people skills, but in the context of construction knowledge and understanding. One other is suggested, which takes into account the uniqueness of each construction project context:

- Facilitate appraisal and training regimes, which encourages the development of a defined project culture.

Cornick and Mather [9] suggest that there are four main cultures that the leadership evolves through. *Directing* where very clear instructions come from the leader; *coaching* where the leader prescribes advice and exercises to bring on more autonomy to inexperienced members of the team; *supporting or facilitating* where the leader provides resources and an open door for consultation if required to experienced members; *delegating* where the leader is able to brief the team, agree tasks and become a member of the team. Partington [11] refers to the difference between

- transactional leadership as working within the explicit and implicit contracts to achieve the task;
- transformational leadership, which is the charisma needed to change the status quo and unite people to the achievement of a new vision or 'higher purpose'.

In the latter situation the team is most productive and a good degree of trust, respect and understanding has been developed. The output should exceed the sum of the parts and room for further improvements should be planned. There is also a sense of supporting each other so that bottlenecks around the leader can be eliminated by interchanging roles. More team leaders may also be created.

The motivation aspects of management have been well documented, but the main motivational principles in the workplace can be categorised as economic reward or other extrinsic incentives, intrinsic satisfaction gained from the work and social relationships connected with the workplace. In the context of project teams, any of the three types of motivation might be applied. At a professional level, it is almost certain though that there needs to be a degree of satisfaction from the work itself and where there is a close performing team this is also likely to provide satisfaction and social stimuli. Care should be taken with the notion that a happy team is a productive team as it might just be comfortable and complacent or into group think. From the individual's point of view the psychological contract, which consists of the expectations between the individual and their employer and between them and the project leader, should be balanced. Frustration of the balance of personal and organised goals can be demotivating; on the other hand problem solving any issues leading to desired goals leads to constructive behaviour. Maslow [12] introduced a theory of satisfying needs and self actualisation, Herzberg [13] modified this by distinguishing between hygiene or basic level factors and those that motivated, introducing the idea of dissatisfiers.

McCelland [14] described needs in terms of achievements. Other process theories such as Vroom [15], Porter and Lawler [16], Adams [17] and Locke [18] looked at motivation as being dependent on the dynamic relationship between several factors such as expectancy, effort, rewards and performance. In other words if the credibility of these factors is poor, for example, the effort expended did not match the reward offered, then frustration and demotivation set in. It is difficult to 'put money on' any one of these theories, but the instinctive style of the leader should be used to be able to analyse factors that may be causing problems for the individual or by extension, the team. It is also clear that individuals may be motivated in different ways and there needs to be a different treatment of team members to get the best from them. Equally important is the principle of equity within a closely knit team, who will see themselves as needing equal treatment. It is in this context that valuing differences may help to build on the different strengths and manage conflicts and stress.

Conflict and stress

Conflict can be defined as a difference of opinion or a difference in objectives. Some conflict is inevitable and in a healthy competitive form can be beneficial in improving productivity. The very nature of construction creates contractual parties that by definition have differing objectives. In traditional procurement – a predictable conflict is created between the two parties to the contract, but also between design and construction. This is conflict, which is an inescapable consequence of our trading relationships. Other conflict that is dysfunctional – actions which have gone outside of the functional and have significant impact – is best resolved in order to stop escalation and the negative results of retaliation. Sommerville and Stocks [19] have produced a model by which to measure conflict on a three dimensional parameter model. This would grade complexity by the level of certainty, functionality and significance. In particular a significant, dysfunctional conflict with a high level of uncertainty requires attention and is disruptive. Other conflict may be insignificant, or within experience.

Stress is a natural reaction that normally enhances performance and may be used as a management tool, but prolonged stress or excessive levels of stress can produce adverse physical or psychological symptoms, which have the opposite effect. The Health and Safety Executive (HSE) [20] defines stress as 'the adverse reaction people have to excessive pressure or other types of demand placed on them'. Individually stressful situations like dysfunctional conflict may be created, which create de-motivation or absence. In the UK, it is estimated by the HSE that 5m people feel stressed or very stressed at work and that it costs industry £3.7b because of stress. The concept of stress hardiness as a measure of the effect of stress on different individuals is well known, but is also as a way of preparing people to react more positively to stressful situations. It may be possible to increase stress hardiness by means of control, gaining commitment and challenge. In addition, to do this a project manager may wish to introduce organisational help by teaching positive coping strategies, making stress counselling available, training or by matching individuals in teams.

In construction there is a machismo which has propounded as the norm certain levels of stress, which are now being questioned as exclusive practices that far from inculcating toughness and durable productivity, are distracting workers from the work in hand, nurturing unnecessary safety and health risks and creating regressive conditions which put off some from joining the industry. This may also be affecting the ability for creative alternative technology and methods designed to increase VFM. What can be done to deliver a non-threatening culture of continuous improvement and help construction to remain competitive? What are the main areas of conflict that are adversely affecting construction projects and how can they be resolved?

Our central thesis then for productive working is to psycho-engineer the environment by providing individual and team motivation, aware of the need to resolve dysfunctional conflict and prolonged stressful situations. The Belbin team roles and Myers Briggs personality indicator are just two methods that have been used to measure attributes and to use this information to match individuals in teams and also to allow them to have the information to self-develop (see later in this chapter). The following section identifies the particular causes of stress on construction projects.

Conflict and the life cycle

Thamhain and Wilemon [21] collected data on the frequency and magnitude of conflicts found in a range of projects to produce a measure of conflict intensity for different categories of conflict. This allows the possibility of heading off conflict at an early stage. Their findings categorise conflict under seven general conflict categories of project priorities (e.g. the use of a tower crane by many operatives), project procedures, technical problems, staffing

Table 7.1 Main conflict causes compared with project life cycle

Project life cycle stage	Conflict cause (significantly exceeding average conflict intensity)
Project formation	Priorities
	Procedures
Pre-construction (build up phase including design)	Schedules commitments
	Priorities
Construction programme	Schedule commitments
	Technical issues
Finishing stages	Schedule commitments
	Personality conflicts
	(on average, but significantly more than other phases)

Source: Thamhain and Wilemon [21]. Adapted from report as in Meredith and Mantel [22].

needs, costing issues, schedules commitments and personality conflict. These were measured by the degree of average intensity for each conflict type at each stage of the project life cycle. All conflict types were present in each life cycle stage, but some went above the average conflict intensity significantly as shown in Table 7.1.

Table 7.1 clearly shows the importance of managing time, cost and other schedule commitments throughout the executive stages of the life cycle, but differentiates project procedures, project priorities and technical issues in the first three stages. We must also differentiate between the implications of each conflict type for each life cycle stage. The conflict over priorities indicated by the solutions at the formation stage are making decisions in the strategic planning, whilst at the build up stage it is the allocation of resources in support of these priorities. Schedule conflicts develop from commitments in the formation stage to the breakdown of packages in the build up stage, slippage at construction phases and reallocation and completion at the final stages. Personality clash is listed as under average intensity in all stages, but is worst and on average in the frantic finishing stages of the project. However by implication, interpersonal skills are needed to resolve the conflicts that are occurring as a prime solution. Meredith and Mantel [22] list perfectionism, motivation and conflict as being the main behavioural problems facing the project manager. Others have done similar work for construction projects and have shown how conflict type and intensity varies between projects with different procurement types.

Meredith and Mantel have gone further and identified three generic categories of the reasons for conflict which are:

•   different goals and expectations of individuals or organisation;
•   uncertainty about authority to make decisions;
•   interpersonal conflict.

They identified the main parties in conflict as project or senior management versus client and project management versus senior management. Further, conflict takes place between contractors and between project management and contractors. The matrix nature of supply chains means that resource priority conflicts result from contractor involvement in several projects. Unless this is managed by the project manager, there would be different goals and expectations for project-based staff giving uncertainty about whether project or employer authority would influence their decisions the most.

To resolve conflicts of authority and objectives, a project manager is likely to use soft management skills, so that *project* loyalty is engendered. Disputes between client and project about schedule, cost and time objectives and possibly about authority in some of the technological decisions need to be made clear at the start. These are connected to the degree of involvement the client has in the project team and the procurement method employed. In more recent years, there is a push towards partnering objectives where the more experienced client is more fully involved in technological decisions that increase value. It is important to note that partnering is unlikely to lower the conflicts encountered, which are an attribute of the contracting conditions, but as attitudes are changed so the intensity of conflict should be lowered and more productive working conditions experienced especially in the area of personality conflict. They also give incentives for contractors to be involved earlier and to make suggestions for technical improvements. Partnering also involves 'soft' management skills and encourages approaches which produce a 'win–win' situation and tries to leave behind a competitive win–lose scenario resulting in residual resentments, but care needs to be exercised where one party is a dominant size or may exercise undue influence on the agreed terms of the pact [23].

Personality conflict may appear at first to be the source of all conflict. Meredith and Mantel [22] suggest that these are mostly created as a subset of technical conflict, the methods used to implement project results or the approach to problem solving. To resolve them care is required to determine underlying problems first and personal animosity may well be resolved as a result. Knowledge of personality type and behavioural strengths and weaknesses will also be key to the solution.

Principles of negotiation and conflict resolution

In a traditional approach to construction conflict it has been established that, many disputes are put off only to resurface at the final account stage and this is because of the heavy investment by all the parties in the project. If the project were not to complete especially by the time the implementation stage comes around then all parties would incur heavy losses. Conflict resolution at these stages, as Meredith and Mantel [22] point out, is based

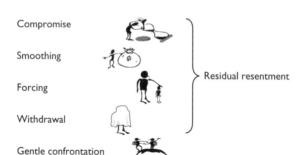

Compromise

Smoothing

Forcing

Withdrawal

Gentle confrontation

Residual resentment

Figure 7.3 Different ways to resolve conflict.

on 'allowing the conflict to be settled without irreparable harm to the project's primary objectives' i.e. finishing the building on time and to cost. Other disputes which involve determination of contract, unfinished buildings happen, but are not the norm in spite of what the press would have us believe. Escalated disputes causing failed objectives are not the main point of this section. Collective bargaining may have some similar characteristics. There are five generic ways of approaching dispute resolution and Figure 7.3 gives the essence of each, and although it points to one method, are all the other methods all bad?

In forcing, a dominant partner insists on a 'win–lose' solution which a weaker partner is not in a position to refuse. In compromising, two positions of last resort are established and a halfway point is agreed, commonly applied to financial agreements. In smoothing, persuasion is used in order to make it comfortable to accept a win–lose solution, often a short-term solution with long-term problems. In withdrawal, neither side refuses to admit that there is a conflict nor a conflict is put aside for a time. Where withdrawal, is one sided it may be similar to forcing in that psychological pressure forces one party not to face up to the implications of disagreement. Finally gentle confrontation is the opposite of withdrawal, where both sides agree that more effort should be taken to get to the root of the problem to bring about a 'win–win' solution.

The technique of principled negotiation developed by Ury and Fisher [24] is one of gentle confrontation and is a useful one for a project with an open culture, seeking to attain a 'win–win' solution. The four steps are set out as follows:

1   Separate the people from the problem.
2   Focus on interests and not positions.
3   Before trying to reach agreement, invent options for mutual gain.
4   Insist on using objective criteria.

A typical construction problem on poor information access may have created abortive work because a contractor did not receive their copy of the revised drawing on time. Tempers are running high with accusations flying in both directions and tools have been downed until 'some one sorts this mess out!'

Step one means sitting down around the negotiation table and allowing an element of time. In order to calm down emotions that may be running high it is important to define the actual problem and collect facts. This encourages each party to work on the problem rather than the feelings. Neutral facilitators are useful here.

Step two could establish that one side feels that this has happened too often and they are expecting it to happen again. The other side has particular concerns about the knock on effect on the programme, because it will highlight responsibility without one-sided blame and refutation. The matter of abortive work will follow. In focusing on positions pre-judgements have already been made. 'I can't possibly finish that in the time allocated' becomes a barrier if there are heavy liquidated damages. In focusing on the main interest, which is to finish on time without demotivation, an agreement is possible by removing fears.

Step three is the creative one and involves lateral thinking. A more water-tight system to avoid this and related problems in the future is clearly the first focus of the negotiation. If more than one viable solution is proffered then the other party is offered the dignity of choice. 'You must remove this architect from the job', may be countered by a separate proposal 'they are staying, but they will complete some design health and safety training'. Other solutions for mutual gain may emerge by comparing solutions from both sides and taking the best from each. This is not the same as compromise as residual resentments arise.

Step four looks at the principles to be used in coming to a financial agreement. It is clear that financial compromise is often the starting point for thawed relationships, though paying for abortive work is not a solution that deals with the cause.

Conflict and communication

Information supply is key to the successful construction project. Any delays disrupt the whole production process. Information supply is a fundamental prerequisite to raising productivity CPN [25].

CIRIA [26] indicates the need for systematic written communications (or a joint project extranet) on larger projects in order to bring about a co-ordinated understanding of requirements across a broad range of organisations and individuals. Typical communication problems are no allowance for geographically distant partners, poor communications in the early stages, more changes being made because right first time is not achieved, objectives that are divergent. There are also particular issues for communication and conflict in the use of fast track construction where construction is started at an earlier stage by progressively finishing areas of the detail design. It is most important that the design team have fully considered the interfaces between the different packages. Particularly helpful is

- face-to-face communication and the confirmation at meetings of action taken and to be taken;
- a record of formal approvals and changes to approvals and briefs;
- open and honest communication to take place, particularly when discussing changes, which so often produce dispute. Co-location, or use of IT tools for 'virtual co-location and better visualisation';
- regular open workshops, routine debates, a good ideas notice board;
- a well-differentiated and thought through circulation of material, with easy access to further material and involvement of the supply chain in all or any of the earlier mentioned.

CPN [27] have also compared the benefits for more overlapping models of communication between design and manufacture indicating earlier involvement of the contractor to increase communication and again to give more chance of 'right first time'.

A communication model

The model in Figure 7.4 illustrates the basic elements of communication, which are:

- a sender;
- a transmitter choosing a medium such as speech, document or phone;
- a receiver.

The difficulties that may be experienced are indicated as distortion or noise, but other psychological factors are in play as well. It is likely that communication in the construction industry suffers from some of the built-in prejudices that exist between different disciplines in the built environment and may also be influenced by the effect of varying education styles, for example, between the architect's studio, the engineer's laboratory and a manager's office. Individual situations and personal circumstances may affect the climate of relationships and clearly subconscious signals, often called body language, play a more powerful role than words themselves. Many different media can be used and the correct one should be used for the type of message sent and the impact required. More than one medium may reinforce the message and also provide feedback to the sender as to how accurately the message was received.

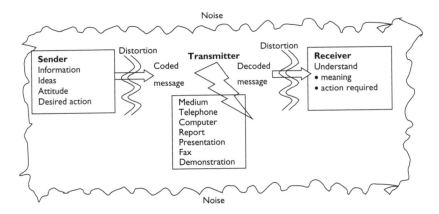

Figure 7.4 A general communication system.

Source: Adapted from Lucey [28].

Essentially, it is the responsibility of the sender to choose an appropriate medium and to make reasonable checks that a message is received and understood accurately. When using non-verbal language, different cultures may interpret these in different ways. For example, a 'put off' may become a 'come on!'

Clarity of language, the inconsistency in action compared with words, non-verbal signs, past experience and the relationship that exists between the sender and the receiving audience create noise. The 'non verbals' speak louder than words; in spoken and face-to-face communication the tone and the noted facial expressions or involuntary body movements provide an opportunity for rich interpretation. For example, folded arms may be taken as a sign of keeping uninvolved, but can also indicate a desire for privacy. Covering your mouth whilst listening can indicate a lack of belief in what is being said. Facial expressions and tone make words like 'no' mean entirely different things! Responsive 'non verbals' are also likely to influence what you say and the way you use language.

Perception is an important part of communications and this concept is well introduced in Mullins [29]. In communications, it refers to the unique way in which an individual views and interprets the world. This means that we may all get a different message from the same memorandum or instruction and it may be influenced by our personality, our attitude and our reaction to the current environment or by our past conditioning. To avoid this in key messages some sort of feedback should be encouraged on receipt. Meetings are an opportunity for interaction and getting an equal message across to everyone.

The project manager may attempt to converge individual interpretations by introducing a project culture, an inspired and inclusive communication system and encourage respect and attention to making us understood by others. This is harder if each new project has a new team, or if the same team works for a different client. It can even be difficult if assumptions about previous behaviour are wrong with the same client and team in a different environment and location. In other words, the dynamics of the situation vary on each occasion and there is only a limited project period to develop and understand culture. It is logical then to carry over a consistency, which is connected with the client or the project manager or company style, and to provide some training and feedback in the early stages of the project.

Noise is the element that impacts on communication as an external element. In simplest physiological form, a loud noise outside the meeting window means that only two-thirds of the words are heard or a headache affects concentration. In a more subtle form, it is referring to the other things that are going on for the receiver or sender outside the project. An unresolved domestic row is likely to loom more importantly than accurate interpretation of work documents. Work on another project that is going wrong may affect the tolerance of a subcontractor negotiating price changes on the current project. Severe time constraints may provide problems in ensuring safe environments.

Psycho-engineering the team

Psycho-engineering in this context, refers to the knowledgeable use of psychometric methods and motivation to build a robust team which is then able to form good working relationships and outputs that have been productively enhanced by the team. It is well known that teams may introduce waste (witness four work persons watching one person dig a hole), but a new team needs to create a synergy that is more than the sum of its parts to arguably justify its existence.

A construction project team is formed by appointing a number of different consultants, the project manager, the contractors' managers and possibly the client. The resulting organisations choose suitable members on the basis of availability and experience. It is possible that a client may specifically request a named individual, but this is not the norm so the team is not usually selected on the basis of personality, or with any other psychometric tests, that are common to standard recruitment. The job of team building by the project manager is vital especially in respect to working relationships. What can a project manager do if there is the wrong mix of team roles? How can the project manager deal with a personality clash, given that they have limited influence on the choice of personnel?

The Belbin hypothesis was based on observation of the interaction of management teams during training exercises at Henley College of

Plant	Creative and able to produce ideas
Resource investigator	Operates a good network of contacts and good at resourcing the ideas
Shaper	Able to whip up enthusiasm and to motivate the team
Chairman	Co-ordinator and listener. Good at drawing people out and summarising ideas for agreement
Team worker	Willing to support and develop other people's ideas and encourage working together towards common goals
Monitor/evaluator	Able to objectively appraise ideas and to keep control on critical parameters, for example, cost
Finisher	Good at finishing off and ensuring the final details are in place to get and maintain a working solution
Specialist	The starting point for many technical teams, relevant where specialist knowledge is required. Not really a generic role

Figure 7.5 Belbin roles.

Management by Meredith Belbin [30]. He noticed that teams that were stuffed full of clever people were not helpful to the solution of problems. He also noted that where a team was allocating roles and complementing each other, they were usually better than those teams where team members had similar roles and were competing against each other. The roles that he identified as important after further research connected with the smooth maintenance of the team. They are explained in Figure 7.5. He identified eight roles, in which some have changed names over the years, but have not changed in character. In order to assess these reliably, he developed a self-assessment inventory to help him to identify the role that each individual would find themselves most comfortable with. In reality individuals are comfortable with two or three roles and can cover roles where there is not a strong team representation of the role. This can work well in smaller teams.

Belbin also recognised that the team needed to determine a common goal. The main problems occurred when there was an overlap of two or more dominant team roles such as the shaper and the plant as these members were less likely to tolerate other ideas, or being 'shaped' themselves, unless the team could be subdivided. A set of pure team workers without any leaders may also be ineffective, but in the absence of team workers and a chairman the team was tetchy, indecisive and fragmented. The omission of a finisher meant that in complex problems, solutions often missed out on vital ingredients and without monitors there was a common problem with schedule overrun. Over time, Belbin remixed the teams to prove a startling improvement in the group dynamics concurred with his theory.

Belbin can be applied by trying to pick and mix a team, based on the inventory results as Belbin did to optimise performance. However if the team is already there, the team can be made *aware* of its weaknesses and agreement to cover the roles as discussed earlier. This awareness of the team working, together with a desire to work together can be as powerful as having natural role players. Roles are reinforced by training and practice.

In addition, the occasional or regular use of an outside facilitator or a creative workshop may reinforce these roles if they are lacking. The main reasons for removing persons from the team are:

- they are not team players;
- there is a deeply entrenched personality clash, which has been unresolved;
- there is a failure to accept an open culture;
- unwillingness to submit to certain team parameters and agreement.

This may be described as removing oneself from the team. There are occasions when a client has requested that a core member of the project team is moved on. Project disciplines are relevant in construction projects, but functional contribution has already been covered by appointment. However by themselves, they do not add to the synergy of the team, or overcome human problems.

Myers Briggs

Another approach is based on Jung's psychology of in-built personality type. A template has been developed by Isabel Briggs Myers and Katherine Briggs to provide an ethical approach to the identification of personality preferences in four dimensions. In-built preferences are polarised on each of the four scales of:

- extroversion(E)–introversion(I);
- sensing(S)–intuition(N);
- thinking(T)–feeling(F);
- judging(J)–perceiving(P).

A self-assessment questionnaire has been developed over a period of 50 years and has a high reliability factor. The combination of letters differentiates 16 foundational personality types (the Myers Briggs Type Indicator or MBTI™) indicating a starting point for an individual. Used ethically, they can point towards strengths and weaknesses and can guide individual choice and help to identify areas of personal development. Ethically they should not be used for selection, as they do not define behaviour. Behaviour is a function of additional factors, such as environmental conditioning, experience and choice. Used in teams for problem solving a mixture of personality types in teams should complement each other to arrive at better solutions. This gives a much richer understanding than Belbin about why clashes may occur and how they may be avoided between types. One of the more powerful applications for a project manager is to use an understanding of someone else's type to adapt his/her own natural preferences to improve individual communication with them and to tailor development

ISTJ	ISFJ	INFJ	INTJ
ISTP	ISFP	INFP	INTP
ESTP	ESFP	ENFP	ENTP
ESTJ	ESFJ	ENFJ	ENTJ

Figure 7.6 Myers Briggs type table.

Source: Myers Briggs Type Indicator is a registered trademark of Consulting Psychologists Press Inc.

programmes to suit learning styles. This also has implications for improving the outcomes of negotiation in an open culture. Personality type does not explain individual motivation, but it can give clues to understanding individuals' motivation.

Figure 7.6 indicates the 16 types possible using a letter from each pole of the four dimensions. These combinations represent the foundational preferences of an individual and do not seek to stereotype individuals as Jung acknowledges a natural desire for the development of the personality to master other areas, which are not immediate preferences. An example of this from the table is that the eight types represented in the first and last columns, which have a thinking rather than a feeling preference, may as project leaders see the benefits of types which recognise the use of harmony (F) rather than step-by-step logic (T) in the management of the project team.

An ESTJ (a common type found in Western cultures) may be roughly interpreted as loyal, practical, with a logical step-by-step approach and the ability to close out projects and make effective and systematic decisions. In many cases, they make excellent leadership material in a practical industry like construction. Their limitations will be the tendency to over control, a resistance to change where it 'ain't broke' and a pragmatic rather than a creative outlook. In taking on leadership, it may be necessary to develop more creative approaches and team-building skills, initially using a deputy who has complementary skills that you can work together with and gain confidence in strategic decision making.

Jung recognises the role of environmental conditioning to individual personalities. For example, socially there is need for expressing oneself publicly and this is not the natural preference of the eight types at the top of the table as they gain their energy from within and not from others. This produces a particular twist in that those with an 'I' preference unconsciously present themselves publicly as something they are not and others easily misinterpret their preferences. This stated, it is possible to interpret the behaviour of different types with some degree of training. It is dangerous to guess the type from the behaviour! Environmental conditioning and conscious personal development provides depth and uniqueness to individuals' personalities and contrary to popular opinion has the opposite effect to stereotyping. MBTI™ can be used for developing team working, ensuring balanced problem solving, improving communications and resolving conflict. It is often used for identifying leadership style, but it is not a measure of good or bad leadership.

Behaviour can also be altered by severe stress and there is a possibility to cross over from one's preference to one's opposite in these circumstances. This can be recognised when a team member is perceived as acting out of character. For example, a normally placid ISTJ will take on the extroverted characteristics of an ENFP and will move from present to future possibilities in assertive fashion warning of doom and gloom. It may also be apparent when certain personality types under stress are asked to go through major organisational change.

Myers Briggs and conflict

It might be thought that by having opposite types working together that conflict might result with a natural instinct by the leader to keep certain types apart, but this is not the case if progress is to be made in existing teams in standing in each other's shoes. The key is to use the clear vision and blind spots of different personalities and to resolve conflict by an understanding of conflict generators for that particular personality type and what they need from others and in the longer term promote personal development. The most dangerous situation is when the team is working under significant stress. In this situation, what you see is not what you always get.

The research done by Killen and Murphy [31] have indicated that the two last letters contribute most to conflict. They call these the conflict pair. In the fourth dimension, judging types have most difficulty adapting to creating space for decisions to be made and vice versa; the perceiving types have most difficulty seeing why there is such a rush to make decisions and need more time for information to check it out. In the second dimension indicated by the poles of feeling (F) or thinking (T) there is a major conflict between the Ts who want to fix what is wrong and the Fs who want to make sure that everyone is heard and respected. This creates four possible combinations TJ, TP, FJ and FP, each showing up in four of the personality types (Figure 7.7). The thinking types' objections tend to arise out of hurt pride in a challenge to their authority or trust. The difference between the TJs and TPs is that one wants a defined process and the other wants closure and conflict sorted. Both detach themselves from emotion. The feeling type objections arise when people are not listened to. The difference between the FJs and FPs is that one wants intact relationships and the other wants open exploration and for all to be heard. They both accept the role of emotions.

By mixing types it is possible to get a good mix for solving problems and the model used is to create space using the Ps, add value by using both the T and F approach and to bring about closure by using the Js.

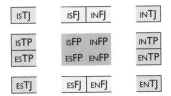

Figure 7.7 Killen and Murphy conflict pairs (explored Myers Briggs type table).
Source: Adapted from Killen and Murphy [31].

Myers Briggs and communication

Communication may be improved by the sender being aware of the receiver's tendencies to understand the message. According to the communication model, the ability to choose the right media and to encode the message with reference to the language understood by the receiver would help. Myers Briggs personality types might give further clues to this process where known. According to Brock [32], this is best determined by the use of the Myers Briggs middle letters. This has been researched with reference to effective selling by understanding your buyer and further applied by Allen [33] to the health service, which has a reputation for poor patient communications. This approach helps to communicate points clearly and persuasively in a negotiation. The technique will be most useful in aiming to understand mutual needs and producing a win–win result. Used as a tool for domination they are not suitable for the sustainable relationships needed in projects.

The Mobius™ [34] communication model is another application which can be linked with Myers Briggs. This model equates acknowledgement with perceiving, responsibility with sensing and capability with thinking, judging with commitment, possibility with intuition and mutual understanding with feeling. The main reference to communication is the connection of extroversion with blame or praise (faults or strengths first seen with others) contrasted with the connection of introversion with worry or claim (faults or strengths first seen with self). The communication approach for Mobius gives a negotiation and problem solving sequence as follows:

1 Mutual understanding – of likeness and differences, this is designed to create an atmosphere of well-being (F).
2 Explore the possibilities – recognise the common ground (N).
3 Commitment – where choices made within the framework of possibilities get firmed up (J).
4 Capabilities – indicating an implementing strategy for deciding on resources and skills (T).

5 Responsibility – this commits the who, what and when to the project so that people are involved in doing it (S).
6 Acknowledgement – is a monitoring and feedback action which provides a basis for assessing success (P).

It can be seen this is also useable as a win–win negotiation framework and its connection with Myers Briggs suggests that the use of a team of negotiators, each engaged at the relevant part of the process to use their natural preferences, might be beneficial. The first three steps bring the negotiation to agreement and the final steps ensure the efficient working of the agreement before. Applying this to a PFI negotiation indicates the difference between selection of the preferred bidder stage and financial closure.

Appreciative inquiry

In the area of feedback, appreciative inquiry (AI) is a methodology for giving positive feedback by breaking people out of their typically negative mindset and habits. It looks to the concept of learning opportunities rather than the more negative concept of problem solving. The main drive is to motivate a no blame environment, which encourages learning by doing in which mistakes are seen as the tools to teach. Cynics may call it the art of ignoring the disgruntled, but this may be countered where there is a widespread change in culture developing.

The steps include identify, appreciate, inquire, envision, focus, affirm and sustain. These steps allow a neutral discussion of the issues which have the potential to be improved, by looking at what you do well and analysing whether generic processes may be applied to improvement of this and other areas. A brainstorming 'what if' scenario might be useful in the envisioning step, which then needs to identify areas for action and maintenance of better systems.

An example of AI in customer care is to inquire 'Describe a time when you went an extra mile for the customer – what made it possible?' This provides an opportunity to explain something that went well and analyse it for the innovative features that could be reused.

Change and developing a project culture

Part of setting up the project organisation is the desire to influence the nature of relationships. Traditionally relationships operate on a 'them and us' basis between the consultants and the contractors and between the consultants/contractors and the client. The closed culture is also likely to have infiltrated the supply chain relationships and the contractor organisations themselves. AI might help the latter in the context of overall change. The discussion here relates to the nature of an open culture and how to move away from a culture of closed and regulated relationships.

The CIRIA C556 [26] report describes a culture of openness as a blame free environment, with an admission of errors on the basis of agreement on the best way to move forward. It also provides a single system of reporting to client, where a single understanding of the time and cost safety margins is tendered and change is acknowledged by all parties. This system allows the formation of a true team and incentives which encourage waste saving improvements.

The CIC [35] emphasises the need to identify and ensure standards of specific project management competencies that enable continuous improvement and the appointment of a project manager with experience fitted to the particular needs of the client's project. This may be enshrined in an integrated team of people, which they list as the project manager, design manager, construction manager and trade contractor managers. These competencies are listed in 25 areas under the five main heads of:

- strategic;
- project control;
- technical;
- commercial;
- organisation and people.

This raises the issue of extending the culture to supply chain relationships (including consultants and construction managers) and the need to integrate the culture through different life cycle stages. It may also put some responsibility on the client, to appoint a team that can work together well in an open culture.

It depends upon competent management to keep costs down and an open book tendering system which allows the client to see where costs are incurred in the event of change which realistically is part of the deal in rapidly changing markets. In the case of cost certainty then extras need to be balanced by savings and a claims culture should be avoided by agreeing who takes on the risks and who pays for contingency. The main challenge of an open culture is the sustained commitment for fundamental culture change of the whole supply chain and perhaps some clients' reluctance or inexperience in becoming closely involved. A number of contracts are suitable for this type of contracting, but they need to be used in a spirit of partnership to be effective.

Traditional contracts encourage a closed culture with familiar, but complex rights and conditions which have grown up with modern contracting, putting the client on the 'other side of the fence' and providing them with a 'fait accompli' that asks for specification in return for product and keeps the client away from the project team decision making. This may be what the client wants, but in many cases greater flexibility is required with client choice to work the system knowledgeably and not by trial and error.

The benefits of a closed culture theoretically lie in the experience of professionals in operating the contract, the 'fixed price' where little change is likely and the ability of the project team to proceed without interruption. In reality, these advantages only exist in some market conditions and over short gestation and construction periods. Track record and conditions should be vetted by the client to ascertain that these advantages exist for their project.

The fixed price contract is an erroneous title as changes and redesign are common and may create extra cost. In a traditional contract, even with a bill of quantities, revised prices for change are rarely fully defined prior to go ahead. This makes it difficult for the client to predict final cost and in the case of disagreement at final account one or both of the parties is likely to lose out, because an agreed common path was not taken at the original change position. An additional complexity is the role of third party design consultants, who may have had a role to play in the creation of additional costs. In short any one of changing requirements, competitive tendering conditions, insufficient time and agreement for the design, information flow, newly formed project teams and less than competent management can spell disaster.

The NEC is people rather than contract centred as it sets out early warning systems, which allow the two sides to get together and to firm up the effect of change on programme and budget and requires a regular resubmission of an agreed programme and formal approval of revised budgets to proceed (Case study 7.1). It has also been criticised for its lack of legal clarity if tested in a court of law. Is that good or bad?

Case study 7.1 A best practice case study of the use of NEC contract Heathrow T5 [36]

The electrical installation of the £4b Heathrow Terminal 5 contract is worth £72m and is being installed by EDF Energy and includes a 365 day, 24 h maintenance of the occupied airport high voltage (HV) and low voltage (LV) supplies. The design is being carried out simultaneously with construction with a varying element of supplier design. It presents many challenges to a traditional form in providing cost certainty over a long period, with interfaces with a large number of sub packages which are on a supply and install basis, mostly by suppliers who are not used to being contractors also. It was claimed that the NEC family of contracts was chosen because of its virtues of flexibility and clarity and that it embodies the principles of effective communication. Because of the wholesale lack of experience of this type of contract and because of the limited contractor experience, a joint residential training course was set up to focus on the practical application of the contract. The project team accepted that the open culture of the NEC contract helped the resolution of conflicts at package interfaces and enabled cost certainty because of the quick settlement of compensation events. The learning curve though steep, was achieved more easily because of the simpler structure of the contract, which suited those who had no prior expectations of other contracts.

The conclusions were that the initial difficulties in understanding were more than overcome and offered excellent reporting and visibility. Visibility also enabled substantial additional works to be carried out at a fair price to the client.

(CIOB [36])

Conclusion

The industry has gained in the past from non-bureaucratic relationships which are a characteristic of project work. However, now the construction industry has only paid lip service to new ways of changing the confrontational culture which has been created by the contractual forms commonly used over the last 60 years. There is a long way to go in catching up with other industries in developing integrated ways of working which present continuing opportunities for reducing waste and adding value through better understanding of the benefits of an open culture. More is needed in engineering in the form of a conducive open environment, moving away from blame and building relationships that recognise the importance of trust and openness. One of these initiatives is 'respect for people', where people are clearly indicated as a valuable resource.

There are opportunities also in the area of personality profiling and team building in order to deal with conflict positively and to stop it becoming personal and dysfunctional. This is not another alternative dispute resolution with additional mediators, but a wholesale change in manager attitudes throughout the supply chain who

- believes in the bottom line benefit of a 'type Y' approach to management;
- cultivates the mutual benefit of win–win negotiation and understands the importance of personality in all types of conflict;
- convinces clients of the benefits of collaborative type contracts such as NEC in breaking down contractual attitudes and developing where appropriate long-term relationships.

There is plenty of scope for improving the quality and value of one-off construction projects by training and motivating the workforce with these attitudes, so that clients respond and repeat work is generated by their recommendation. The challenge for implementation is 'up front' investment in training people and sticking with it until it happens. In a wider adoption of good practice, human resource development is changing ingrained attitudes which have been diametrically opposed.

References

1 Likert R. (1967) *The Human Organisation*. McGraw Hill, New York.
2 Tuckman B.W. (1965) 'Development sequence in small groups'. *Psychological Bulletin*, 63: 384–9.
3 Argyle M. (1989) *The Social Psychology of Work*. 2nd Edition. Penguin, Harmondsworth.
4 Argyris C. (1964) *Integrating the Individual and the Organisation*. John Wiley & Sons, Chichester, UK.
5 Allcorn S. (1989) 'Understanding groups at work'. *Personnel*, 6(8): 28–36, also in Muller J. (1993) *Management and Organisational Behaviour*. 2nd Edition. Pitman Publishing, London.
6 Hiley A. (2004) 'Transferring to teamwork'. *CEBE News*, Update issue. 10 June.
7 Walker A. (2003) *Project Management in Construction*. 4th Edition. Blackwell Publishing, Oxford.
8 Gunning J. and Harker F. (2004) 'Building a project organisation'. Chapter in *Architect's Handbook of Construction Project Management*. RIBA Enterprises.
9 Cornick T. and Mather J. (1999) *Construction Project Teams: Making Them Work Profitably*. Thomas Telford Publishing, London.
10 Construction Industry Council (2000) *Construction Project Management Skills*. CIC, London.
11 Partington D. (2003) 'Managing and leading', in R. Turner (ed.), *People in Project Management*. Gower Publishing, London.
12 Maslow A.H. (1943) 'A theory of human motivation'. *Psychological Review*, July, 50: 370–96.
13 Herzberg F. (1974) *Work and the Nature of Man*. Granada Publishing, London.
14 McClelland D.C. (1988) *Human Motivation*. Cambridge University Press, Cambridge.
15 Vroom V.H. (1964) *Work and Motivation*. John Wiley & Sons, Chichester, UK.
16 Porter I.W. and Lawler E.E. (1968) *Managerial Attitudes and Performance*. Irwin (Richard D.), Homewood, IL.
17 Adams J.S. (1965) 'Injustice in social exchange', in Berkowitz L. (ed.), *Advances in Experimental Social Psychology*. Academic Press, New York.
18 Locke E.A. (1968) 'Toward a theory of task motivation and incentives'. *Organizational Behavior and Human Performance*, 3: 157–89.
19 Sommerville J. and Stocks J. (1992) *Psycho-Engineering the Productive Environment*. Strathclyde University.
20 HSE (2004) http://www.hse.gov.uk/stress/index.htm (accessed 30 November).
21 Thamhain H.J. and Wilemon D.L. (1975) 'Management in project life cycles'. *Sloan Management Review*. Summer, also in Meredith J.R. and Mantel S.J. (1995) *Project Management*. John Wiley & Sons, Chichester, UK.
22 Meredith J.R. and Mantel S.J. (1995) *Project Management*. John Wiley & Sons, Chichester, UK, p. 175.
23 See the CBPP/CIB (1999) Model Project Pact. May. Also comes with guidance notes from http://www.ciboard.org

24 Ury W. and Fisher R. (1983) *Getting to Yes.* Penguin Books, Harm

25 Construction Productivity Network (CPN) (1996) 'Design mar industry', CPN Workshop Report 21. 17 October.

26 Lazarus and Clifton (2001) *Managing Project Change: A Best Practice Guiac.* CIRIA/DTI C556, London.

27 Construction Productivity Network (CPN) (1996) 'How others manage the design process', CPN Workshop Report 20. January.

28 Lucey T. (1991) *Management Information Systems.* 6th Edition. DPP Publications, London.

29 Mullins L.J. (1993) *Management and Organisational Behaviour.* 3rd Edition. Pitman Publishing, London. pp. 129–49.

30 Belbin M. (1981) *Management Teams Why They Succeed or Fail.* Heinmann, London.

31 Killen D. and Murphy D. (2003) *Introduction to Type and Conflict.* CPP, Palo Alto, CA.

32 Brock S.A. (1994) *Using Type in Selling: Building Customer Relationships with the Myers Briggs Type Indicator®.* Consulting Psychologists Press, CA.

33 Allen J. and Brock S.A. (2000) *Health Care Communication Using Personality Type: Patients are Different.* Routledge, London.

34 Stockton W. (1994) *Integrating the MBTI with the Mobius Model.* Mobius Inc. mobiusinc@msn.com

35 Construction Industry Council (2002) *Construction Project Management Skills.* CIC, London.

36 CIOB (2004) Contractors struggled with ECC at Terminal 5 but discovered its benefits in the end. 21 July. CIOB International News. http://www. ciobinternational.org

Chapter 8

Managing risk and value

with Tony Westcott

The chapter:

- tracks the development of risk and value techniques and the influence they have on today's practice;
- considers the concept of risk and value, people's values and attitudes and their connection with decision making and business planning;
- demonstrates how value, risk and opportunity are assessed, managed and reduce uncertainty in construction projects and evaluates the tools and techniques;
- investigates the influence of procurement routes on managing risk and value and supporting best practice in the appointment of the supply team;
- critically appraises procurement guidance as proposed by private, public and EU sources for managing risk and value and deriving an ethical approach;
- considers the effect of and response to change in the client brief.

Ten years ago the application of RM and VM in construction projects was very limited and the industry has lagged behind others in recognising how critical they are to building client confidence and in meeting their requirements. As clients have become more sophisticated they have made more demanding targets that have left less room for manoeuvre and these have led to tighter budgets, more innovative technology and less tolerance of time slippage, which is often connected to more onerous penalties for lack of performance. In response, the industry has looked at enhancing value for clients and managing their risks better by the use of more advanced techniques. It is now recognised that the scope for changes in design to give significant improvements in value come mainly at the feasibility and strategy stages and that it becomes more difficult as the design is progressed. This comes about as efforts are made to understand and incorporate client value systems better.

Traditionally, risk control in the construction stage was based on implicit heuristic assumptions and allowed for in the budget or the programme as a pragmatic contingency or, in the case of the contractor's tender, as a risk premium. Risks which make assumptions about external factors and arise when early design decisions are made also have to be managed. These require better predictive models to cover the range of probability in response to factors outside the control of either the client or the project manager. Simulation may also be used to model scenarios where a number of factors may all change.

Enhancing value is often connected with risk analysis, through the assessment of alternative methodology and functional value. They are both used to weigh up the operational, as well as the immediate capital costs. This inter-dependency of risk and value is recognised as critical and the two separate techniques are integrated in this chapter to allow a more holistic discussion. This is not an 'how to' chapter on techniques, but argues the case for an holistic, integrated approach in order to support effective managerial decision making. Previous chapters have identified risk and their fair allocation amongst the parties to the contract.

Discounted cash flow (DCF) models may be used to assess the viability of medium to long-term projects and to optimise the balance in life cycle costs between the capital expenditure and operating costs. Thus, the DCF model provides a method to model the risks and maximise the value. They are particularly valuable in DBFO and related forms of procurement where the sustained risk of the provider and their interest in the operational costs of the building combine most strongly.

Historical approaches to RM and VM

An over-arching concern of clients in procuring construction is to achieve added value from the project. The value of a construction project is determined by the benefit it creates for the client. Typically, this value is measured in monetary terms and therefore the VFM is the primary concern. In the 1960s, when construction cost-planning methodology was still in its infancy in UK, James Nisbet identified VFM as the most difficult of three facets of cost control to achieve. The other two facets of budgetary control and appropriate balance of expenditure over the construction elements are much easier to demonstrate and deliver through effective cost management and cost planning, but VFM requires understanding and evaluation of the client's values as part of defining the project brief. This is no simple matter when the client and/or end user may be a complex organisation or diverse group of stakeholders with disparate interests and requirements.

The principle of VFM in public procurement has been a long-standing principle espoused in HM Treasury procurement guidance; failings in construction procurement have been a long-running criticism of the NAO, particularly in capital projects for the construction of nuclear power stations, motorway construction, hospitals, the Thames Barrier and most recently for the Scottish Parliament. VFM in government projects has traditionally been measured against cost targets set by reference to historic average costs and by competitive tendering where the contract is awarded to the lowest bidder. Since the mid 1990s, the emphasis has changed to VFM measured against the project business case, where life cycle cost considerations can be as important, and even more important, than initial cost in determining VFM.

Evaluating the business case for a project implies an assessment of the degree of certainty in the project outcomes. Project uncertainty in terms of completion to the expected project functionality, cost and programme are the main cause of stress for construction project clients; unexpected changes to any of these key parameters are the primary cause of client dissatisfaction with the construction process. The process of gateway reviews at key milestone points are designed to identify and manage project uncertainties, and to allow projects to be deferred or abandoned if project uncertainty is unacceptably high.

VM seeks to maximise value in relation to function and cost, whilst RM seeks to minimise the uncertainty of it not being realised. Both are, therefore, central to ensuring the successful project delivery. Both VM and RM services are essential features at each stage in the Gateway Review™ process of managing procurement in the Achieving Excellence in Construction procurement guidance issued by the UK OGC [1].

The origins of VM can be traced back to the Ersatz movement in Germany when the country was facing the difficulty of regenerating the industrial and commercial infrastructure in a climate of postwar austerity with a desperate shortage of money and materials. The search for added value was not necessarily about lowering cost; value could be added by raising functionality. German industry, exemplified by names like Mercedes, BMW and Audi, have demonstrated that the quality and reliability associated with *vorsprung durch teknik* has added value, rather than cut the initial purchase costs of these cars.

Value engineering, 'the organised approach to providing the necessary functions at the lowest cost', owes it origins to Laurence Miles and his postwar work at GEC in USA. Early development of the techniques of value analysis, function analysis and the VM Job Plan was primarily for US military projects. These techniques, which are characteristic and essential elements of VM methodology, are incorporated in the VM standards of SAVE International, which describes itself as the premier international society devoted to the advancement and promotion of value methodology.

The economic turbulence of the 1970s, triggered by the 1974 and 1979 oil crises and their associated property and construction slumps, stimulated a period of great interest in UK construction of comparative analyses against other countries' practice, most notably of USA and Japan.

In the mid-1980s, the construction management company of Bovis teamed up with Lehrer McGovern in the USA to introduce American techniques to UK on the Broadgate development, including value engineering, and resulted in innovations such as cladding panels supplied with integral internal finishes and pre-installed engineering services and 'chandeliering' of multiple floors of steel beams into position on one crane lift.

Investigation by Kelly and Male [2] in the early 1980s, identified differences in approach between VM in the USA and the UK. Central to this was the use in USA of a VM team to audit a design at scheme design stage. Several states have made this a requirement for verifying VFM in publicly funded construction projects and an example was given of this procedure for prison capital works in the state of New York.

The VM audit is undertaken by a multi-disciplinary team under the direction of a VM Team Leader and the team of experts are appointed separately from the design team by the client. The additional fee is often justified on the basis that it will represent some 10% of the savings identified and the savings will be at least 10% of the overall project budget. The difficulty is that these 'savings' are implicitly a criticism of the original design team. Who will pay for and take design responsibility for the required design changes?

By contrast, the UK has tended to promote VM as an additional service provided within the original design team as a way of improving the understanding of the client's value system, business process and the project brief; a way of testing the VFM of the design as it develops and, in the process, a major contribution to team building and encouraging collaborative interdisciplinary teamwork to improve integration of the design and construction process. Fragmentation of the design and construction process in the UK has been a major criticism of UK government reports for more than 40 years, as evidenced for by Emmerson [3], Banwell [4], Latham [5], Egan [6] and Fairclough [7].

Research by Stuart Green in the late 1990s focused on the contribution of VM as a group decision support method, or a method to improve strategic decision making by project teams, which was differentiated from the SMART or MBO approach to one which is defined by Connaughton and Green [8]: 'a structured approach to defining what value means to a client in meeting a perceived need by establishing a clear consensus about the project objectives and how they can be achieved'. This gives VM a social value in improving the way the project team work together in defining criteria for project success and then delivering to these criteria and in this respect furthers the argument for the provision of VM and RM services in consort. However, Green identified the problem of a lack of an established theoretical underpinning to the softer approach to VM by comparison with the recognition given to the more established quantitative techniques within the SAVE International and other existing VM certification schemes. Davis Langdon and Everest [9] presents success as a fusion of managing value, risk and relationships, with the conditions for success being a number of things, but including:

- project objectives that are clearly established and understood by all;
- reconciliation of stakeholder needs;
- reduced risk;
- improved communications and understanding;
- innovation and creativity;
- enhanced team working;
- effective use of resources;
- elimination of unnecessary costs and waste;
- optimised whole life costs (WLC).

Thus effective VM at the strategic project stage can substantially reduce the potential uncertainty in a project and increase the chance of success.

VM and RM as decision-making tools

VM can be used as an early decision making tool if the client's value system is known and is harnessed by the involvement of the client and RM can be used as a decision making tool if the client's attitude to risk is known.

Client values

A project has many stakeholders who may have different values. A client's value system is derived from value chains from each stakeholder group which set the strategic criteria and from a project value chain which develops in tactical detail as it moves through the phases of the project life cycle and may be exposed to different stakeholders in each stage. Various stakeholder interests are related to the sector such a social housing, school or road construction project. Figure 8.1 shows different emphasis through the life cycle.

Understanding the Client Value System is crucial to successful project management and is central to function of VM. The essence of a VM study is to question convention or preconceived ideas by a rigorous process of creative

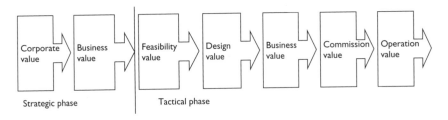

Figure 8.1 Individual project value chain.

Source: Based on Bell (1994) and Standing (1999) in Male (2002), citing [10].

and constructive analysis. This is followed by dedicated VM workshops when the client and the project team – designers and constructors if possible – are brought together. It is vital that the VM facilitator encourages participants to share their ideas as equals in a creative process. Decisions may be made more and more by the project team as the life cycle develops, but the client values still remain at the centre of their ongoing approval of distinct stages.

Risk attitude

Risk attitude has been classified by Flanagan and Norman [11] as either

- risk loving;
- risk averse; or
- risk neutral.

It is popular to cite bold risk taking as a feature of successful business, but in the context of a project there is a greater tendency to pass risk down the line. Many clients are taking a risk averse view that as amateurs in construction and property they should pass related building risk down the line and resist the lines of risk allocation established in traditional contracts where many building risks are shared. This can be successful where the real estate and technical decision making is also delegated. In PFI, DBFO and BOOT a facilities solution is offered and agreed for a substantial period of time so that the client can concentrate on their core service and/or profit making.

A client who views their building as an investment option will apply a risk premium to their returns, if they feel that they are taking more than normal risk. A client will carry out an investment appraisal to look at the affordability and returns of a real estate investment and the returns, which are likely given the market predictions. This is carried out on a similar basis to the appraisal techniques covered in Chapter 2, which illustrate discounting techniques. A risk premium can be applied through a higher discount rate, which will emphasise the importance attached to a short payback period and uncertainty in the future. This will raise the hurdle return on investment by which a project's viability will be judged.

In a similar way, the contractor will lower the tender price at the tender evaluation stage in order to make themselves more competitive for a project, or raise the price to cover the perceived higher risk. Risk loving contractors may actually see an opportunity in the higher risk project to make more money. The danger with a traditional form of construction procurement is that the client perceives the burden of risk has been moved to the contractor on a lump-sum fixed price tender, whilst the contractor perceives that those parts of the building not fully detailed provide the opportunity to bid low and recover costs on later variations.

A risk neutral approach is an approach that could be interpreted as a zero sum in discussing risk allocation to benefit all parties and put risk where it can best be managed. This requires an open attitude and the development of an associated VM approach which in particular makes the early appointment of contractors in the development stage of the contract essential. There is nothing easy about this type of approach; it requires more effort and more early commitment, often at a stage prior to finalising contracts.

The choice of procurement systems introduces its own risk allocation and this will be discussed later. This may also be associated with VM as a process tool.

VM in practice

To illustrate the process of VM, consider a VM workshop for students who were asked to develop a brief for a national Millennium project. After some initial ice-breaker and team-building exercises, the students were allocated a range of different stakeholders to represent, covering potential investors, developers, designers, constructors, operators, residents, tourists and sponsors. This immediately raised the issue of who were the most relevant stakeholders, how should they be represented and in what proportion, but clearly the developer was going to need public support and acceptance of his proposal. Next the question was raised as to whether the project was necessarily a building or several buildings or an event, rather than a building.

A mission statement of 'Provide a national focus for the celebration of the new Millennium' was agreed and this was brainstormed for performance criteria to help formulate a strategic brief. Part of the function tree diagram is reproduced in Figure 8.2.

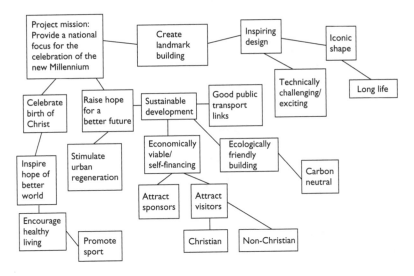

Figure 8.2 Function tree diagram for Millennium project.

An example of a strategic VM study to clarify the project definition was a study undertaken in Ireland (Case study 8.1) for a developer proposing to build a new hotel.

Case study 8.1 VM at the early concept stage

> The combination of a tax-free industrial area and the growth of international flights connecting Ireland with America had created a strong basis for economic development and particularly a strong growth in international tourism. As the VM workshop explored the drivers for growth in tourism, the importance of an international golf and conference facility was realised, to the extent that the project brief became a much larger regional and international attraction than a hotel.

Case study 8.2 is a more detailed example of the use of VM at the early stages of the project life cycle where there is the biggest scope for developing the brief to build value. It also illustrates how a multi-disciplinary approach to problem solving can produce more VFM. It also illustrates a situation where spending more produces higher added value by a factor of two or more.

Case study 8.2 VM at the early concept stage (office refurbishment)

The situation involves an office refurbishment of approximately 2500 m² of office block behind a listed Victorian façade, which has to be retained. The five floors of the building are to be gutted internally and a new steel frame office is to be built inside the retained façade. This façade will have to be retained during demolition and rebuilding; an existing party wall will help to stabilise it during this process, but in this wall there is also a major central chimney as in Figure 8.3(a).

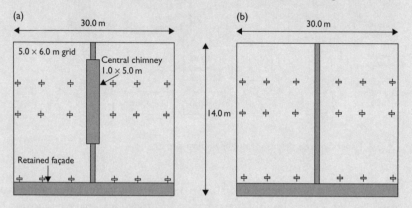

Figure 8.3 Office refurbishment: (a) option A (retain chimney) and (b) option B (chimney removed and more floor space).

The structural engineer advises that if the chimney is retained (as in option A) it will save on demolition costs and provide stability during rebuilding but also reduces the structural bracing required in the new steel frame with a possible saving on the construction programme of one week. The QS assesses the construction costs of the two options and estimates that option B, with the central chimney removed, will cost an extra £15 000 for demolition plus an extra £8000 for structural bracing to the steelwork. On the basis of construction costs and programme there is therefore, a clear argument to recommend option A, but during the value engineering appraisal the commercial property surveyor raises the issue of the loss of net lettable floor space in option A. Option A has some 21 m² less net lettable floor area, allowing 300 mm for the party wall thickness, than option B and the developer values this at £71 000, based on a rental of £20 per square foot or £220 per m² and a yield of 6.5%.

Calculation
This equates to a years purchase of 100/6.5 = 15.38;

$$15.38 \times \text{floor area } £220/m^2 = £71\,055.60.$$

After deducting the additional demolition and bracing costs, this leaves

a net additional value of £48 000 for taking option B.

By the same means, we can evaluate the benefit to the client of saving a week on the construction period, which is

$2474 \, m^2$ at £220 per m^2 per annum/52 weeks = £10 466.

Assuming an immediate let, option B still provides

a net added value of some £37 500.

Functional analysis and FAST diagrams

Function analysis is a fundamental precursor to the creative or innovation stage of VM. Once the VM team have received a full briefing on the project, it involves the VM team in brainstorming to identify the required functions of the project. The discipline of describing functions by a combination of a verb followed by a noun helps to sharpen the clarity of thought. Figure 8.4 illustrates how the required functions of a door might be expressed.

These would typically be collected by team members individually brainstorming the project goal or focus of the VM study for the required functions and writing these on post-its, displaying these anonymously on a team board and then discussing how these can be structured into a FAST diagram using the convention outlined in Figure 8.5. The higher order function or focus of the VM study is set on the left of the diagram; primary functions relating to essential needs are moved towards the top left of the diagram; lower order functions or wants are moved towards the bottom right; the assumed functions which will indicate potential solutions then emerge on the right. The model is tested by iterations back and forth across the diagram: moving left to ask why this assumed function is right and moving right to ask how the required function can be achieved.

Internal doors
- Enclose space
- Give separation of functions
- Resist fire
- Reduce noise transmission
- Separate environments
- Support signs
- Provide access/exit
- Resist vandalism
- Aid security
- Aid vision
- Provide architectural feature

Figure 8.4 Functional analysis of a door.

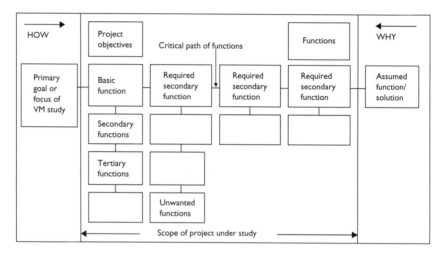

Figure 8.5 FAST diagram.

Examination of the model against the client's value system will begin to identify the relative importance of the various functions, and the relative costs of functions over and above the basic functions will help to identify the extent to which secondary and tertiary functions are relevant.

As the model is developed and examined, a range of potential solutions will emerge. After an initial sorting of the most promising solutions, these can be examined by more detailed life cycle costing to identify the optimum solutions to present to the client.

It should be clear by now that the function analysis can be undertaken at different levels, usually determined by the stage in the project life cycle at which the study is undertaken, that is:

• Tasks, for example, What is the primary goal of this project?
• Spaces, for example, What rooms and where?
• Elements, for example, What form of building enclosure?
• Components, for example, Do we need marble from Italy?

This depends on client values discussed earlier.

Understanding the client value system is crucial to successful project management and is central to function of VM. The essence of a VM study is to question convention or preconceived ideas by a rigorous process of creative and constructive analysis. The following questions illustrate this:

• Have you thought of ... ?
• Is this better?

- Can it do this?
- Is this cheaper?
- I've got an idea...!

It is vital that the VM facilitator encourages participants to share their ideas as equals in a creative process. This is done in dedicated workshops.

VM workshops

To be part of a complete integrated system, VM needs to interface with the project life cycle at key events, Kelly and Male [12] describes a continuum of staged events, including those in Table 8.1.

Woodhead and McCuish [13], bemoan the stereotyping of VM in UK construction by short 1–2 day workshops, or even half-day workshops, which they believe do not give enough time for detailed value and function analysis, and claim they become structured cost studies focusing on cost reduction. They propose VM studies facilitating common understanding of project objectives, shared values, innovation and creativity over a longer period.

The Job Plan in the SAVE International toolbox is commonly described as a 5-day workshop, covering the following phases, which have been merged with Woodhead's proposals for the steps of the initial VM workshop:

- *Information*   Senior management determine metrics. A base case is determined where the project team develop an explanation of their current thinking about the project (i.e. how they are currently proposing to deliver the project and its results). The base case is translated into the functions that the system (i.e. the elements of the project) must perform. The functional representation frees us from any particular method (or solution) so that we can later consider alternative ways of performing the functions.
- *Creativity*   Brainstorm different ways of performing the functions that leverage value. That is, target specific functions and ask: 'How else can we perform this function?'
- *Judgement*   Sort out the ideas on which we are going to spend quality time writing up considered proposals.
- *Development*   Take the considered proposals and create a menu from which we calculate the business results of each scenario and pick the ones we will present to senior management for approval.
- *Recommendation*   Develop our presentation, where senior managers from day one, return to hear the team's recommendations and make a decision as to what the new base case should be.

The practical difficulties of time tabling a multi-disciplinary panel of experts, stakeholders and senior management to meet mean that a shorter event is often held incorporating many of the aspects.

Table 8.1 Continuum of VM workshops through the life cycle

Description	Duration (days)	Focus	Stage
Pre-brief workshop	0.5–1	The strategic brief	Clarifying the business need
Brief workshop	1–2	The project brief	After the initial business case has been established
Brief review workshop (Charette)	1–3	Examination of the brief already prepared by others (as an alternative to the two workshops earlier)	At the end of briefing, before concept design
Concept design workshop	1–2	Outline sketch design	Before submission of planning application
Detailed design workshop	1–3	Final sketch design	Examination of the design by functional elements
Implementation workshop	0.5–1	Are recommendations being implemented?	
Contractor's change proposal	1	Buildability studies	After appointment of contractor or in second stage of two-stage tender with performance incentive
Post construction review	1	Feedback, especially where ongoing programme if capitalworks	

Source: Based on Kelly and Male [12].

By virtue of utilising techniques such as brainstorming, option appraisal, life cycle costing and buildability studies, VM has had some difficulty in the UK being accepted as a separate discipline. Again, 'Why should the client pay extra for something that the design and construction should be doing anyway?' is still a typical response. To some extent, value engineering has been seen in the UK as the lifeboat that is brought in to bring a project at tender stage back within budget, and the danger is that participants see this as no more than a multi-disciplinary cost-reduction exercise, with little rigour in the focus on value and function analysis, with little creative thinking and the adoption of solutions based on previous experience. Alternatively, VM may be viewed as a contractor-led exercise to improve buildability with a cost-sharing clause added to the contract as an incentive, and yet the textbook approach proclaimed over more than 30 years has been to emphasise the benefit of early strategic VM exercises. An interesting question to pose of companies offering VM services to the construction industry is whether they use VM internally to improve their own product and processes.

Definition and evaluation of risk

Risk arises out of uncertainty. Risk in the context of project management is a realistic approach to things that may go wrong on the site and it is used in the context of decision making and in answer to the question, 'What happens if...?' Once a risk has been identified and defined, it ceases to be a risk and becomes a management problem. In this context, it needs to be analysed and a response made – usually to accept the risk, mitigate it, reduce it or transfer it. People reactions are important and the response to the same risk varies according to who is effected, how many and who is responsible for it. This may also lead to the ethical question of acceptable risk. It is important then to classify risks so that appropriate action is taken. It is also important to separate the source of the risk from the consequence of the risk happening:

- Edwards [14] classifies by *source* distinguishing between *external* risk that is beyond project manager control, for example, interest rates, *internal* risk that is caused by the activities on the project, for example, accident, damage, design or methodology impact and *transmitted* risk which has an impact on others for example, environmental damage.
- Burke [15] classifies risk as to the degree of uncertainty, connecting it to the level of available *information* about the project. He introduces a risk continuum from total risk to no risk (unknown unknowns > known unknowns > all knowns).
- Flanagan and Norman [11] also distinguish pure and speculative risk. The former typically arises from the possibility of something going wrong and has no potential gain. The latter is connected with a financial, technical or physical *choice* and has the possibility of loss or gain. This type of risk can be quite creative and may be evaluated as worth the pain for the greater gain.

The consequence and probability of the risk happening help to decide the priority and assessment of the risk, which is discussed later. The unpredictability of the risk is going to lead to less control, or to more sophisticated assessment methods to increase predictability. Many current techniques used in construction projects, do not analyse the degree of unpredictability and therefore, do not estimate probability correctly. External events tend to be less predictable and their consequences are likely to be greater, so they cost more when they occur outside the planned schedule. They also affect a broader range of activities and is therefore, potentially more disruptive. Ignoring or reducing risk by lowering probability of an unknown is foolhardy. A stochastic simulation or sensitivity analysis approach makes sense in this situation. Another approach is indicated by Goldratt's [16, 17] theory of constraints which provides a framework for identifying critical gatekeeper bottlenecks/risks for focused priority control.

Risk can be evaluated by quantitative, qualitative methods or both. Quantitative methods allow a tangible value to be attached to the risk and are suited to weighing options such as investment choices or assessing the greatest probability as in the time taken for a programme. These often use tools such as expected monetary value (EMV), Project evaluation and review technique (PERT) or Monte Carlo analysis to provide a range of sensitivity. Qualitative is more holistic and provides a systematic evaluation of a range of factors and their interaction. Tools such as decision trees, heuristic analysis and 'fishbone diagrams' provide a logical framework. In both cases some subjectivity is involved in assigning priority or a value for probability or financial impact as in any of the systems mentioned earlier. Classification then provides a basis for the proper treatment of risk.

The consequence of risk in terms of its likelihood and severity may help to decide what to do with risk. If it is a very frequent risk, but does not create a huge financial impact it is likely that the risk may be retained and managed better to minimalise the risk. If the impact is larger or harder or more expensive to manage, then insuring it provides a way of spreading the cost of the risk. If the impact of the risk is very large such as an environmental disaster and is more than rare, then it may be expensive to insure and a decision may be made to move away from that activity or to specialise in managing it. The case of asbestos removal is an example of the latter case. Another issue in consequences is how closely it is possible to predict the impact and frequency – is it known? What track record is there in its control?

Response is usually in terms of reducing the risk by management, spreading the risk (as in portfolio management between several unrelated projects), avoiding the risk, transferring the risk to others or insuring against it. It is important to note that in the uncertain and fast moving climate of projects that there will always be a residual risk which needs to be managed. Transferring risk to someone who has no particular experience in managing that risk is likely to be expensive. Not all things can be insured against including the result of illegal activity and fines. There are many other areas where insurance costs become prohibitive.

Some insurance especially for third parties is compulsory or required and further insurance may be requested by clients to protect their interests and property and these can be provided on a level playing field. Other insurance has been brokered by professional organisations, as a condition of fair trading for members to provide an ethical and professional service which protects clients and the reputation of its members. PII insurance to cover the impact of human error and negligence is an obvious example of this. Other insurance may be taken out on a voluntary basis, but the cost of insurance and the limitations of compulsory excess and upper limits needs to be balanced sensibly with the cost of managing the risk in order to remain competitive. This blurs the boundary between pure and speculative risk. In cases of rare catastrophic loss and as part of the management solution to limit liability for loss, a subsidiary insurance company known as a captive insurance company may be created [17].

Integrated approach

Any risk management (RM) system will cost money to administer and this cost also needs to be monitored with the cost of the consequences. Sticking with familiar tasks and investing in training for new ones reduces the exposure of a company and keeps them competitive. This does not preclude entrepreneurial activity to move into new markets, but marks a note of caution. New markets such as DBFO or BOOT create risk and as the track record of managing that risk improves, so funders feel happier to reduce the cost of finance and to make the market more competitive. New ways of working that ultimately reduce risk, such as partnering and repeat working in prime contracts have the potential to reduce costs. Here the key aspect is the building up of trust between partners not only to value manage, but to eliminate wasteful double coverage of risk and reduce or eliminate the risk of dysfunction conflict. It should also provide the opportunity for synergistic team working and of joint insurance coverage in the core team. Unproductive risk transfer may be reduced by the application of better supply chain management and risk retention can be used on a voluntary and not on a compulsory basis. Subjective delayed payment can also be outlawed more effectively where teams are prepared to work together better with automatic payment of suppliers, therefore reducing the risk of non-payment. This broadens the application of RM beyond the application of quantitative techniques, because the measures to work together comprise a managed risk to trust each other. However, where it is agreed it is likely to reduce many risks for the project team and the client.

RM process

The process of managing risk starts at identification often by brainstorming ideas as well as referring to any standard list of risks. A risk register records the significant risks and tries to categorise them in order to try and assess them. Significance is assessed by evaluating predictability, probability (frequency) and impact. There is then, a response to reduce significance by eliminating, reducing risk levels or transferring risk and, finally, managing the residual risk. The process is shown in Figure 8.6.

Figure 8.6 The RM process.

Risk assessment

The process of assessment involves finding the level of significance to provide some guidance to the seriousness of the risk. This is usually carried out on some sort of matrix or scale. Significance is a product of impact and probability. Figure 8.7 is a simple way for assessing a hierarchy of importance. A rating which is 2 or over would indicate the need for investigation. A rating of four or over (i.e. frequency or impact high) needs attention.

Risks become more risky when they are the more uncertain. And this can reflect itself in a subjective rating of probability, as indicated in the EMV method in Table 8.2. EMV takes each risk and provides a monetary value for impact and reduces it by multiplying by the degree of probability. Monetary value is another way to quantify the impact and probability is rated 0–1. If this is carried out for all items in the risk register, a total risk transfer value is achieved by summing all the extended values.

The bottom-line shows a total valuation of risk which, if the risk rating is consistently prepared may also be used to compare the risk of other options and thus support the decision making in option appraisal. Risk assessment in this context particularly varies with the procurement option chosen. Later occupancy or whole life risk impacts will be better shown as discounted as this impact is later.

From a ranking point of view it must be remembered that these figures might be subjective in both the financial assessment of the impact and the probability and can therefore be unreliable, or biased according to the compiler's subjectivity. This can be especially be the case where the compilers want a business case to be successful. Objectivity can be built in by forcing compilers to use standard ranges of values for given common situations, or

Impact Probability	Low	Medium	High
Very unlikely	1	2	4
Occasional	2	3	5
Often	4	5	6

Figure 8.7 Rating significant risk.

Table 8.2 Risk assessed by monetary value

	Probability £'000	Impact 0–1	Extended = impact × probability £'000	Ranking
Client scope change	200	0.9	180	1
Planning qualification	100	0.4	40	2
Planning delay	200	0.1	20	3
Higher cost of construction	600	0.4	24	3
Industrial action	500	0.05	25	3
Latent defect	100	0.2	20	3
Heating failure	50	0.05	2.5	4
Total monetary value			311.5	

interrogated through a KBS. A project however is unique and an audit of such a method is important where the affordability and VFM are close to the bone. It is, however, an easily understood way of assessing risk and the project team can own the results and easily explain the impact and probability figures. Probability figures may be backed up by stochastic forecasting, but the final probability should be adjusted for the riskiness as discussed earlier. For example, predictability is affected by the distance of the event away from the time assessed and the experience of the contractor or client in dealing with that risk.

Risk attitude is important in how the response is managed as we have seen. A defensive attitude to risk is to add a risk premium to the tender, if the risk is sensed to be abnormal to normal business, or to hold a retention on a contractor if the contractor is perceived to be untrustworthy or incompetent and/or to require bonds. Risk transfer may be defensive or good business. Reducing risk may be considered an expensive add on or it might become part of the business of adding value to the contract. The narrower view involves managing up to socially acceptable risks and failing to use the opportunities of positively increasing value. Speculative risk might provide innovation and value for the project or it might be one risk too far.

The opportunities to change risk attitudes come with integrating RM and VM, so that there is a positive motivation factor over and above the level playing field of hygiene factors. The impact of this is to encourage sensible risk in the area of innovation and to build in systems that properly plan the risk and other issues such as quality and environmental response. This provides the opportunity to see RM as a money spinner rather than a cost to the overheads.

The assessment then, is the basis of the executive decision making of the project manager, affected by their attitude and their ability to influence the client and the project team. In DBFO and BOOT projects and other negotiated tenders, there is a chance to develop a common approach to risk and to share it in the most profitable way.

Risk response and RM

Risk response occurs to eliminate, mitigate, deflect or accept the risk [18] and logically will reflect the cost benefit of the RM process. Mitigation is action taken to reduce the risk and deflection is action taken to transfer the risk. They are not mutually exclusive, but deflection alone is not a way of reducing the probability. Mitigation may have the effect of reducing probability and impact. Accepting pure risk is a management decision that eliminating it is not economic, or creates too many other risks. Certain pure risks (e.g. health and safety) will have legal or social expectations for mitigating them. Managing residual risk is part of the process of RM.

Flanagan and Norman refer to certain rules for risk taking such as, 'Don't risk a lot for a little, don't take risks purely for reasons of principle or losing face, never risk more than you can afford to lose, always plan ahead, consider the controllable and uncontrollable parts of the risk and consider the odds and what your intuition and experience tells you'.

They also refer to the need to gain knowledge about the risk and to remove ignorance. It is in any case unlikely that a risk will be eliminated completely without creating other risks. Ethical and social considerations should be taken into account as well as the business economics in deciding what level of residual risk is acceptable. The reputation of the client and the business is also at risk.

The difference between pure and speculative risk is important in the response. Speculative risk will be accepted in order to take on an opportunity. Success in this case, will be in the identification of the risks accepted and mitigation of those that are not acceptable. Here a business may build up its reputation by taking the right risks and managing them successfully. The transfer of risk by the client is a factor of the procurement method and further risk may be transferred to the contractors in the negotiated contracts, so that VFM in allocating the risk to the party best able to manage it is gained. It may be further transferred through passing it down the supply chain (as in asbestos management), by insuring against it or by raising a bond. If the client ultimately wants reliability, the use of inclusive maintenance contracts should also be considered.

Contractors, designers and clients will also put different priorities on the same risk. For example, the risk of an overrun in budget because of design changes is a risk that will most affect the client in a traditional contract who is likely to pick up the bill for any design and scope changes. Design changes can mean extensive reworking of design for limited remuneration to the designer too. It also affects the designers reputation, if there are too many unexpected changes for which there is little to balance against elsewhere. The contractor will also have to suffer disruption if the scope is changed late, with many knock on effects for resource management and abortive work. These disruptions may be absorbed, or a decision made to

Table 8.3 Risk assessment methods for speculative risk

Title	Risk approach	Example/application	Connection with VM
Investment analysis	The methods are described in the business case chapter and may be assessed on payback, NPV or through building in a risk premium into the client's 'discount hurdle rate' using IRR or breakeven. More confidence is achieved by also carrying out a sensitivity analysis	Assessment of financial payback associated with balancing CAPEX and OPEX, with income received from a building over a period of time	Excellent use for assessing financial risks involved in comparative schemes or assessing costs of an alernative maintenance régime for WLC
Sensitivity analysis	This is a mechanism for testing if a small change in one of the expected variables has a disproportionate effect on the overall result. More sophisticated analysis can also be applied such as multiple regression and multi-attribute value theory	If the capital cost of energy insulation was to go up by 20% then it might facilitate against the additional insulation by cancelling out discounted savings in heating costs or making the payback too long. Each of the variables can be plotted on a graph of life cycle cost and percentage variation in parameter	This has obvious usage in WLC and may help to establish the evidence for the proposed value for money and the optimum balance between CAPEX and OPEX
Portfolio management	Portfolio management is diversification of investment to mitigate the risk of loss if one market collapses. By definition it has less potential for premium returns. It is applicable for managing the risk of poor returns and emerges from financial planning. Diversification does not work if the whole industry is affected by adverse conditions such as poor demand for buildings. Niche markets are another variation in looking for opportunities and exploiting them to balance poor returns	Portfolio management may be used by the contractor to spread their risk of loss by taking on a range of projects so that if work dries up in the leisure market then there are other income earners say in housing. Businesswise this can cut both ways as projects are unique and an even wider diversification of skills cuts down efficiency and hence margins may be effected	This is not really a VM tool, but it may be useful to the client in choosing between projects

Table 8.4 Risk assessment methods for RM during the project

Title	Risk approach	Example/application	Connection with VM
Expected monetary value (EMV)	A risk register is compiled and a value is attached to the consequences of things going wrong. This is illustrated earlier. A decision needs to be made as to what premium is added to the project	Use in a public sector comparator (PSC) in a PFI contract and in comparison of different methods of procurement	Alternative schemes can be compared or methodologies to cut down the degree of risk or the probability on the risk register
Decision analysis	A flexible technique for structuring decisions into 'stages' and to apply weighted probabilities for alternative choices. It provides a method which takes into account related decisions and gives flexibility to allow for risk attitude and subjective impressions. There must be a similar number of stages for alternative decision chains.	Use to look at probability and cost of say alternative types of river crossings such as bridge, ferry, tunnel, do nothing. Stages might be 1 Which one? What type? 2 When to start? How long? 3 How much?	This is very valuable for giving an integrated VM/RM comparison for each of the expected monetary values that have been devised. Discounting can also be applied if there are ongoing costs
Theory of constraints (TOC) Goldratt	Applied to project management Goldratt based his project application on identifying the weakest link where there is most risk and focusing all management effort on these in what is called the critical chain. This prioritisation is reached after project wide workshops which link to VM and stakeholder management	This has been used holistically to identify major risk of the weakest link in a critical chain of programmed events for major rail infrastructure projects that are logistically demanding and fast track [19] and to manage it closely on the basis that last past the post is the critical factor for completion of the whole	This method brings value in itself by saving on management time and releasing them for value adding activities
Monte Carlo simulation	Stochastic probability model. Generates sets of random numbers for each of the variables within a range and compares the cumulative outcome (e.g. cost, time) on a frequency basis for the total of each of the random models created. This is most powerful where the number of iteration (simulations) generated reaches a large number, making it statistically significant	The PERT model below is a Monte Carlo simulation making a model for project time. The highest frequency of project time estimates is represented as the best estimate of time	Simulation may be limited because it is difficult to quantify functional requirements or clients values, but it has a role in assessing WLC's

| Project evaluation and review technique (PERT) | PERT is used on schedule planning. It is a probalistic model based on an evenly distributed frequency distribution curve with a weighting of most likely, m = 4, optimistic, a = 1 pessimistic, b = 1. The equation expected time/cost = (a + 4m + b)/6 is built into a computer programme based on a range of values for each activity that are usually weighted towards the pessimistic end. There is a level of subjectivity in choosing the range and its distribution around the mean

| Pertmaster have developed a risk anyalysis which is able to produce a Monte Carlo simulation generating a large number of combinations of activity time or costs within the range for each activity. The most probable overall cost or duration is given together with the range of values for a given standard deviation. Pertmaster generate this on a triangular distribution | This is mainly a scheduling tool, but could be used to test the programme, or cost certainties of different designs |

| Multiple estimate risk analysis technique or root mean squared | This method is based on calculating a risk free base estimate and an explicit calculation of a risk allowance rather than an arbitrary percentae for contingency just based on experience. The risk allowance is calculated by taking the difference between the maximum (90% probable not to be exceeded) cost of each particular risk and the average likely (or 50% probability) cost of that risk, taking the square of that difference in cost for each risk, adding the squares for all risks and taking the square root of the total. The result is a base estimate, an average likely estimate including risks and a maximum likely estimate | A project approval process might require the maximum likely estimate does not exceed the average likely estimate by more than 15% to receive project approval to proceed to the next stage. The method has the benefit of identifying the extent to which risks have a fixed or variable total cost as well as their probability and cost impact | The process of identifying the risk improves project understanding and clarity of the brief. The likelihood of successful delivery is improved by making potential risks more explicit |

claim for additional payments. Likewise weather conditions are mainly a contractor risk, but an extension of time affects all parties.

RM techniques

Many textbooks have been written covering different risk techniques and here it is proposed to summarise the techniques, which best fit in with a combined risk and value approach. An example of its application is given, but this is not an exclusive use of the method. Table 8.3 is split into speculative and pure risk methodologies.

The given table indicates methods used by clients and contractors in assessing their exposure to risk in assessing whether to take on a project and is concerned with the risk of costs including overheads exceeding the returns that may be had from investing the money in bonds or less risky ventures and needs to show economic profit over and above that. It requires entrepreneurial skills over and above the RM techniques to maximise opportunities, so on their own these techniques might help to compare options or to support a business case or a key market evaluation. Tender adjudication depends heavily on the market conditions and their predictability is often assessed by speculative risk assessment techniques. Major innovative methods and life cycling risks may also be judged to come in the speculative category as they present opportunities, which over the extended period of the life cycle are less predictable.

In short Table 8.3 shows other methods of RM which can be used at the project definition and implementation stages of a project in order to differentiate between different alternatives. They have more relevance to the value engineering approach and can also help in managing the project in the cost and time control areas.

Table 8.4 deals with RM and assessment techniques during the project that are connected to a feed forward approach to control.

Risk, value and procurement

We have already referred to procurement types in the chapter on strategy. The continuum of risk and value assessment is carried out at feasibility, for example, development appraisal, design, tender, construction, commissioning and fitting out, occupancy and life cycle. One of the key factors is the allocation of risk by determining the procurement system to be used. This should match the client's preferences and risk represented by the procurement type is most easily represented by Figure 8.8.

The Table 8.5 indicates that the client has a wide choice of risk strategies, but the golden rule is to allocate risk responsibility where it can be managed best. The headline is that clients are looking to deflect risk because of their

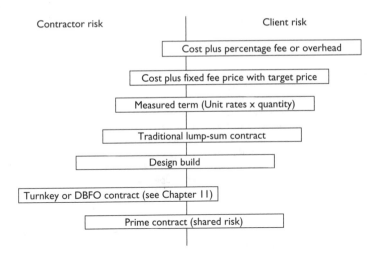

Figure 8.8 Procurement and the allocation of risk.

Source: Adapted from Flanagan and Norman [11].

experience of having to pick up the 'tab' on traditional lump-sum contracts, but by less involvement this may take away more opportunities to create value. Another option is to use a shared risk approach like prime contract that has a built in pain–gain incentive, which is associated with a no blame culture. The DBFO contracts often have a chance to negotiate risk at a very early stage of the development life cycle and the relationship is extended (especially in the PFI format over a long post construction occupation phase).

A professional judgement should be made on the tender price. However, the client chooses the degree of risk they take, by the degree to which they require the contractors to guarantee their final price. Changes are inevitable both in detailed design changes and in scope and the risk here is to be managed as discussed later.

The latter is the reason why some experienced clients have chosen to go with construction management, or design and manage, bringing in contractors early and bringing the whole team round the VM table. Clients have more control over their projects, but take more risk. Design and build has taken most of the risks away from the client as there is a clear contractor responsibility for the whole design and construction process, but risk arises from change. In turnkey projects residual building management risks and capital funding and sometimes operation of the facility risks have been handed over to others. This has not taken away the VM process as it has all happen up front around the negotiation table. Prime contracting represents a risk-sharing approach where there is a particular concern for innovation

Table 8.5 Table to show comparative financial risk to client of different procurement types

Type of contract	Client financial risk	Measures to control budget
Cost plus	High – any factors difficult to track expenditure and so control it except by limiting extent	Daywork rates
Measured term	Medium–high – rates control prices, but many rates may not be covered. Work can be limited if budget exceeded	Measured rates + inflation allowance
Traditional lump sum	Medium – quite a wide range of additional rates, e.g. design changes, unforeseeable events and scope changes inevitable push up the 'lump sum'	Contingencies, provisional sums and dayworks
Design and build	Low if no changes – contractor takes on most risks for unforeseen changes and or aspects of design development not to do with scope change	Re-price any scope changes on the basis of known rates
Construction management (Construction management or Management contracting)	Medium – specific managed budget and package procurement. Unforeseeable events still applicable, but more integrated design and construction than lump sum so opportunities for initial VM.	Incentive to share savings
Design and manage	CM takes no risk, but may have an incentive contract to share savings	
DBFO or PFI	Low – best certainty once figure is agreed. The risk is that the negotiated financial close with the preferred bidder may be higher than expected and client is committed. However this figure is regulated for 25 years	Affordable service charge paid yearly by the client, but may limit facility in the long term
Prime contract and partnering contract (PPC 2000) or NEC	Medium – better culture for working together to contain budget, but no guarantees against poor planning and unforeseeable events	Use of compensation events to avoid culture of blame. Full access to contractor accounts. Culture of continuous VFM improvements

and developing a value solution. This is often associated with long-term strategic partnering where relationships are built up, synergy established and the learning process is not started all over again for each project. However, a client may make different decisions based on the type of project and the degree of control that is perceived necessary. Clients may have a risk avoidance attitude and this may be tied up tighter in the type of contract chosen, or opened up as in the case of the NEC contract and the partnering contract.

Risk attitude and behaviour

Risk behaviour is an interesting subject and a lot of research has been done in the area of health and safety risk, for example, driving safety and the area of gambling risk. The latter perhaps being more akin to the RM of projects as a whole. Greenwood's [20] has measured a difference between the degree of risk we take personally and that, which is taken on, on behalf of others. This influences decision making in the area of investment, project choice, the degree of innovation and planning and the response to external risks.

The context of the risk is very important to the behaviour displayed. Bidding theory identifying the success of 'work hungry' contractors [21] indicates the risk of the lowest price being a loss leader to get into a market or to retain a minimum workload. This in its turn may lead to contractor claims to recoup loss. At the other end of the scale, contractors may collude where there is much demand to share out the work at higher prices. Client risk is increased in both of these dysfunctional events for time cost or quality and RM needs to be aware of the context.

'Group-think' indicates a greater willingness to ignore risk in coming to group or committee decisions where an established group converges their members beliefs by familiarity and starts failing to see the wider picture. For example, the decision to use the nuclear bomb in the Second World War could have emerged as a 'group-think'. It might be thought that projects are unlikely to fall into this category due to the formation of fresh teams, but industry or contract cultures are created that reinforce contractor claims, project overruns and less than excellent performance, creating risk for not meeting project objectives. These are generally by assumptions that 'there is no other way'.

With health and safety Wilde [22] uses risk homeostasis theory (RHT) which suggests that people compensate if there are technological improvements to machinery or environment to experience the same level of risk (target risk) indicating that it is not easy to change someone's risk behaviour without changing their attitude. Familiarity with a task is likely to induce less caution than the approach to innovation. Safety training can, for example, have a negative effect on behaviour if it induces more confidence and therefore a greater willingness to engage a risk, because of supposed familiarity with the outcome and a loss of fear. Attitudes on the other hand are created and are associated with say a greater willingness to use RM methods, or to avoid certain things altogether. Their correlation to behaviour can be weak as other factors such as norms and peer/client pressures will have an influence on behaviour. However, these attitudes can be iterative with behaviour and make significant changes in behaviour over a period of time, if there is motivation. Reinforcement such as incentives to change, or the opportunity to experience benefits

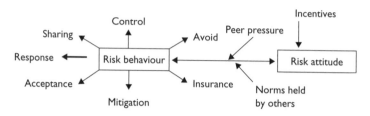

Figure 8.9 Relationship of risk attitude to risk behaviour.

can change beliefs/values. If this motivation reaches a critical mass for a project the ingredients are in place to make a change in culture. For projects, ongoing relationships will help to establish norms and positive peer pressure.

Risk attitude is particularly important in how a risk is responded too as shown in Figure 8.9.

The challenge for RM in construction is therefore to challenge and encourage collaborative working, to develop a 'no-blame' culture of openness and honesty and to work within procurement strategies that identify and share risk according to the ability to manage and accept those risks, rather than passing the risk 'parcel' down the line to the weakest member, often the subcontractor. The difficulty for RM is that much of its techniques are based on mathematical analysis of probability and come from the 'hard' engineering side of the industry. Whilst techniques such as Multiple Estimate Risk Analysis have a mathematical rigour, their application and the presentation of their results is often viewed by clients and fellow consultants as confusing and overly complex when they are looking for clear direction and certainty in their project advice. But a similar problem has been experienced by the medical profession in trying to explain risk in surgical treatment, vaccines and susceptibility to diseases. The awareness and understanding of risk needs to be raised in education and professional training and, therefore, how to develop this 'soft' area of managing culture change with the acceptance of 'hard' engineering based techniques is a fertile area for research.

Another issue is how to achieve the balance between the independent RM facilitator, who may be managing risk in an area outside his particular technical expertise and needs to charge a separate consultancy fee, with the in-house risk manager, where RM is provided as part of a wider project management service. Is RM an additional service to be paid for by the project client or part of the service of successful project delivery by the project design, construction and possibly operation consultants and contractors? Attitude certainly has some bearing on it.

Whole life costing

Part of the business plan in commissioning a new building is the consideration of a wider set of costs, which occur over the life of the building. These costs are often greater than the capital cost and so may influence the initial expenditure in order to save larger amounts of money later. The later costs are usually discounted in order to bring to present value. These combined discounted costs are described as WLC. Public procurement clients are insisting on the WLC approach.

Life cycle costing (LCC) is defined by the RICS [23] as the present value of the total cost of an asset over its operating life including

- initial capital cost including site costs and fees and mitigating grants;
- occupation costs, including energy, letting and rates;
- operating costs including, maintenance, repair and replacement, security, insurance;
- cost or benefit of the eventual disposal of the asset at the end of its life.

The term WLC also include revenues which are generated by the building [24] in the NPV figure. Kelly and Male [2] define it as 'the economic evaluation which accounts for all relevant costs during the investor's time horizon and adjusting for the time value of money'. This is useful as it makes the term relevant to all building clients whether they will retaining the building or not. It helps also to make a universal connection with VM. One of the problems with WLC is that apart from public clients the process may be too comprehensive and provide the wrong information for strategic client decision making. For PFI, DBFO or BOOT schemes however the client and the operator have the same interest.

The advantages of WLC are the better evaluation of the business case in considering the whole picture. There are however a number of *barriers* to its use, as noted earlier by Pasquire and Swaffield [24], and Robinson and Kosky [25], which include

- lack of reliable data on components and maintenance life and energy consumption before building occupation;
- the separation of capital and revenue budget management which stops an integrated approach so that companies want to reduce capital costs to impress their shareholders;
- a short-term interest by the client who sells or leases the building on;
- taxation and fiscal policy encourages allowances on capital expenditure, but this does not include developers for whom it is 'stock in trade';
- business running and maintenance costs can be set off 100% against the corporation tax immediately it is incurred.

Table 8.6 Whole life elemental breakdown

Relevant cost	Proportion of operating costs	Possible value measures
Maintenance	Offices 12% Homes for the aged 7% Laboratories 30%	Increasing component lives or easy access to replace
Energy and services	20–35%	There is more scope now to tender competitively and special bulk rates may be possible with direct suppliers. Water is not yet truly competitive. Energy saving methods are popular and environmental
Cleaning	10–20%	Access is a key issue in cost such as access to windows. Owners may also choose easy clean options such as self-cleaning glass at greater capital cost
Security	Varies	The potential for a secure building depends directly on the design. Significant revenue costs in security staff can be saved by the use of passive devices. There is a trade off between insurance premiums and extra security investment. Intrusiveness is a factor for productivity as employees may resent control
Business rate and capital tax allowances	15–45%	The principal determinant for rates is location. Grants, capital allowance or rate relief is available in regeneration areas. Short-term relief is possible from councils trying to attract employers

Some interesting figures for the relevant importance of the life cycle elements are given by Kelly and Male [2] in Table 8.6.

Life-cycle costing uses a number of techniques to determine costs, mostly connected with DCF techniques. These techniques assess the present value of later operating and maintenance costs and annualise them in what is called an annual equivalent which can be added to the capital cost. Capital allowance tax, depreciation and inflation/replacement costs can also be considered in the model by the use of a sinking fund calculation and years purchase can be used to calculate how much is needed to be invested now to provide an income for an annual yearly cost. There are many textbooks that deal with these calculations in detail and, tables are available to support the calculation.

One of the key issues is the reliability of data for the life of components in the building and for the cost of maintenance, energy costs and insurance. Where these comply to a trend historic figures are available, but step changes in costs can occur creating new taxes and grants such as energy tax, fiscal changes, introducing new statutory requirements such as accessibility for disability and updating to keep up with new technology imperatives such as IT networking. In any case new buildings may be designed on for a completely new operating régime and internal figures will not be available. Trusting external figures for costs needs to be properly adjusted to take account of unique business and location factors. Other risks are the level of inflation and interest rates and the possibility of the discount rates not being adequate or being out of step with inflation.

This creates the need for a sensitivity analysis on each of (or a combination of) the assumed factors, to understand those which create the greatest sensitivity to the overall calculation. There is then a need to factor in contingency to allow for some unknowns on the basis of probability. Table 8.6 indicates some of the global ranges of cost for different elements of operating cost and is helpful in arriving at where to concentrate the sensitivity.

Appointing consultants and contractors

Consultants are appointed by the client or the project manager. Traditionally construction consultants are selected on a quality basis for a given fee and contractors are selected on a price basis against a detailed and prescriptive project specification and drawings or pricing document. The trend towards VFM means that both could be considered on a quality and price matrix. The quality issues may be selected by the client, but there is a responsibility on the part of the client to have a selection process that is clear, consistent and complete. This ensures good competition, a basis for a reliable selection process and builds trust with the service providers. The award of a project to a consultant or contractor needs to be transparent and fair and as such standard rules of appointment and tender are available for public and private tenders. The rules allow for a choice of procurement, but lay down certain basics. In the EU, compulsory purchasing rules have been standardised for public works projects in the context of breaking down trading barriers and creating a level playing field. Private clients will benefit from using standard procedures as these will attract competition, but may also make known bespoke rules, which fit their situation.

The EU Purchasing rules identify open, selective or negotiated tenders as three possible approaches to the appointment of contracting or consultancy services.[1] All public tenders over a certain size need to be advertised in OJEC and are listed on the Tenders Daily service to ensure maximum exposure. The open tender is not recommended in construction because of the extensive work involved in preparing tenders and evaluating them. So the advertisement provides outline information about the project and invites expressions of interest. This creates the facility for pre-qualification of the tenderers. The restricted or selective tender method can then be operated with a shorter list and tenders invited. Good practice requires that these lists are restricted to no less than three up to no more than six. In the case of consultants, the CIOB [26] recommends no more than two consultants on the short list. However in any one situation, a justification may be given for expanding this list. In negotiated tender the aim is to carry out a second stage to the tender which develops the project VFM. This provides the best of competitive and negotiated systems, by ensuring more involvement of the contractor in the development of the project definition. Private clients may use the NJCC Conditions for single or two stage tender or may prefer to use the later CIB [27] Code of Practice for the selection of consultants and main contractors. The DTI Construction Line [28] is an example of a pre-qualified contractor and consultant list.

Consultants are also classed as service providers and are appointed according to the standard industry codes (e.g. RIBA, NEC, RICS or ACE). Selection may be entirely competitive (fee bidding), but it would be unusual not to appoint key consultants with reference to quality factors such as their reputation and track record. This brings in an assessment of their experience and ability in the specifics of the project. Selection techniques may include design competitions, viewing previous work, or proposals presented during an interview with the team expected to provide the expert service. This provides a better forum to assess expertise, compatibility with client objectives and the suitability of consultant experience. Project managers, managing contractors and possibly main contractors also fall into this category.

Clients have to develop their strategy in appointing consultants by deciding:

- Do they want one appointment who contracts with the other consultants and co-ordinates feedback as through a project manager or do they contract with a circle of consultants who will provide them with direct advice that they manage?
- At what level do they delegate authority for decision making?
- Is the appointment for the whole project life cycle, ongoing phases or on a call off basis?
- What scope of duties is required of each consultant and do all the consultant briefs cover all areas without overlapping duties?
- Do they pay consultants on a time basis, lump sum or *ad valorem* – according to a percentage of the value of the project? (Clients are asking whether this latter method in simple form does in fact motivate VFM).
- What level of professional indemnity or collateral do you require? This relates to the level of risk which they carry in the case of mismanagement of the project and poor advice.

The procurement strategy is influenced by these decisions. Other issues such as partnering, dealing with conflict and integration overlay the procurement method.

Selection process

We have already spoken of the need to have clarity, consistency and completeness in the selection procedure and this means that the project brief and scope must be unambiguous and where there are queries that a process of providing answers available to all parties simultaneously is provided by having regular mailings or a mid-tender briefing session, to deal with queries and emerging issues. Best practice requires that a quality threshold is established especially where a large number of bidders are likely to be interested. This provides the clear criteria for a select tender list and for short listing. According to context (public or private), this may be pre- or post-interest qualification. Quality thresholds may include client values such as a wish to see certain environmental friendly project policies, the existence of a third-party quality assurance scheme, technical suitability, resource availability, financial stability, whole life costing and performance in partnering are just a few. It is necessary to be client organisation specific, to reflect client values and past experience.

The price, of course, is important and provides an awarding mechanism once a minimum quality threshold has been achieved. It is obvious that where the quality threshold is only judged after appointment on the basis of price, the appointment is subject to risk of greater conflict as it depends upon contractual interpretation and possibly expensive litigation. In order to choose between price and quality it is important to determine the weighting between the two. The procurement guide three [29] gives guidelines on the basis of project type and whether it is a consult or a contractor. It suggests a range 90:10 to 10:90 quality/price with complex and innovative projects having a much greater emphasis on quality and 'straight forward' and, interestingly, repeat projects having the lightest emphasis on quality factors. Surprisingly though, it suggests no greater than a 40:60 quality/price emphasis for even the most innovative project in the selection process for contractors and 20:80 at the bottom of that range. This puts the emphasis on consultant quality factors, but firmly puts quality into the frame of all contractor selection processes.

The balance of the WLC factor against capital cost has already been discussed, but is much more dependent on what type of client is going to use the building. Another factor in selection is the cost of tendering, which is often associated with the number of tenderers and the complexity of the tender information asked for. If this includes substantial architectural design elements, such as in design and build, then the cost goes up and contractors will choose to bid more strategically in order to boost their chances to win and recoup more of their tender overhead. Clients on the other hand may induce competitiveness as well as design control by appointing a concept designer who novates[2] to a design build contractor at the scheme/detail design stage. More care in the selection stage is needed to match the design and build contractor and the designer.

Award of contracts

The award of a project on price and quality involves choosing the best value contractor or consultant and needs to have a fair and tangible final selection criteria. One way of doing this is to use a balanced scorecard, comparing contractors by providing a weighted score against each of the criteria for the quality scores. The weights much add up to 100. These criteria are relating directly to the tendered information and interview processes and are scored on a percentage and multiplied by the weighting of the criteria. Thus, if functionality was scored 80 and weighted as 50% amongst three other criteria at 25%, 12.5%, 12.5% respectively, then it would receive a score of $80 \times 50\% = 40$. The others would then score as the table to give an overall quality score out of 100.

Table 8.7 provides comparative data which combines the quality and price ratings. Much depends on how you rate the criteria, quality and price to provide a weighted score for the comparative prices. This should be standardised perhaps, according to the tightness of the budget. The more criteria, the more likelihood that no criteria will make much difference in the quality scores.

It is therefore important to check out the calculated outcomes with other methods such as a face-to-face meeting and to ascertain the validity of any risks that you have specifically asked the tenderer to price, which are not to be expected. These will give further clues to the validity of a tenderer with experience. You may also differentiate between essentials and desirables in the scoring.

Another way of dealing with a dual criteria is to use the two envelope system which means that quality data is evaluated prior to the price data in order to assign additional weight on those contractors who have made excellent quality before the price creates a bias. This illustrates the importance of changing cultural norms to bring about VFM over the life cycle of the project. A public body will have to work harder to demonstrate objectively that VFM has been achieved where the cheaper short-term solution was not VFM.

Notification and terms of engagement

Most consultants expect to be appointed on standard terms and conditions and would price differently for the risk involved in accepting other conditions that might be standard for the client. The basic conditions should refer to the scope of the work and the responsibilities of the consultant. It is good practice for separate consultants to be aware of each other's conditions and for there to be consistency of style between them and to ensure that the team of consultants can work together CIOB [26]. Post appointment meetings need to negotiate the practical outworking of conditions need to check the PII and/or all risks insurance, bonds, warranties and other formalities, which are part of the risk allocation procedure. Further checklists

Table 8.7 Illustrative example of a balanced score card

Quality weighting 60%
Price weighting 40%
Quality threshold 55%

Quality criteria	Criteria weight (%)	Organisation A			Organisation B			Organisation C		
		QT reached	Score	Weighted score	QT reached	Score	Weighted score	QT reached	Score	Weighted score
Quality scores										
Functionality	50	yes	65	33	yes	70	35	yes	90	45
Programme	25	yes	45	11	yes	50	13	yes	60	15
Risk manage	12.5	yes	45	6	yes	30	4	yes	60	8
Environment	12.5	yes	50	6	yes	85	11	yes	65	8
Totals	100			56			63			76
Price scores										
Tender price		20 000 000			27 000 000			22 000 000		
Price score (average £23 m = 50)		80			30			60		
Overall scores										
Quality weighting × quality score		60% × 56 = 33.6			60% × 63 = 37.8			60% × 76 = 45.6		
Price weighting × price score		40% × 80 = 32			40% × 30 = 12			40% × 60 = 24		
Overall score		65.6			49.8			69.6		
Order of tenders		2			3			1		

Source: Based on Procurement Guidance 3. Appointment of Contractors and Consultants [29].

and detailed information about appointment can be researched from the CIOB code, the Treasury/OGC [30] Procurement Guides and the CIRIA guide [31] on competitive value-based procurement.

In some cases private clients will be only interested in selective lists based on recommendation, or may wish to further partnering arrangements, which share work between a limited number of framework consultants or contractors. Partnering should be clearly delineated from pre-qualification, which provides opportunities for pre-qualification for each tender, whilst strategic partnering pre-qualifies for a period or series of contracts, to build up collaborative working.

Change management

Change suggests that there should be some flexibility built into a project, management suggests that this should be well controlled. Change management can be classified as elective and required change, CIRIA [32]. The required change comes as the result of an unforeseen event such as additional concrete in foundations. Case study 8.3 is an example of managing everyday change to benefit the project.

Case study 8.3 Design development change

> The project is a specialist research facility. The building is circular and involves the use of specialist materials, for example, heavy concrete. A special company (the client) has been set up for the purpose of running the laboratory and commissioning the works. The contract is inspected by technical staff from the company which has also appointed a project manager who is responsible for vetting the budget and co-ordinates the design and construction sides at a high level. The programme time is 55 weeks + enabling works + a link bridge.
>
> Changes were managed that developed the design. A no-blame culture is incorporated to account for up-to-date knowledge of the project with a clear definition of project scope. In this case, packages were either tendered on a bill of quantities (ground works) or on drawings and specifications. Risks need to be properly allocated to parties in the contract and grey areas adjudicated. The Architects Instruction (AI) was to be the official instrument for instructing change. AIs were used to record all changes, whether financial or non-financial and were also the procedure used for issuing drawings.
>
> However, in many cases change is identified by Requests for Information (RFIs) where there is a lack of clarity in the drawings or specification or a clash between them or the BOQ. In these cases a contractual approach is rarely helpful and contingency planning is

employed which seeks to focus on the key objectives of operational programme imperatives, which are:

- quality in the sense of fit for use;
- overall rather than elemental budgets;
- other key client values which may be effected such as sustainability.

This is not an approach which provides a 'contingency sum' as this is unfocused.

In the case study there was a late change in the foundations, in order to cope with the ground conditions and this pushed out the programme by three weeks. In order to retrieve the programme loss, the drainage layout was revised with main runs outside the foundations so that the drains were taken out of the critical path and other works could be commenced. This meant that key operational programme objectives could be obtained without a major budget inflation. For success it depended upon an immediate assessment of the impact of change.

This shows the close relationship between design, methodology and cost. In order for contractors to cope with their own contingency, they may build in early programme targets to allow for some time contingency if the programme slips. This is becoming less easy where programme times are already tight. Phased handover is another possible option where scope changes have expanded the work, but are not desirable to affect handover of critical sections of the work.

The elective change is a choice and may emerge from the need to do things better. This may be as a result of a VM or RM exercise or as the result of changing circumstances such as market forces. In either case the change should be managed so that the impact of change is well known by the whole team, including the client, and effective choices are made on the basis of full information. CIRIA [32] mention three objectives for change management:

- Avoid the disruption and associated cost escalation that arise in the absence of the management of change.
- Give clients both control and choice about how their money is spent.
- Avoid clients being committed to the consequences of change without their prior knowledge and approval.

The disruption of change is greater if there is change to information which is already agreed (fixed), because resources may already have been abortively expended. Design development is excluded from this. It may also disrupt the flow of work affecting productivity and confidence in the decision-making process. On the other hand, an open culture is prepared to work on the causes of what CIRIA calls 'post fixity' change and limit its disruption by not apportioning blame. Formal approval through gateways are important places for reviewing the integrity of what is being approved to limit the disruptive element of change. A VM and RM system both contribute not only to reducing the chance of disruptive change, but also to promoting timely elective change. Another reason for an escalation is a poor match of the contract and procurement process to the clients needs. For example, an IT client will need the flexibility to adjust capacity and process to match rapidly changing market conditions. If they are fixed into a single stage design build contract it will be very expensive to change after the original contract has been let, as both they have committed to a detailed design right from the start – any minor or major changes will be considered disruptive and expensive. If communications are poor, then the brief may have been misunderstood and again major changes made late on in the project. Lack of co-ordination can also cause poor design interfaces and unworkable tolerances that the client has no control over, that have to be resolved with budget and delay consequences.

It could be argued that required change at a later stage is inevitable and acknowledgement of this risk, requires a change management system to be put into place that meets the principles outlined earlier. Case study 8.4 indicates how this system might work within the principles.

Case study 8.4 Scope change

In the development of a postgraduate education centre, a procurement system was set up which required the individual approval of the client of each of the packages as they were let and the use of the NEC form of contract by which compensation events were advised on an early warning system by either side as soon as they were perceived as a problem. The client had an absolutely fixed budget on the basis of an agreed mortgage and any additional costs had to be offset by a saving so that the project remained in the same budget limit. A contract period was agreed with a single handover.

In the event, the client recognised the necessity to occupy part of the floor area at an early stage to decant another department who became 'homeless' due to late completion of another project. It was agreed with the contractor under an early warning compensation event to accelerate completion of the top floor and the lift and one set of stairs so that occupation could take place two months earlier. In order to compensate for acceleration costs and a protected access to the floor, a shell and core finish was agreed on part of the ground floor to be fitted out, in a separate contract by the eventual tenant of that area. The NEC system also requires a fast turn around for costing and reprogramming the change prior to approval of the change. The contractor reprogrammed to show a phased handover without affecting the final handover date. Design changes were also necessary, but where these were not a change in scope they became the responsibility of the contractor. One such change involved the resizing of masonry to suit the lift door size. This was disruptive, but mitigated by agreed changes to the lift design.

These type of changes cannot be envisaged, but flexible systems and contractual procedures had been risk managed to meet the stringent restriction of a totally fixed budget and to deal with any contingencies and design changes.

Pre-fixed change opens up the relationship with the client and allows the client choice and control over the design process. Alternatives may be offered with a rationale for choice, so that savings can be offered by the contractor and possible pain gain incentive schemes. Alternatively the client will be invited into the VM process in order gain optimisation of the functionality of the building. Agreed risk sharing to suit the abilities of each party to manage that risk will be less costly than transferring unknown risk, which a contractor prices at an excessive rate to cover or to insure the risk. Elective change, as an outcome, is controlled and fully evaluated before implementation. VM works best for elective change where there is an open culture and a capacity to bring in a contractor or construction manager early to give a fuller perspective. In repeat contracts or strategic partnering arrangement this is a natural move. The open integrated culture has been discussed more widely in Chapters 6 and 7.

Conclusion

The various tools and techniques that underpin both VM and RM are well established in textbooks and industries, where the business case has been demonstrated and where collaborative working within an integrated supply chain is established or has been strongly promoted. The traditionally fragmented and rather adversarial nature of the UK construction industry, particularly with its separation of design and construction, has both been an opportunity and a constraint on the development of VM and RM in the UK. Conversely, using these approaches within project management can also provide a means to encourage and facilitate change. However, both are still largely paid lip service, rather than being applied as distinct project services with their own distinctive and robust methodologies.

In one sense RM and VM is about better procurement and this is the angle taken in improving procurement in recent reports. The Atkins (1994) Secteur

Report [33] for the European construction industry summed up the malaise of the industry with the recommendation for competitiveness and VFM:

> improve the relationship between quality and cost and improve the reliability of the service provided to customers in terms of the achievement of quality, cost and time targets and worry free construction.

Recent writings[3] in the UK and Europe, particularly for public consumption, have been about urging the use of more sophisticated techniques in assessing risk and value and in recognising risk priorities in sustaining VFM. It is also clear that RM has developed a two-edged approach of looking at the opportunities and the threats. This has been clear in the more recent editions of leading RM guides. This has helped to bring it closer to the concept of VM and it now makes sense to view the techniques as two sides of the same coin. In considering value, WLC has always been a consideration, but more development of long-term procurement relationships such a DBFO and PFI has given the incentive for use of the technique and better accessibility to reliable life costs. Wider use of benchmarking for operating costs has given confidence in its use, for long-range decisions. The use of readily available IT has also made more sophisticated techniques such as probability and sensitivity analysis viable, but to better support and not make the decisions.

Neither RM or VM is new, but they are being seen as a key way of assuring *quality* in construction, so that promises made at tender stage are backed by evidence that value is not going to be compromised with cost cutting in the later stages. VFM is seen as identifying a broader range such as competency of consultants and providers, sustainable and environmental design, ability to finish on time, better life-cycle costs and energy consumption, promoting considerate contractors and better and safer products. These can all be built onto a matrix for considering contractor risk and functional value by bringing the client, the designers, contractors and other stakeholders together at an early stage of the life cycle.

A higher capital cost due to these factors that indicates significantly greater inflation and excludes a client's ability to build, is also a VM problem. PFI or 'sell and lease back' are examples of solutions that spread the cost of building over the life cycle when significant related savings can also be made. The issue of longer term RM then becomes critical too. For example, in assuring that the cost of financing does not exceed the value gained.

The pitfalls of RM and VM arise out of the late or inappropriate use of these techniques, or the wrong interpretation of client values. Risk evaluation depends upon subjective judgements of impact and probability, VM may have subjective selectivity. The level of predictability is often ignored as a significant risk, but is generally associated with the unreliability of long range 'guestimate' material. Mechanisms for removing or testing bias for whatever reason (client or consultant) need to be in place, but non-exclusion because the factors are financially intangible is more dangerous than holistic, systematic, experience based decision making. These decisions can be justified by a constraints analysis which identifies the 'gatekeeper' risks and values which are critical to determining the global risk and value. For example, a 30% cheaper museum is going to be doubly ineffective if the visitor figures realistically will still not support the income to pay the finance and extra running costs, which might arise. More fundamental VM and RM is required, perhaps involving the securing of grants, moving location and cheaper entry charges. More time and effort can be allocated proportionately to gatekeeper value and risk.

Two other drivers could also encourage wider use of VM at the earliest strategic stages. These are the increasing push for sustainable development and corporate social responsibility. Both require engaging with a wider stakeholder group, where identifying and agreeing mission statements and strategic goals will require more robust and more transparent decision support mechanisms, including identifying value systems to inform both strategic and more tactical option appraisal to arrive at best value solutions. Developers are increasingly being asked to justify their planning applications in terms of a sustainable development assessment. This requires considering community impact as well as economic and ecological. As well as multi-disciplinary team input at the design stage, clients are increasingly wanting engagement with stakeholders from the wider community, rather than just the planning officer. The challenge for VM is will its methodology be robust enough to rise to this strategic role, or will the management consultants step in and subsume it under other services?

At the heart of the M4I and Constructing Excellence is the use of KPIs to measure key quality criteria and drive a process of continuous improvement. Continuous improvement is the required benefit to the client as part of the new partnering agreements. Will VM provide the mechanism to facilitate this continuous improvement and will this culture of partnering encourage more 'buy-in' by clients and contractors to the concept of using VM to improve their service, whether or not initiated and paid for by the client?

These things point to a more flexible approach and earlier involvement for the use of the techniques, but also sound follow up in implementing and monitoring the early 'poor predictability' assumptions made. This is why a hierarchy of VM and value engineering meetings is a good system for ensuring VFM. As it crosses over with a business case, the consideration and/or involvement of a wide range of stakeholders is critical.

On small projects the costs of such techniques is costly in terms of person hours and the cost of specialist advice is prohibitive. Here the use of simple, but effective heuristics may be a better approach with the direct involvement of stakeholders. This is why EMV has been illustrated. A culture of monitoring and control should also be in place as failure costs have more impact on a smaller budget.

Notes

1 Contractors come under The Works Regulations SI 1991/2680. Consultants come under The Services Regulations SI 1993/3228.
2 Novation is the process of the concept architect employed by the client to do the scheme design is transferred to a design contractor as part of the contract conditions to do the detail design.
3 The Atkins [33], Latham [5], Egan [34] Reports have spawned the development of the European Procurement Rules, CIB Value management guide and the Treasury Procurement Guides, The OGC Achieving Excellence in Construction guidance, Building Down Barriers, Local Government Task Force (LGTF) Toolkit, Gateway Reviews, Prime contracting, 4P's, The National Strategy for Procurement (2003), The Improvement and Development Agency's (IDEA) e-procurement software. MOD's SMART procurement, The Strategic Forum for Construction Integration Toolkit and demonstration projects in the M4I.

References

1 Office of Government Commerce (OGC) (2004) Constructing Excellence Procurement Guides. http://www.ogc.gov.uk
2 Kelly J. and Male S. (1993) *Value Management in Design and Construction: The Economic Management of Projects.* E & FN Spon, London.
3 Emmerson H. (1962) *Survey of the Problems before the Construction Industries.* HMSO, London. A report prepared for the Ministry of Works.
4 Banwell H. (1964) 'Ministry of public building and works'. *The Placing and Management of Contracts for Building and Civil Engineering Work.* HMSO, London.
5 Latham M. (1994) *Constrcuting the Team.* Final Report of the Joint Government/Industry Review of Procurement and Contractual Arrangements in the UK Construction Industry. HMSO, London, accessed from www.constructingexcellence.org.uk
6 Egan J. (1998) *Rethinking Construction.* Department of Environment, Transport and the Regions. London, accessed from http://www.dti.gov.uk/construction/rethink/report/
7 Fairclough J. (2002) *Rethinking Construction Innovation and Research: A Review of Government R&D Policies and Practices.* Department of Trade and Industry, London.
8 Connaughton J. and Green S. (1996) *SMART Technique.* CIRIA, London.
9 Davis Langdon and Everest (2003) Guide 'Creating the Conditions for Success'. June.
10 Male J. (2002) *Building the Business Value Case.* In J. Kelly, R. Morledge and S. Wilkinson (eds), *Best Value in Construction.* Blackwell Publishing in conjunction with RICS Foundation.
11 Flanagan R. and Norman G. (1993) *Risk Management and Construction.* ©RICS. Blackwell Publishing, Oxford.
12 Kelly J. and Male S. (2002) *Value Management.* In J. Kelly, R. Morledge and S. Wilkinson (eds), *Best Value in Construction.* Blackwell Publishing in conjunction with RICS Foundation.
13 Roy Woodhead and James McCuish (2002) *Achieving Results: How to create value.* Thomas Telford Publishing, London.

14 Edwards L. (1995) *Practical Risk Management for the Construction Industry*. Thomas Telford Publishing, London.

15 Burke R. (2003) *Project Management: Planning and Control Techniques*. John Wiley & Sons.

16 Goldratt E. (1997) *Critical Chain*. The North River Press Publishing Corporation, Great Barrington, MA.

17 Goldratt E. and Cox J. (1989) The goal: a process of ongoing improvement Gower.

18 Burke R. (2004) *Project Management: Planning and Control Techniques*. 4th Edition. John Wiley & Sons.

19 Gregory A. and Kearney G. (2004). *The Theory of Constraints. Association for Project Managers Yearbook*. APM, UK.

20 Greenwood M. (1998) Risk Behaviour. Unpublished Master's Thesis. University of the West of England.

21 Harris F. and McCaffer R. (1985) *Modern Construction Management*. 3rd Edition. Butterworths, London.

22 Wilde G.J.S. (1994) *Target Risk: Dealing With the Danger of Death, Disease and Danger in Everyday Decisions*. PDE Publications, Canada.

23 RICS (1999) *The Surveyors Construction handbook – Life Cycle Costing*. Royal Institution of Chartered Surveyors.

24 Pasquire C.L. and Swaffield L.M. (2002) 'Life cycle/whole life costing'. In J. Kelly, R. Morledge and S. Wilkinson (eds), Chapter in *Best Value in Construction*. RICS Foundation/Blackwell Science, Amsterdam.

25 Robinson G.D. and Kosky M. (2000) Financial Barriers and Recommendations to the Successful Use of Whole Life Costing in Property and Construction. CRISP, UK.

26 CIOB (2002) *Code of Practice for Project Management in Construction and Development*. 3rd Edition. Blackwell Publishing, Oxford.

27 Construction Industry Board (1995) Selecting Consultants for the Team: balancing Quality and Price.

28 DTI. Construction Line. http://www.dti.gov.uk/constructionline

29 HM Treasury. Procurement Guidance No. 3 Appointment of Contractors and Consultants (undated). Replaces CUP guidance 13 and 26a and b.

30 Office of Government Commerce (OGC) (2002) Procurement and Contract Strategies. Achieving Excellence toolkits.

31 CIRIA. Guide 117 Value by Competition. A Guide to Competitive Procurement. CIRIA, London.

32 Lazarus and Clifton (2001) *Managing Project Change: A Best Practice Guide*. C556. CIRIA research project.

33 Atkins W.S. (1994) 'Secteur strategic study on the construction sector final report (1994)' – In *Association with the Centre for Strategic Studies in Construction, University of Reading, Euroquip, Dorsch Consult, Prometeia, Finco Ltd; Agrotech Inc and ESB International*. F2347.050/SDR/ECC.6A. For the ECC 111/4173/93.

34 Egan J. (2002) *Accelerating Change*. Strategic Forum for Construction, accessed from www.strategicforum.org.uk/pdf/report_sept02.pdf

Chapter 9

Design management

Design management has been heralded more for the increase in the design and build procurement route and many contractors have set up training courses for their personnel creating a career path alongside the project and the procurement manager. This chapter is intended to investigate the design process and the role of any professional in the management of the design stage of the project. This will clearly be of interest to architects and engineers as well as contractors.

The objectives of this chapter are:

- To have a basic knowledge of the design process and to demonstrate the challenges of improving its efficiency.
- To understand the importance of urban design in design quality, sustainability and managing the project stakeholders.
- To generate skills in the flow of design information from the various project parties and to manage its integration between design and construction in the supply chain.
- To set up an effective design change model, which recognises the need to reduce waste (in the spirit of an open culture) and in the context of different procurement frameworks.
- To appreciate the impact of the information technologies on the integration of the project team including client approvals, fast tracking and fast build.

The CIC [1] defines the role of design management as,

> ensuring that the right quality of building design information is produced at the right time and conveyed to the right people.

With reference to its management, it goes on to say that,

> the increased number of specialist designers and subcontractors contributing to the overall package of design information to practically

realise a project has created the need for a defined approach to their direction and co-ordination with a new single point of responsibility. Design management responds to this need and attempts to improve the efficiency of the design process and its integration with the construction process.

Nature of design

Gray *et al.* [2] define design as a process of human interaction and consequently the outcomes contain the interpretations, perceptions and prejudices of the people involved. Another definition by Hirano [3] defines it more dynamically as,

> The learning creative-developmental process can be represented as a spiral. Viewed from the top, a spiral is a moving circle constantly expanding in scope. Viewed from the side upward movement is evident...the addition of experience and understanding.

Design is an *iterative* process with the designer producing a working solution which is tested and improved or redesigned for further review as greater mutual understanding is achieved in the 'spiral' effect of the process shown in Figure 9.1. This is related to the traditional stages of the RIBA Plan of Work [4] and the deepening detail of the life cycle progression.

The role of value analysis brings together client and designer on a regular basis and is a vital part of ensuring design meets need without excess at an early stage. It may be formal (outside facilitator) or informal (design review with the client).

Several related parts of the design could be, being developed in parallel, subsequent to the concept design such as services, structure and sound. These will impact on each other and need to be co-ordinated. In addition there will be other constraints like planning, statutory requirements and regulatory control such as fire, building control and health and safety. In building projects, these are specific to location and context.

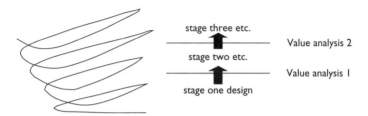

Figure 9.1 Iterative design process.

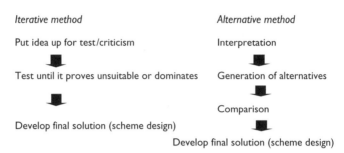

Figure 9.2 Comparison of design development methods.

Alternative design solutions may have been identified post-concept stage, partly dependent on location as well as other factors such as access. This method of design is called option appraisal and the choice will be based on client business needs and values as well as efficiency. The two methods are compared in Figure 9.2.

Maguire [5] suggests that design comes out of three inputs – the designer, the problem as given and the data about the problem. He admits that not all available data is found or used and also indicates later the constraints represented by the site and the budget/time given.

Engineering design

Gray *et al.* [2] talk about the difference between craft-based construction with traditional materials and engineered technology with its tighter quality and fit requirements. Thus a traditional house design is well known by those who have to build. However, the use of engineered components such as steel framing or prefabricated non-cavity walls needs much more design co-ordination, research and supervision. This complexity in the detailed design stage increases with the move from standard (well known) components to specially developed components and of course with the size of the project. This leads to a need for

- revision in the concept and scheme in the iterative cycle of design development;
- a larger and more specialist design team with better communication;
- many specialist inputs from designer contractors;
- many drawings of a manufacturing detail;
- many designed component interfaces.

This stage needs much more management due to the greater co-ordination required whilst the concept and scheme stages are likely to be in the hands of one or two people. It could also be argued that the earlier involvement of the construction manager is appropriate where the methodology for construction is changed significantly.

Implications for design management

The following are identified in Gray et al. [2] and should be good practice:

- Allow the designer time for reflection.
- Choose designers of relevant experience and provide support to problem solve.
- Keep objectives in focus through a framework which allows cycling.
- Provide access to the client and review provision of information.
- Allow time for the design implications of change.

Design management is based on an awareness of the key milestones of the master programme such as value analysis meetings, outline and full planning submission, key procurement dates (supply and contractor) and BAA have a system of process gates which provide joint targets which go beyond design to construction and user acceptance. These gates are client approvals which recognise the coming together of parallel processes (simultaneous engineering), but should not be allowed to become bottlenecks.

They further listed nine essential steps for successful design management:

- Recognise the inherent complexity of design.
- Carefully manage the designer selection process.
- Recognise the changing design leadership role as the design progresses.
- Integrate information supply with construction need.
- Obtain agreement at key decision points.
- Actively manage the integration of contributions.
- Plan at each stage.
- Manage the interfaces.
- Control design development.

Design management at the design stage

The RIBA [4] sequentially only fits the traditional procurement method as covered in JCT 98, but is acknowledged as a valuable framework for understanding the various stages of outline design, scheme design, detail design and production drawings. Design is often carried out by contractors at the later stages with different degrees of input depending upon the procurement system, so with this proviso design management can still be overlaid. The key issue in determining leadership at this stage is the form of procurement.

There are two models:

1   a lead designer;
2   a manager with a knowledge of the design process.

The key areas for management need to be defined project by project. Gray *et al.* [2] again, helpfully identify quality as the primary driver for the management of design and it is this concern which will drive the need to plan in detail, to manage the interaction of the team and ensure the timely production of the information for its use by others, to meet targets for client approval, to tender and production and to work within the key cost parameters. This is followed closely by the need to ensure that health and safety issues are considered, which does not take the liability away from the individual designer for the safety of their design, but there is a complication here in the overlapping role of the planning supervisor to co-ordinate.

PFI and design and build procurement is easier to manage due to the executive single point of responsibility of the provider/contractor over design. Design and manage contracts are the same, but in this case there is a consultant responsibility similar to an executive project manger. In construction management, there is a tighter client involvement and the design and construction management and the contractor inputs need to be reconciled, by setting up project guidelines to give the design manager authority.

In traditional procurement, the architect or engineer has a contractual role in design management in conjunction with the cost planning role of the quantity surveyor and liaises closely with the client. It is usual however to appoint, or more informally use, the design contractors for their specialist inputs.

An ideal situation for improving communications and quick decisions is to bring the core design team together in one office, but this co-location has not often been used in the past due to its expense and the detachment of designers away from their core information and their need to be available for other project work. Case study 9.1 gives an example.

Case study 9.1 Office fit out

This £27m London office fit out was carried out under construction management procurement. Representatives from each of the contributing consultants were housed in an open-plan office with the construction manager. This proved very effective in managing the production stages of the design and dealing with client and design change. The specialist package contractor managers also had open-plan office space allocated on the same floor of the building, making their design and management services accessible. Lead contractors were responsible for managing the interfaces of other contractors in a particular zone or area, for example, the floor or the ceiling. Monthly meetings were held, although some client related decision making was delayed due to the remote location of the client in New York. The project itself has been nominated for a management award, coping with significant client changes without altering agreed targets.

Design and construction co-ordination

One of the key areas of design management is the interface between the design and construction stages. Traditionally these are separated so that, there is a need for a robust communication system to be employed to ensure that information arrives when and how it is needed. Particular stress is put upon this system when a fast-track approach is used requiring construction to take place on certain elements before the whole design is complete. Some of the areas for co-ordination are as follows:

- health and safety risk exposure and hazardous materials;
- specially developed or technologically advanced components;
- sequencing and timing impacts;
- tolerances;
- constraints which affect construction methodology;
- material lead in times;
- available or required mechanisation levels;
- prefabrication methodology to maintain benefits and timing.

As a contractor is not often available at the early stages, unilateral design action is often taken which tries to predict or later to inform, certain construction methods. Contractors should always be aware of the implications for the constraints of the site and the pre-decisions imposed on the methodology when tendering. In an open culture, value may be enhanced where construction management aspects can be dealt with directly with the contractors at an early stage and improve productivity.

In recent reports where it has been claimed that 30% productivity savings may easily be made on current practice, the authors have banked heavily on process savings of this nature. The following list indicates some of the main areas of savings which impinge upon design:

- Design waste elimination by engendering efficient methods.
- Constructability, for example, workable tolerance.
- Standard components and processes.
- New technologies and research.
- Change management.
- Benchmarking.

Case study 9.2 indicates a mechanism for efficiently co-ordinating design and construction.

Case study 9.2 Design co-ordination on specialist research facility

Although it is a traditional contract, the contractor was involved in the enabling works and, therefore, played a forward role in checking the buildability of the drawings.

At the primary level, the lead designer was responsible for co-ordinating the different designers who do a full-performance design. Full-detailed drawings and specification are offered to the principal contractor for scrutiny and tender. He is responsible for assessing the buildability of the drawings and the specialist subcontractor puts in a detailed design which may suggest minor changes to suit economic ways of working. The short-listed contractors are also encouraged to fill in question and answer sheets in order to provide acceptable detailed proposals.

The need to change doors in the dry-lined partitions in order to give a workable tolerance was an example of a change made to benefit the buildability and efficiency of the project package for ceilings and partitions. This would have mutual benefits for the contractor's installation and the client's quality. Value engineering inputs on the design by the contractor are difficult even though the design overlaps the operational phases of the contract, as there is no specific brief or system which defines the responsibilities for design liability where design has been adjusted. This makes it difficult to assess the risks involved. Incentives to share financial benefits with the client may help to cover risks.

This may be less efficient than a full-blown construction management contract with a 'pain or gain' clause, but significantly quicker and cheaper solutions have evolved and good co-operation achieved. Neither is this a formal partnering contract, but the client made use of the availability of the contractor in order to involve them in the design stages of the second phase of the contract. The experience and availability of the enabling contractor on the site was also acknowledged by putting them on the tender list for the second phase construction, which they subsequently won in competition. The informal direct relationship between the client and the contractor project team ended up with a limited formal line of command through the project manager. This was also encouraged by the direct client involvement in the quality inspections. This meant that the project manager has been able to use contractor expertise.

Information flow

One of the key areas of co-ordination is the flow of up-to-date and timely information to each designer and design contractor party and the review of change implications. This requires an efficient change management system.

Design information flows through four levels to production as shown in Figure 9.3. Production information flow is traditionally controlled by an information-required schedule, which provides a date based on the contractor's estimation for the lead time plus a contingency from going out to tender and receiving delivery on site. A procurement schedule programme is helpful. A simple indication of a procurement programme is shown in Figure 9.4.

Design information needs equally critical sequencing and timing for its inputs with the condition that design information from specialist members

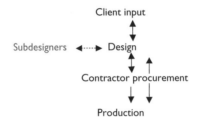

Figure 9.3 Project information flow.

Procurement of	3.3.XX	10.3.XX	17.3.XX	24.3.XX	31.3.XX	7.4.XX
Steelwork contractor	O	△		□	D ▬▬▬	
Roof specialist	O		△		□	D ▬
Windows		O		△	□	D ▬▬

D Delivery
△ Design
O Order
□ Manufacture

Figure 9.4 Procurement programme for construction work.

of the team (consultants and contractors) must be co-ordinated, so as not to clash, be incorrect or to omit areas. Specialist component information must be made known to appropriate parties, so the information flow is two-way, iterative and needs to incorporate client comments as appropriate. Information in the ground or hidden in the structures of refurbishment projects will be revealed from surveys and later, as construction proceeds.

The Taywood Report [6] on the influence of design on site productivity, written from a contractor's viewpoint, suggests that productivity may be improved

- if there was a single person responsible for co-ordination;
- if contractors and specialists were allowed to be involved earlier in the design process;
- if the design process is programmed in detail and co-ordinated with construction, with regular design reviews and feedback;
- if construction only started after detail design is adequately advanced.

From a designer's point of view, flexibility is needed for the creative process and time is needed, if information is to be correct first time. The nature of fast-track projects (overlapping design and construction) contradicts this, but *fast construction* is a concept to maximise design time at the expense of construction time. This becomes a reality when contractors are taken on early to be involved in the design process so that they can procure in advance, in order to pre-plan the construction phase. For example, prefabrication and simultaneous engineering techniques may be applied to reduce time on new build projects with relatively predictable market conditions.

There are opportunities for the client to reduce construction and design time when standardisation is applied and integrated project teams are doing repeat work for the client and can understand the client's business better. These usually involve partnering or framework agreements.

Change management and information flow

Change management is an inevitable part of design development and also of client flexibility, both of which are part of an open culture encouraging creative design environments and building in client choice. Design management seeks to control the boundaries of change without closing down the options for later specialist inputs. The RIBA plan of work indicates three stages of design – outline, scheme and detailed. These stages should have defined parameters which can be fixed and agreed with the client. If this model is used then the following is an example for the boundaries of control.

Outline   Client value system is understood and building location, scope, footprint and orientation is established. Outline elevations and layouts based on adjacency proposed.

Scheme   Clear site layout, elevations, plans and cross-sections are established with textures and colours of cladding materials so that scheme is suitable for detailed planning. Structural and services systems based on performance proposed.

Detail   Layouts are finalised and structure is firmed up with working detail suitable for tender, no major changes should be envisaged. Specialist contractor inputs.

Production   The development of details to suit alternatives and developments to suit on site conditions.

In a traditional role there is a need for a formal freezing of the brief. If a design build position is held then the risk for design development lies with the contractor, but the management role on design development does not go away. In an open culture procurement incentives for value engineering need to be built in. For example, 'pain or gain' value incentives help to produce benefits for all the team by encouraging innovation and discouraging poor management. Full-scale transfer of risk to the contractor for design change often ends up in heavy premiums unless they retain long-term interests such as in PFI/PPP. Case study 9.3 illustrates this.

Case study 9.3 Inappropriate risk transfer

The Ministry of Defence (MoD) prime contract with DML reported to cost £505m and an all up cost of £650m at its start in 1997, is now likely to cost the MoD at least 50% more by its completion (2003) with other possible MoD liabilities for overrun. The contract had planned to transfer all risks for overrun on to DML, but they had ignored the rule that this was limited to one-third of the value of the contracting company which amounted to £20m.

(CIOB International News [7])

The role of urban design

One of the key issues in a project is the acceptance of its impact on the environment by the community. There are many examples of this. Consider the examples in Table 9.1.

Table 9.1 Impact of building on the environment

Project	Impact
Heathrow Terminal 5	Impacting on noise and wildlife habitat
Newbury Bypass	Pollution and wildlife habitat
Nuclear Power Stations	Fear of major accidents and health fears
Green belt housing	Foreshortening amenity such as views, congestion

Table 9.2 The CABE aims

- Campaigning until every child is being educated in a well-designed school and every patient treated in a decent healthcare environment.
- Fighting alongside the public for greater care and attention in the design and management of our parks, playgrounds, streets and squares.
- Helping people who are looking to buy or rent a new home by improving the design quality of new houses and neighbourhoods.
- Building up the evidence that good design creates economic and social value, so that investment in good design is seen as a necessity, not a luxury.
- Keeping every client of a new building on their toes by demanding design quality in projects of all shapes and sizes, in all parts of the country.
- Working across the country with all those involved in the planning, design, construction and management process so that opportunities for design innovation are always exploited.
- Thinking ahead to the new demands that will be made of our built environment in 10 or 20 years' time, through drivers such as climate change, technological advance and demographic change.

These may be categorised as related to the building itself, the construction, or the proposed process and its intensity. Urban design influences relate to the process of incorporating social values and may influence patterns of land use, transport provision, building design, landscaping, environmental impact, typography and orientation. These all have the ability to enhance the experience both for the occupiers and the neighbourhoods of buildings and are normally controlled by the planning authorities, but designed or funded privately. There is often a tension between developers wishing to maximise profit and planners trying to maximise amenity, which results in ugly compromises – more is needed. The Commission for the Built Environment (CABE) has been set up to engender appropriate enthusiasm not just for prize-winning architecture, but for cities which reflect a 'sense of place', maintain historic views and provide a distinctive character or archetype appropriate to the location. CABE [8] formal aims are summed up rather dramatically in Table 9.2.

These aims hint at the connection of good urban and building design with sustainability and with the true success of built environment projects themselves, if ongoing social value is considered.

Loosemore *et al.* [9] challenged architects to take greater interest in the urban design issues, which they believe architects are in a unique position to develop and lead and if not led, could have a critical impact on the project's success. CABE [10] who support this, have discovered through a survey of local authorities that just less than half have an urban designer on their staff to assess planning applications for design and only 26% turn down more than 20 projects per year on the basis of poor design. The reasons given by the authorities is that, there is lack of policy guidance and support from the Planning Inspectorate and there is tremendous pressure to accept borderline design which is adequate and yet not inspiring when they are outweighed by the other economic benefits to the community. Worse still, because of the latter resource pressures, they have not applied pre-application negotiation to improve a schemes design and help it enhance its context. This raises another issue about the ability of the public planning process to filter out poor design and to consider it on a wider context of urban design.

CABE's [11] priorities in the next three years include getting the public more involved in the design and management of public space and more critical of the standard of new housing developments, by giving them information, getting local authorities more committed, trained and supported by the statutory planning system so that they recognise regeneration and neighbourhood renewal as principal strategic goals.

Further work by the Audit Commission in joint publication with CABE and OGC [12] in assessing good value in general, outlines six value drivers that include good design and the positive impact on the locality. Included in the 'impact' value driver is the capacity to create place, the experience of the users and visitors in the internal layout and the enjoyment in use of the building, its looks and quality and 'clarity of composition' in the detailing and its ability to present a distinct corporate image. Interestingly another value driver is effective project management, which relates to the efficiency of the delivery process that for public buildings has recently been under the spotlight.

The metrics which they propose for assessing positive impact on the locality are:

- post occupancy evaluation;
- design awards;
- design quality indicator (a mix of impact delight, build quality and functionality, assessed by client, designers and stakeholders on a six point scale);
- response of the LA.

These metrics seem plausible, but for the reasons discussed earlier require more resources and education, if they are to be driven by the local authorities.

Their value assessment tool chooses weightings for a wide range of criteria, but in an example for a court building which has stunning looks and impact and low level maintenance costs, still only gets a score which indicates room for improvement. It loses out, of course, on the high weighting areas of financial performance, business effectiveness and delivery efficiency.

Case study 9.4 is an example of the complexity and difficulty of this type of assessment.

Case study 9.4 City centre mixed use development

This £300m, 5-year mixed use development using brown field space in the centre of a provincial city in the UK was presented for planning permission by a private property developer with particular interest in city prime residential development. This was part of a larger scheme by the City Council to regenerate the area and regain a valuable waterside site for public enjoyment allowing better usage of the existing docks and access to important historical sites. Any developer would receive partnership money towards the scheme to help the excessive infrastructure costs, but would also have to sell the scheme with less parking as a congestion reduction measure.

The scheme was commercially viable, was potentially highly advantageous to the city and was going to clean up an obsolete gas works, but the scheme took nearly 4 years to get through planning, because of a strong community objection to the first two proposals which gave bad publicity to the developer. The scheme was eventually passed by the dogged determination of the developer to overcome the hostility that had now been engendered and using a neutral third party to redo the Master Plan and to publicly consult to uncover the complex set of variables which might satisfy public objections and still remain viable. Simple suggestions were taken on board by producing better site lines to the cathedral from key points and by providing more open residential views and wide-ranging public access to the waterside and around the residential areas.

This scheme presents an obvious case for urban design that the developers thought they had been taking into account with public access and leisure requirements as well as providing access improvements to the road. Two reasons have been put forward for the strong public reaction:

- a sense of ownership of an area of the city which connected up to its history and the need to enhance this with any new development;
- the strong lobby provided by the Civic Society to organise the latent feeling which existed and the lost city centre parking which would result.

Conclusion

It is important to understand the iterative nature of design before you can manage it and give time for reflection and development. Management needs to optimise creativity and design impact in its assessment of value and performance. Design impact will pay back with long-term benefits. Design may be managed effectively by a designer or a non-designer who may be the architect, the project manager or the design and build contractor. Efficient information flow between designers and contractors is a key aspect of productivity. This may be enhanced by the earlier involvement of the contractor in the design and the use of a single design and construction co-ordinator.

The management of design is most applicable in controlling cost and to a detail stage, which is characterised by the co-ordination of many different specialist designers including supplier input. At this stage, design has a two-way flow with the production process which continues to need the input of design. A design manager has a responsibility for ensuring that procured packages have the correct design inputs and produce their design outputs for approval.

The new challenges for integrated design management are:

- Optimising creativity in reduced overall project implementation (scheme design and construction).
- Working in the context of sustainability and to co-ordinate the backward linkage of the building with urban design for true project success taking into account the final users and the social impact of the building, which does not work properly by depending on the statutory planning process alone.
- Developing of teambuilding by the use of excellent communications on complex projects. This can be done through co-location or through virtual teamwork, by standardising electronic documents to remote positions supported by face-to-face communications wherever possible.

References

1 Construction Industry Council (2002) *Construction Project Management Skills.* CIC, London.
2 Gray C., Hughes W.P. and Bennett J. (1994) 'The successful management of design'. *A handbook of building design.* Centre for Strategic Studies in Construction, University of Reading. ISBN 0 7049 0523.
3 Hirano, Takuo (2000) 'The development of modern Japanese design'. In V. Magolan and R. Buchanan (eds), *The Idea of Design: A Design Issues Reader.* MIT Press, Cambridge, MA.
4 Royal Institute of British Architects (2000) *Plan of Work.* 2nd edition. RIBA Enterprises, UK.

5 Maguire H. (1980) 'A conflict between art and life'. In Mikellides B. (ed.), *Architecture for People*. Studio Vista, London, pp. 130–2.

6 Taywood Engineering Limited/DETR (1997) Report No. 1303/96/9383. *The Influence of Design on Construction Site Productivity*. A partners in technology report. p. 18, table 4.

7 CIOB International News (2003) Article 3999. 'Devonport, what went wrong'. September.

8 CABE (2004) Aims as indicated http://www.cabe.org.uk (accessed on 29 July).

9 Loosemore M., Cox P. and Graus P. (2004) 'Embarking on a project – the role of architects in promoting urban design as the foundation of effective project management'. In Murray M. and Langford D. (eds), *Architects Handbook of Construction Project Management*. Chapter 6. RIBA Enterprises, UK.

10 CABE (2003) Survey of design skills in local authorities. CABE, http://www.cabe.org.uk (accessed on 29 July).

11 CABE (2004) Priorities as indicated http://www.cabe.org.uk (accessed on 29 July).

12 NAC, CABE and OGC (2004) *Getting Value for Money from Construction Projects Through Design: How Auditors Can Help*. March. pp. 6, 8 and 22.

Chapter 10

Project safety, health and the environment

Construction sites have the potential for serious accidents as there are many people close together, many activities are unpredictable and the tolerance towards risk is traditionally quite high, making the frequency and impact of unplanned activities very high. Although there is a poor health and safety record on construction sites the mass of legislation designed to reduce accident reduction has not been as effective as could be hoped. Health conditions are less visible, but just as critical to improving conditions and their recent link with the impact of construction on the environment has given this a higher profile. Accident prevention, good health and safety and environmental management is the key to improving conditions and reducing accidents that occur.

The objectives of this chapter are:

- To understand the principles of workplace health and safety legislation.
- To legislative principles to construction projects.
- To discuss a framework for managing project health and safety.
- To provide a critique for effective health and safety planning and risk assessment.
- To consider environmental issues and their integration with health and safety.

Introduction

Overall the industry has not improved its accident record in real terms over the last decade and rates as the second worst industry in the after mining and agriculture. The construction industry is rated on an accident per 1000 of workforce, as being four times worse than manufacturing for serious accidents. And an EU report produced in 1992 showed a survey of worker perception of work safety in construction was 10–20% worse than other industries. It is easy to blame a 'cowboy' element for the depressed figures, but it is clear that there is still a major problem. Egan claimed that accidents can account for 3.6% of project costs and this also gives an economic as well as a social and ethical reason for improving performance.

In round terms, the HSE figures indicate that there are 30 000 construction related accidents incurring more than three days' absence (under three days are not reportable) and over 4000 are rated as serious injuries. The HSE has reviewed and reported on the hierarchy of accident types with falls from height listed as the most common serious accident (1300 or 42%). For example, this would include falls from ladders, from or through roofs and scaffolds and into excavations. It is also generally recognised that only about a third of accidents are reported and studies seem to agree that there is a proportion of at least 1:3 between serious over three-day absence injuries and under three-day non-reportable accidents and one major report indicates 16 times. Many others need some first aid and there are at least eight times as many again near misses [1], producing an iceberg effect shown in Figure 10.1, where there is a large bulk of accidents which will not be known about, but are not or do not need to be reported.

In addition to accidents there are many concerns for poor health caused by the workplace. In 1997 the World Health Organisation indicated that worldwide there were 160m new cases of work-related diseases caused by various types of exposure at the workplace. There are huge numbers of chemicals that have been identified by them as important health hazards. A well-known killer is asbestos, which kills more than 3500 people each year and is set to rise. This is more than all road accidents in the UK. It is more difficult to tie diseases into particular construction projects, but suffice to say that construction workers are in the front line of handling disturbed asbestos. Health is also a factor where 30–50% of workers in industrialised countries complain of psychological stress and overload. There is a much closer connection here with environmental policy, which is also concerned with the impact of construction on the health of others.

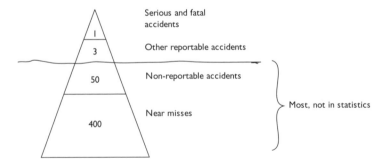

Figure 10.1 Relationship of reportable and non-reportable accidents.

Source: Based on Tye and Pearson [1].

There has been a whole raft of legislation which has hit the industry, mostly generated by directives from the EU.[1] It is important to be aware of the principles of the legislation which were set up in the Health and Safety at Work Act 1974 (HSWA) as the EU directives have not wavered far from these far-sighted guidelines. The Act set up a framework for more focused Regulations.

Principles of modern health and safety legislation and the HSW Act 1974

Prior to 1974, there was a piecemeal industry approach with quite antiquated and often contradictory legislation. In construction the Factories Act 1890 and to a lesser extent the Offices Shops and Railway Premises Act 1964, together with a wide range of supporting Regulations, were used as a basis of Health and Safety on construction sites. In the early 1970s a new committee was set up, chaired by Lord Robbins in order to put more emphasis on preventing accidents, and draw the older legislation together.

Integrated guidelines to provide simpler health and safety legislation with parity across all industries were required and the resulting HSWA 1974 has provided robust forward looking principles which have allowed modern regulations to be produced for all industries under the umbrella principles of the Act. These principles are based on a personal responsibility and not a prescriptive approach:

- The responsibility of the employer to provide a safe working environment to employees.
- The responsibility of the employee to comply with all reasonable provisions for their own safety and to act responsibly towards others.
- The responsibility of the employer and owner of premises being accessed by others to other persons who are not employees.
- The responsibility of designers, suppliers and plant hirers to provide safe products and components.
- It requires every firm to prepare a safety policy.
- It provides for criminal as well as civil liability and set up the HSE to implement and enforce the Act.

The point of this approach is to encourage a risk assessment method which puts the onus on the employer to adapt and develop measures relevant to their context, rather than a blind adherence to prescriptive measures by others. It also allows, in the requirement, for individual identification of responsible person's accountability at different levels for measures which are insufficient to stop accidents. It is quite clear for instance that the CDM Regulations follow this principle of identifying persons who will assess the risks and set up measures to control and reduce that risk.

Enforcement was entrusted to the HSE, which has an advisory and policing role in construction (HSWA sections 10–14). Enforcement can lead to a written notice to rectify specified faults promptly in the form of

- a warning;
- an improvement notice;
- a prohibition notice;
- prosecution.

A prohibition notice would shut down the site until suitable action has been taken to make the site safe again. Prosecution with resulting fines or imprisonment can be imposed in the case of non-compliance to the notice where the employer is proven guilty in court. In the case of serious accidents resulting in death or serious injury an employer or their agent is *criminally* liable to prosecution. There is also a warning notice for action to be taken in so many days.

Reporting

Reporting of accidents through the Reporting of Injuries, Diseases and Dangerous Occurrences Regulations 1995 (RIDDOR) becomes a statutory requirement whenever there is an accident involving more than 3 days' absence to a worker or where a dangerous occurrence on site incurred a near miss, with a safety or health hazard. The latter may cause substantial property damage, but reporting should be wider. This allows the HSE to investigate accidents and provide accident statistics. There is, however, a serious shortfall of reported injuries which is indicated by the discrepancy between, for instance, the 150 000 reported accidents and the 380 000 which were recorded in the annual Labour Force Survey in 1995–6. This indicates that only 40% of accidents are reported.

All accidents however small have to be recorded in an accident book. They do not have to be reported if they are outside RIDDOR, but it is important to review any accidents and near misses to see if there are any patterns which indicate poor practice that can be improved.

EU directives leading to UK regulations

EC directives which are adapted into UK regulations under the umbrella of HSW Act 1974 all have adapted a pattern of risk assessment identifying responsibilities. The way in which this is done is slightly more prescriptive than HSWA.

The Framework directive (89/391/EEC) for 'measures to encourage improvements in the health and safety of workers at work', led in the UK to the Management of the Health and Safety at Work Regulations 1992

(MHSW), revised in 1999. This defines the risk assessment process and prescribes generic measures such as training, planning, health surveillance, organisation and monitoring and escape which might be applied to a wide range of more specific Regulations.[2]

The Framework directive led also to daughter directives applying to any workplaces to deal with such areas as hazardous materials, manual handling, biological agents, use of work equipment, noise, electricity at work, protective equipment, equipment safety, visual display screens, pregnant workers, protection of young people at work, safety and health signs and worker safety representation. Figure 10.2 indicates the framework.

Each of these regulations specifies certain hazards and limits in relation to the areas indicated and expects a risk assessment process with control measures to reduce risk, training as necessary and information given out for employees' information. The Temporary or Mobile Construction Sites' Directive (92/57/EEC) led to two UK Regulations which apply to construction workplaces in particular, these are:

- The Construction (Health, Safety and Welfare) Regulations 1996 (CHSW), amended 2000, taking out more specific areas such as lifting equipment (LOLER 1998) and Working at Height Regulations (WAHR 2004).
- The Construction (Design and Management) Regulations 1994, dealing with client responsibilities, the co-ordination of supply chain risk assessments, health and safety plans, general site welfare, designer responsibility for user safety and co-ordination of designer safety measures. The Approved Code of Practice (ACOP) was amended in 2001 following extensive consultation to provide better guidance for designers.

The main difference in construction is the temporary nature of the workplace with unique new hazards and the need to fundamentally consider the new

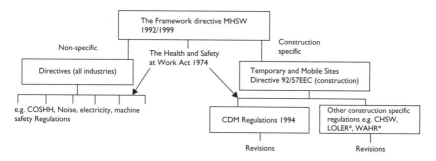

Figure 10.2 The framework directive and daughter directives.

Note

* See text.

risks. For example, the use of foreign workers for a bricklayer gang even on a house of the same design, will mean that safety posters do not communicate so well. In addition the access to another house in the same workplace will have different hazards. There are also physical changes taking place throughout the construction project, which create a dynamic risk environment. The use of competent people permanently on the project becomes more critical here. The CDM Regulations identify responsibilities for the management of safe environments.

The main requirements for health and safety practice are to be found in the CHSW 1996. These together revoked many of the old regulations under the Factories Act 1962 and the Offices, Shops and Railway Premises Act 1963.

The new Lifting Operations and Lifting Equipment Regulations 1998 (LOLER) replace the Construction (Lifting Operations) Regulations 1966. The Provision and Use of Work Equipment Regulations 1998 (PUWER) describe in general terms the safe use of plant and equipment and insist on use only by competent and trained persons. The WAHR supersede parts of CHSW by stiffening up the requirements. This is due to the high proportion of fatal and major injury accidents induced by falls or by being hit by objects falling from working platforms or lifting equipment.

The CHSW together with LOLER, WAHR and PUWER provide a prescriptive framework to deal with specific construction related risks associated with accidents, health and welfare. Typically these are the safety of excavations, falls off or through working places such as scaffolds and roofs, stability of structures, demolition, explosives, falling objects, prevention of drowning, vehicle movements, emergency routes and procedures, fire detection and fire fighting, fresh air to and lighting of work areas, traffic routes, doors and gates and safety in the use of work equipment. Regular inspection and training programmes for competent use are a key part of the operation of these regulations and welfare facilities must be provided and maintained adequate to the size of the workforce.

The Control of Substance Hazardous to Health (COSHH) Regulations 1999 have been regularly updated to keep up to date with European directives which have become tougher on exposure to biological, dust and fibre, chemical and other corrosive or dangerous emissions hazardous to health, some of which have special regulations like those for asbestos.

Liability

The European position on product liability is related to health and safety and the damages caused by product defect. There is a general product liability directive [2] and a product safety directive [3]. In construction this was backed up more specifically with the Construction Product Regulations (1994) which emerged from the need to consider the complexity of liability caused by the interactions of products fixed in the building as a whole. The core of the directive refers to the six 'essential requirements' to ensure the product is fit for the purpose intended when fixed in the building structure. These are:

- Mechanical resistance and stability.
- Safety in the case of fire.
- Hygiene, health and the impact on the environment.
- Safety in use.
- Protection against noise.
- Energy economy and heat retention.

These requirements carry a broad health and safety brief in all six of them and put an onus on the manufacturer to incorporate an auditable 'factory production control' QA system, enabling the issue of a conformance certificate. Labelling is required to encourage correct use where they have a lesser influence once they arrive on the construction site. For example a pre-cast concrete lintel with two reinforcement bars in the bottom must be labelled with the top and also with its safe load bearing conditions. Harmonised standards are part of the drive to use ISO or European (EN) standards rather than national ones. This tries to deal with the hidden problem of products with different standards of safety moving round the EU as well as encouraging the more headline 'free movement of goods'. For example, the performance of a BS476 one-hour fire door is well understood in the UK, but not in Germany.

The liability of services' providers such as designers and contractors lasts for different periods in different EU States ranging from contractual obligation only to 15 years (Spain). Ten years is quite common though not in the UK. The common denominator for most countries would be a five-year period of liability, which the Commission cites as covering 75–80% of the liabilities, whereas 90% would emerge in a 10-year period. There is no agreement on Union wide liability as yet or on its necessity. Insurance rates may be prohibitive for underwriting particularly long liability periods. Damages are the most common outcomes of product and services liability, but there are prosecutions under the Construction Products Regulations that are open to criminal prosecution.

CDM: the responsibilities

CDM Regulations cover the specific management, design and client responsibilities created by the unique fragmented organisation of construction projects. Two *co-ordinators* called the planning supervisor (PS) and the principal contractor (PC) are appointed by the client and are responsible respectively, for co-ordinating the design and construction health and safety processes. They are not directly responsible for the actions of the individuals, but they do have a responsibility for interface management and ensuring good communication between different parties in each of design and construction and have to produce a health and safety plan, which is regularly reviewed and the PS has to produce a Health and Safety File (HSF).

Responsibilities are given to the designers, and the individual contractors to risk manage their work and to be aware of the effect it has on others, so that they provide risk assessments to the co-ordinators who use these risk assessments to compile a co-ordinated approach. Co-ordinators may make suggestions for better amended design or more information, but do not make themselves liable for unsafe design or methods of installation. This liability remains with the designers and contractors who are considered competent.

The *client* is responsible for appointing competent co-ordinators and ensuring that a co-ordinated health and safety plan has been developed by the PC before the construction work starts. In practice, this is done on the advice of their professional team as the client may not be experienced in construction. The client is liable for not passing on information about hazardous substances or conditions which exist before the site is handed over, and they are also responsible for ensuring that sufficient resources are available to cover the costs of sensible health and safety measures. This will be covered by appointing a main contractor who has priced and is committed to proper health and safety measures as indicated in a properly developed contract health and safety plan, which the client must ensure is in place before the work commences.

The *designers* have responsibilities to produce a safe design and are expected to produce evidence of risk assessments for each stage of the design process. They are liable for accidents or health problems attributable to designs which contribute to unsafe environments both for the contractors, the users, the maintenance team and those who demolish buildings.

The *planning supervisor* is not a term mentioned in the original Directive and was only introduced by the CDM Regulations as a way of adapting to the separated design and construction process, which predominates in the UK. Their responsibility pre-contract is to co-ordinate the health and safety aspects of all the designers and to ensure that a pre-contract health and safety plan is available to go into the tender documents ensuring that health and safety requirements are made clear and information is full enough to give 'a level playing field' for each contractor. The client may appoint the lead designer or any other 'competent' person with the knowledge of design processes and experience of the particular type of work. They also collect and present information from all the designers and contractors, collating all drawings as built, safe operating procedures and use of materials and components guidelines into a HSF.

The *principal contractor* is normally the main contractor who formally adds health and safety co-ordination on site to their responsibilities. In practice, this requires all contractors to produce risk assessments before they start work, which are given formal approval or amended to suit the needs of others. Site rules are established to create safe overall conditions with escape routes and measures to manage access to areas, police the wearing of protective clothing and to ensure induction and training with updated skills. A card indicating a level of competence in the worker's trade or manager's supervisory skills is now compulsory on many sites. Updated skills are ongoing and should include safe ways of working.

Health and safety management

Overall health and safety management, including risk assessment, is addressed generically by the APAU [10].[3] The approach reflects total quality management (TQM) and tries to engender a top management commitment to change and a culture that encourages bottom-up participation which is motivated towards improvement. This makes review an important element. The Figure 10.3 indicates HSEs framework for successful management.

The key management issues are a basic policy framework, accountable persons, safe designs, comprehensive planning and organisation, good supervision, segregated or restricted areas, availability of personal protective equipment, careful sequencing of activities which impact on each other and training and induction processes with monitoring and checking. The CDM Regulations 1994 and its ACOP are also important for a good

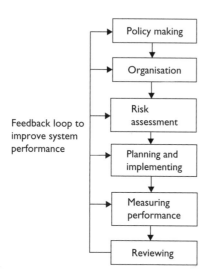

Figure 10.3 A management framework for the management of health and safety.
Source: Adapted from APAU [10]. Used with permission.

project health and safety management system. Implementation of the system is often harder than developing the system as it depends on developing a better culture of safe working than many construction projects have been able to develop.

Safe design means information about the hazards of materials specified and revealed in surveys carried out and the risks involved in special methods implied, for example, manual handling and contaminated land or asbestos.

Supervision cannot be total, but it is connected with knowing what is going on. One way of dealing with this is to require permits to work for particularly hazardous activities such as deep excavations, hot work, confined spaces etc. Other activities will not need to be supervised, but spot checks will need to occur to ensure competent persons are doing their job and that co-ordination with other activities is working.

An effective health and safety culture involves the whole work force in the planning phase and ensures that there is a commitment for ongoing training and updating from all the contractors. To achieve this top management support is required so a budget for a quality improvement approach is committed.

Risk assessment has been added as a separate phase to planning in order to emphasise the importance of this activity at the start of all planning activities. The concept of risk and its assessment in order to take acceptable action is discussed later. Liability in health and safety law can emerge from a number of angles, but the chief of these is design, product safety, negligence in reducing site worker risk, poor site organisation, health impacts which emerge later and public safety. The review stage is an active stage, which for time limited project work needs to apply at regular intervals and particularly when changes are made in the original planned design or methods.

Health and safety policy

All businesses require a health and safety policy in written form where five or more people are employed. The purpose of the policy is to have a clear statement of commitment to the health and safety of employees and to direct the accountability for safe systems of work to named people. A director or partner must be named with executive responsibility for the policy being carried out. Where the firm is larger, there is a separate health and safety manager who is responsible for monitoring health and safety systems, receiving feedback, arranging training and communications and seeking to improve and maintain systems up-to-date. Smaller companies often share an agency who provide a majority of these services for them.

The three main areas to address for any organisation, including a construction project are general policy, organisation for carrying out the policy and safety arrangements Fewings and Laycock (1994).

The primary purpose of general policy is to set out an action plan for health and safety. It consists of

- a general commitment by the project manager to the health and safety of the project by developing safe systems of working;
- stresses the importance of the co-operation of the employees to using these systems;
- setting out disciplinary procedures applicable for negligence and contravention of safety rules;
- establishing communication systems for informing project workers of health and safety matters.

In order to carry out the policy it is necessary to assign various responsibilities.

Organisation

Overall responsibility lies with a director or senior partner of the organisation(s) carrying out the work. In a larger organisation, they are likely to delegate the development of co-ordinated arrangements to a senior manager to provide continuity and parity across contracts. It is also usual to have safety experts who visit on behalf of the main contractor and sub-contractor organisations doing work on site and produce reports for improvements. The preference in the MHSW Regulations 1999 is for the use of competent employees rather than external sources, for health and safety advice. Figure 10.4 indicates an organisation structure.

On a construction project overall project responsibility will lie with the project manager or other named person, who will have health and safety training. He/she is responsible for implementation of the company health and safety policy and co-ordinates the project health and safety plan. He/she monitors the system set up for the project, reports and investigates

Figure 10.4 Organisation to show responsibilities.

accidents and near misses and receive feedback from safety representatives and specialist contractors. Under CDM, the principal contractor has special powers to make site safety rules, recognising the co-ordinating function of a main contractor.

The Safety Representatives and Safety Committee Regulations 1977 also recommend the appointment of workforce representatives who may be the 'ears and eyes' in order to report on unsafe practices on site. This does not take away from the responsibility of all employees and visitors to report dangerous or unsafe structures or practices which they come across.

A construction project is a unique workplace and it is compulsory to set up a health and safety plan within the tender documents, which is developed by the chosen contractor during the mobilisation stage in order to set out the strategy and policies, which are relevant to those unique circumstances. It is also important to keep this under review as the interfaces between different contractors working on the site is dynamic and so new risks arise. One of the commonest causes of accidents is when the sequence is altered and there is an unplanned and dangerous mix of activities going on alongside each other. For this reason the CDM Regulations were developed to establish a transparent set of responsibilities for the client, the designer, the principal co-ordinating contractor and the individual contractors. It is important for these to all work together to provide information, set up training and induction procedures and to co-ordinate the design and construction process.

Risk assessment

Risk assessment is a concept present in the HSW Act 1974, but is more definitively defined in the MHSW Regulations 1992. It is important to define the difference between hazard, significant risk and harm.

A *hazard* is the potential for harm, but the context and the outcome are not defined, for example, the use of hot bitumen. If this is used by competent people, who have proper protection and is not accessible to others this is unlikely to be a significant risk.

Harm refers to accident or longer term health problems. For example, the bitumen might scald skin, cause a health problem through breathing the fumes or gets overheated and causes a fire.

A *significant risk* is the likelihood of harm occurring and follows all the possible harms inherent in a hazard by reference to the severity and probability of that harm occurring. This will be affected by the context, for example, near to the public who are more vulnerable, or being carried out unsupervised, or at height or in bad weather will change the rating of severity and impact. It is these conditions and context that should be addressed as well as the inherent harm of the product.

For this reason the risk assessment should not be generic and transferred from site to site, but if possible reference should be made to the methods which are uniquely used by the workmen.

The risk assessment process consists of three steps:

1   Identify the hazards, assess their significance and then institute control measures to reduce significant risks.
2   Implement these arrangements by

- setting up a written plan in order to establish the procedures;
- train personnel;
- organise the staff and assign responsibilities and communicate the information to all those affected;
- monitor and feedback;
- monitor and control, making sure the procedures are understood, work and are being carried out.

3   Review risk control procedures regularly to make sure they are relevant and effective.

Risk controls

There should be some concern for the way in which a residual risk is controlled and who it might affect and what priority should be given to the actions prescribed. Control of risks is very similar to the system in Chapter 8, but more focused on pure risks.

A risk may be eliminated, reduced, given overall protection against or personal protective equipment provided in this order. It may be possible to eliminate a risk altogether by, for example, eliminating the use of a particular hazardous material or redesigning or banning certain methods of working, but the alternative material or method is also likely to create a level of risk that will need to be assessed. If new materials or untried methods are proposed this is almost certainly the case.

One contractor's system advocates a deeper hierarchy which resonates with the different risk types of elimination, substitution, enclosure of hazard, remove, reduce exposure time, training, safe system of work, personal protective equipment.

The key method then is to reduce significant risks to acceptable levels and to be aware of the residual risk in managing the workplace. Acceptable risk may also vary with improved knowledge and technology and so the control measures should be reviewed for effectiveness.

Those at risk are those carrying out the work, those who work with them (other workers and managers), visitors to the workplace who are less familiar with the workplace dangers and vulnerable groups such as young people, trespassers on the construction site, the general public and foreign or

Table 10.1 Risk priority grid

Frequency	Probability		
	Low	Moderate	Serious
Seldom	1	2	3
Occasional	2	4	5
Frequent	3	5	6

inexperienced workers. The access to risk of each of these parties needs to be considered separately and not lumped together if proper controls are to be devised. For example, holes which are fenced, but not completely covered are likely to be a risk to young trespassers, but not to workers; signs which give written warnings may be misunderstood by immigrant workers newly arrived.

The control of risk especially on construction sites is likely to produce a need to prioritise on the basis of the seriousness of the risk. This is generally dealt with by measuring the severity of the risk and the probability of the risk occurring. This could be measured as an overall impact by using a grid or graph as in Table 10.1.

Normally any risk above 2 would be dealt with as a matter of some urgency with some residual protection given. For example, a nuclear risk would be seldom and serious so, although only a 3 on this system would be important to deal with urgently. A trailing cable across a doorway would be low severity and high probability and should not be left for long as many are exposed to such a risk and it becomes a critical nuisance. The graphical effect is also useful in picking up priority actions. The number of people at risk is partly subsumed into the frequency axis, but the risk is heightened if the risk was affecting an unsuspecting public. A secondary population also increases the number of people affected such as other workers/public affected by noise or dust. Case study 10.1 of such a system is given.

Case study 10.1 A risk assessment pro forma

A contractor is likely to develop their own risk assessment systems, which they feel comfortable with and diagram Figure 10.5 is part of a pro forma risk assessment sheet. A separate sheet is used for each activity or system (e.g. delivery of hazardous materials) on the site. Critically there is a responsible qualified person to write out the assessment and take responsibility for implementation. In order to ensure supervision is in place for hazardous processes a permit to work certificate needs to be issued.

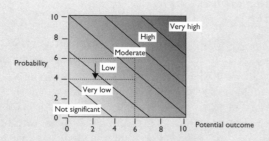

Figure 10.5 Alternative risk priority grid.

Source: Acknowledgement to Pearce Group.

The graphical effect is useful in picking up priority actions. The scores of probability range from unlikely (2) to 'likely to happen at any moment' (10). The scores of potential outcome range from 'minor injury' (2) to 'multiple fatality' (10). A description of all the actual risks and the proposed measures to bring the risk to an acceptable level; length of exposure and the type of people at risk are separately noted on the risk assessment sheet. The risk would be heightened if the risk was affecting young people, or an unsuspecting public.

For example, in the case of tower crane operations an accident's outcomes are likely to be classified as serious, because of the height and wide exposure of employees and possibly the public to falling objects, giving a score of 6. Probability of likely to happen occasionally would give a score of 6 (dotted line in Figure 10.5), making it a moderate risk. This score would catalyse actions to reduce this risk to at least the level shown in the diagram. In this case actions to work on reducing the probability to 4 which is improbable or unlikely brings the risk down to the low level shown. This would be achieved by having strict protocols to supervise loading, test the crane and exclude people from working under loads. It would also be sensible not to slew over external property and highways whilst loaded, because the general public are not inducted into looking out for site dangers, so are more vulnerable.

Research has indicated that many accidents take place because of two normal, but unco-ordinated activities taking place at the same time. So additional attention is needed to measures for overall site management and to any likely conflicting activities – in Case study 10.1 the use of a mobile crane at the same time would cause additional hazards. Any changes to the programme or the original intentions also require further risk assessment. For example, a change of the crane model may affect the exclusion zones. Other things which might be exacerbated are unpredictability, length of exposure, the nature of the risk and the nature of the controls.

Improving the effectiveness of accident prevention

Accident prevention is a primary tool for management effectiveness. The statistics above indicate a limited success in reducing accidents so what are the main principles. The HSE Annual Report 1988 suggests that management action could have prevented seven out of 10 construction site deaths, that is, approximately 70 deaths. The same figure was given in 1985 for deaths that occurred in maintenance. The latest figures for 2002/2003 do show a drop in fatalities and major injuries in construction, but the major injury rate is thought to be low because not all accidents are reported. As discussed earlier, injuries and fatalities from falling or by being hit by falling objects consistently account for 50–60% of the total. Management and lack of organisation and planning has been consistently blamed by reports on health and safety:

- 'Management is infected with a disease of sloppiness'; Sheen Report in 1988 on the Zeebrugge disaster.
- 'London Transport had no system...to identify and promptly eliminate hazards'; Fennell Report 1988 of the Kings Cross disaster.
- 'Significant flaws in Occidental's management'; Cullen Report of the Piper Alpha disaster.
- 'Frightening lack of organisation and management'; the Hidden Report on Clapham Junction crash.
- '35% of accidents in construction in Europe have occurred because of design and organisation faults and a further 25% due to the planning and organisation stages'; EC Report 1992.

The latter report was the basis for introducing the Temporary and Mobile Sites Directive 92/57EEC which spawned the CDM Regulations 1994 which was an attempt to improve the effectiveness of management and to involve the design process more closely. Other approaches have been from a TQM perspective mainly espoused in the APAU [4] and adapted later. The system depends on an adequate culture and commitment from top management to the strategic importance of health and safety management systems. HSE distinguish between active or reactive measurement.

Reactive measurement is normally done by the use of KPIs based on past measurement of accidents, ill health, close misses and other evidence such as past accident reports which informs future planning.

Active measurement measures the standard of planning and organisation procedures and not the failures. It is based on the achievement of objectives and targets for accident prevention and will include the number of health and safety training days and the a test of manager and worker knowledge of safe practices, so that the development in competence can be assessed. One of the key aspects of the active model is the ability to measure performance.

An important point that HSE makes is that a good track record in health and safety does not necessarily mean that the framework for avoidance in the future is in place. Typical activities in active monitoring equate to preventative maintenance by the use of inspections and replacement of components before failure.

In construction high risk operations such as scaffold and lifting equipment are inspected before use. A more extensive check is signed off in a weekly register. Mobile plant and protection of excavations need to be supervised continuously, either by the use of banks persons or by complete covering. Low risks are dealt with by general inspections.

Reactive measurement can be accused of being too little too late, but if it is based on an analysis of minor injuries and near misses and the patterns that occur, specific areas for improvements may emerge as a proactive action. For example, consistent small injuries in one trade may point to a short fall in training or a poor attitude to safe working or inadequate supervision. HSE points out that an observant approach on the behalf of workers with a non-blame, responsive action by management encourages open and honest communication.

The active model is a feed forward model with the use of both types of measurements and aim to prevent accidents by planning ahead and encouraging self supervision rather than intervention. This requires a change in culture discussed later.

In terms of investigation, *underlying causes* are important and these can occur in

- the person (poor attitudes, behaviour and knowledge);
- the organisation (insufficient procedures to ensure a safe working place);
- the task (inherent danger in the process or materials/components).

A risk assessment needs therefore to identify the context of the task with the person doing it and not remain generic. The question is not what actually happened, but what is the worst situation that could have happened. Case study 10.2 is a reactive review on a near miss.

Case study 10.2 Staircase collapse (hypothetical)

An escalator with a cantilever dogleg (Figure 10.6) was to be installed from a floor one storey above a multi-storey atrium to the floor of the atrium. On the way down it passed through a free standing media wall providing electronic advertisements to escalator users.

Initially the escalator was to be made in one piece and a risk assessment determined a suitable method statement for the fitters. In the event the escalator was redesigned and delivered in bits to get it through existing doors and was bolted together in situ. Cranage and

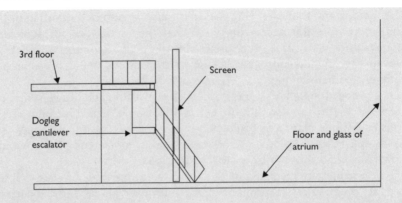

Figure 10.6 Diagrammatic of staircase.

frames were used to lift and support the parts until they were completely bolted together. At completion the frame was taken away and the equipment commissioned by the manufacturer. During its first major usage the assembly slumped out of line hitting the media screen which crumpled and could have hit escalator passengers. The escalator was immediately shut down. The design or the installation had failed.

The subsequent investigation indicated the shearing of two bolts holding two of the sections together. These were not the specified friction bolts or tightened to the right torque. Potential dangers were the complete collapse of a full escalator with subsequent injury or death. The accident was the result of a workmanship problem, but with a narrow margin of safety on the design. The risk assessment was carried out by the installer who was a subcontractor to the designer. They were unaware of the design parameters and a hurried change in the design to accommodate the access restrictions of a nearly finished building had weakened the design safety checks and created a shortfall.

- What can be learned from this near miss incident?
- Who was primarily to blame? Would anyone have been prosecuted?
- How could this have been avoided?

Safety representation and safety committees (SRSC) regulations

These are set up in the Regulations of 1977. There are an estimated 200 000 trade union safety representatives in UK industry and organisations and safety committees have a function to investigate accidents and near misses and inspect premises where they work or others they have been appointed to represent. They may also consult with and receive information from enforcement officers and require meetings with employers and receive information from the employers such as risk assessments, accident reports and hygiene reports. Members are given a basic 10 day union training and represent their colleagues.

In construction, this representation would be supplied through the main unions and the various specialist unions. Companies have a history of resisting training and workplace inspections as it requires paid time off. In construction, except on large sites, many are no longer in the union as they work in small disparate groups for specialist contractors. Clearly they miss out on the benefits of safety representation, which according to a TUC survey[4] was a more important function for the unions than workplace conditions and pay.

Safe design

A designer has responsibility under the CDM Regulations to provide a safe design and to risk assess their design. The three stages in the design process are outline design at the feasibility stage, scheme design leading up to a scheme for planning approval and detail design leading up to the tender documentation in traditional procurement. Risk assessment for a healthy design is best at the earlier stages of design as more opportunities are given to design out hazards (the potential to cause harm). The HSE [5] has a code of hazards called red, amber green, which prioritises under red the hazards that should be designed out. Amber recommends reducing the hazards as much as possible and green refers to design practices to be positively encouraged. The designers' role then is to look to reduce hazard or to avoid hazard at source. For example, heavy beams can have lifting points designed in to facilitate the use of lifting gear and discourage manhandling. This is far preferred over the use of protective equipment. Many hazards are health related and five common areas affecting construction practice are:

- manual handling;
- noise;
- vibration;
- respiratory, for example, asbestosis and asthma related disease;
- dermatological complaints.

Loughborough University now has a website 'tool' to aid designer risk assessment called D4h©, which gives lots of practical advice with case studies to cover the areas of outline and detail design and to eliminate poor occupational health caused through unhealthy design [6]. CIRIA [7] has also produced some guidance notes which support web site tools like these.

Buildings in use

The designer accountability is not for construction alone, but also for a building which is safe to occupy and maintain. This consideration is not new, but the safety of maintenance access has been reinforced. This should include access to services in ceiling and roof spaces, access to change common maintenance activities such as changing light bulbs and clearing glass areas. HSE [8] has produced a Guide for Designers in applying the CDM Regulations to achieve a safe design both for construction and maintenance. It also includes case studies. It should be noted that revisions in the guidance for the CDM Regulations tried to be more specific about the responsibilities of designers. There is also a move to revise the CDM Regulations at the end of 2004.

Particular areas of danger are:

- cleaning windows over the top of conservatories;
- cleaning windows or curtain walling in multi-storey blocks;
- cleaning the underside and top of glass atriums;
- cleaning out of gutters;
- changing light bulbs in tall spaces;
- moving around on fragile roofs, or near roof lights;
- moving above fragile ceilings where there are service access points;
- accessing heavy plant and equipment safely for periodic maintenance.

The creation of an *inclusive environment* is important for the health and safety as well as compliance with the requirements of the Disability Discrimination Act (DDA) 1995 for those with special needs. It considers various needs such as mobility issues, partial sighting, hearing loss and dyslexia and provides where possible access and provision without making anyone a special case. This might include raised braille on signage, tactile paving or flooring for crossings and entrances, contrasting colour schemes and furniture, wheelchair reachable handles and switches and fire doors that are easily manoeuvred or automatically openable, that everybody uses. Fewer special cases means less confusion or frustration for users and this is a safer culture.

The *role of the planning supervisor* as co-ordinator of the health and safety at the design and handover stage has the same range of accountability as the designer, but is generally restricted to the project life cycle. This has often resulted in criticism about the authority of the role, as it is, their responsibility ends upon the proper handover of the HSF and ongoing liability accords to individual designers for the adequate safe functioning of the building in use. It is therefore seen as dependent on the co-operation of other parties to gain access to information. The choice is whether to make this position independent of all other parties or to put it in the hand of the lead designer, client or even the project manager, who can also operate with more power to enforce decisions and design compliance.

Safe construction

The minimum requirements of the PC are shown in the appendix at the end of the chapter and this section looks at the human factors for improving safety on sites. The key thing is that a health and safety plan is not only available with the planned procedures, but that this plan should be effectively implemented and co-ordinated, taking into account all the human factors and the practicality of ensuring good quality safe systems are a fundamental part of individual members of the workforce and management team. This includes training and communications. The issues of competence, inspection, culture, supply chain and communications are discussed.

Competence

A key issue in the maintenance of health and safety standards is the achievement of competent workers. Competence is the ability, the experience and motivation of the individual. This will apply in the recruitment procedures exercised by all contractors, but should not be exclusive of those who are willing to be trained and this will incur temporary greater supervision and feedback (Case study 10.3).

Schemes for ensuring management and operative competency are important moves and there has also been a move by the HSE to target special areas of concern and make themselves available as a training resource during this target period. Health and safety training is part of the competency certificates. This is a necessary addition to site induction, which highlights known hazards, but does not specifically test competency and understanding or provide updates.

Managers need to remain updated if they are to exercise an effective role. Information is passed on to suppliers via their supervisors. The regular development of a health and safety culture is important by keeping health

Case study 10.3 CSCS cards

Many large contractors now require all workers to have passed a test of competence that is associated with a series of compulsory training sessions including health and safety. In the UK competence is recognised by issuing a Construction Skills Certification Scheme (CSCS) card. Critically this is required of all managers and operatives representing any of the trade organisations on the site. The card also confirms that the holder has gone through health and safety awareness training. The cards are graded and recognise competence equal to NVQ level 2 or completed apprenticeship for the trades, NVQ level 3 for supervisors and NVQ level 4 or a professional standing for managers. This important association of competence with health and safety gives a better indication of an individual than a signed disclaimer. Site-based health and safety induction is still relevant for specific site hazards.

and safety in the spotlight through toolkit talks and prizes for examples of safe working. This also has a direct connection with management and shows their commitment.

Supply chain

Most production work is now carried out by package contractors and the self-employed. As indicated they are responsible for safe working methods by carrying out a risk assessment and devising control measures to reduce significant risks caused to their employees and eliminating harm that their work may cause to others. The submission of a method statement allows proper consideration by the principal contractor of sequencing activities to avoid dangerous situations caused by a combination of work activities by different contractors. They may also vet the method statement to protect others and to achieve better standards of safety.

Clearly, there is a supervisory role for both the specialist contractor and the principal contractor. There is also the likelihood that there are sub sub-contractors or self-employed who will need to be controlled further down the supply line and may have their own employees on site. Particular issues to watch are:

- the quality of the risk assessment;
- the receipt of the method statement before the contractor start date;
- the incorporation of changes into the main programme;
- proper training and competence in post;
- competent self-supervision;
- adaptations and project changes to method are properly planned for impact.

The unplanned response to change is the cause of many accidents, because change needs a regular review of health and safety, which may not be appreciated where risk assessment is carried out on a 'told to do' basis at the beginning of the contract, or on behalf of those who actually do productive work.

Competency and health and safety are linked and it could be argued that a quality product is a safely produced one. This has led to many main contractors insisting on CSCS cards for all workers on site including the subcontractors in the belief that they will be able to reduce defects and accidents on site. With the use of a wide range of contractors and self-employed, this target depends on being able to influence suppliers to comply with the training and testing requirements that this will entail. It will be much harder for smaller contractors to afford this investment without help, so a more formal supply chain management procedure should mean more intervention. Health and safety training is a requirement of the Management of Health and Safety at Work Regulations and this gives some leverage for compliance.

Inspection and supervision human factors

Traditionally workplace inspections are carried out by those who manage the site and have accountability for health and safety plans. They include mandatory scaffold and excavation support inspections, spotting those with their hard hats off (trying to make sure they do not come off immediately your back is turned) vehicle checks and formal investigations after things have gone wrong. This puts the onus on a management led system where the burden of proof lies heavily on relevant competence or experience and time available to check. Due to other commitments, it can either end up in the hands of trainees or become a reactive rather than preventative activity.

Dalton [9] notes the importance of a broader inspection procedure due to the downsizing of the HSE enforcement and inspection role. He describes the new approach as a management system approach, firmly linked to QM and environmental systems. These systems would

- pull in a wider view on safety including experts (fire officers, insurance assessors), workforce (trade union or safety committee representation and getting employees to mark up a hazard map);
- note evidence of management involvement and planning;
- take a general view and impression of the efficiency and openness of the implementation of the system – is it clean and efficient in appearance, what are the risk assessments like and are they accessible;
- make specific third party inspections and make reports to management with copies to senior managers with recommendations.

Culture

Developing a blame free culture where unsafe practice is reported in order to maintain a quality product allows every pair of eyes to inspect and encourages ownership of the problem and a responsible workforce by peer pressure. This culture may be engendered where a work area is owned by a particular contractor, with access and controlled by them. This workplace control determined in Chapter 5 applies because of health and safety reasons as well:

- a formal handover between different trades working in that area;
- a clear connection between method statements where work overlaps work to be programmed to take account of this;
- a quality improvement attitude that includes training in health and safety planning;
- a drive towards self supervision to equal or exceed project standards.

APAU [10] put forward a case, shown in the adapted Table 10.2, for encouraging a culture of improvement by contrasting the polarised scales of a culture for deterioration and a culture for promotion of health.

The table indicates the belief that, health and safety is not just a drag on the project budget, but that financial benefits accrue from health and safety in terms of productivity to pay for systems that are continuously raising the health and safety standards.

The same publication also discusses the benefits of self-regulation when workers take responsibility and pride in their work. This in turn will offset the costs of close supervision against the costs of training. In a project with many different employers, it will be necessary for all contractors to buy into the same approach otherwise the system will break down around the weak link. This will need to be driven by the project manager and built into the conditions for the procurement of package contractors. They point out that there are certain underlying beliefs that must be understood to make this culture possible. These are:

- all accidents, ill health and incidents are preventable;
- all levels of the site organisation are co-operatively involved in reducing health and safety risks;

Table 10.2 Spectrum of occupational health culture

Problems No response = Deterioration of health	Benefits Positive response = Promotion of health
Ill health and injuries	More commitment
Damage to property	General health and efficiency rising
Disillusionment	Improved job satisfaction and motivation
Absenteeism and increased liability	Reduced absenteeism

Source: Adapted from APAU [10].

- trade competence includes safe working risk assessments;
- health and safety is of equal importance to production and quality;
- competence in health and safety is an essential part of construction management.

In this context, there is a much more fundamental preparation stage than a remote health and safety plan. Inspection and supervision are shared and discussed and workforce representatives are proactive with management. The system is not unsupervised as building up experience is important and training needs to be ongoing, but it sets in place a wider accountability.

This approach is more to do with smartening up management and providing the benefit of third-party views.

Supervision is also involved in team building APAU [4]. This can be used to achieve team targets for health and safety and to motivate, to coach and to mentor the team in recognising the degree of risk and the importance of risk control.

Communications information and training

One of the most difficult things to achieve is a fail-safe communication system that provides a clear delivery of the requirements of the health and safety plan to the whole supply chain. This must be communicated to supervisors and their workforces. It will be achieved through co-ordination meetings and through regular workforce briefings, posters and signs. Communications are also about ensuring that there are amendments made to method statements to suit design and sequence changes and that amendments to improve procedures are taken on board. Absences or changes in personnel need to be covered.

Basic information about evacuation and site wide hazards and procedures is delivered through induction. Site rules need to give a stable environment and be sharply focused so that they are perceived to be relevant. Different employees will have different training needs and access should depend on competency.

The requirements of different company health and safety systems and specific client requests for health and safety systems need to be co-ordinated.

Consultation and informal employee reporting is an upward channel of communication, which needs to be received responsively if the system of improvement is to be maintained and new systems are to be introduced.

DDA 1994 will also require greater efforts not to impose procedures, physical or mental requirements that cannot be proven critical to core performance requirements. Health and safety is the most used excuse for discrimination. Under DDA from 2004 it will be a requirement to make reasonably practicable adjustments to the workplace to incorporate special needs.

Protecting the public

A project health and safety system is an open system that is affected by its environment and will also impact on that environment. The construction workplace being open is a particular danger to the public who are affected by dust, noise, smells, falling objects, vehicle movements, unprotected excavations, collapsed buildings or scaffolds (to mention a few things). The public is a particularly vulnerable party as it has no extra perception and is not served with any information about the risks to their wellbeing or health. Contractors are also liable for accidents occurring as a result of trespass onto the site.

What is the correct response to protecting the public? Is it an exclusion zone and warning notices or full protection? How do we balance severe restrictions on construction work with reducing nuisance to the public? What are the maximum perceptions of risk that are acceptable when walking past a building site? What level of information do the public need to be safe? The key measures are:

- protection by an effective barrier around the site and any other public areas such as road works and drainage connections, including security guards in busy areas. Scaffold fans or gantries to protect from falling debris;
- information about the hours of work, programme length and activities and irreducible levels of noise or dust;
- viewing windows to satisfy curiosity and limit trespass;
- cut down pollutive levels, for example, dust and emissions, especially exclude any likely to have biological or allergic reactions;
- control noise as public have no PPE. Use exclusion zones fairly and limit hours;
- strictly control any toxic or hazardous materials with secure storage and delivery;
- agree a traffic plan with the LA for safe egress and access and temporary traffic controls;
- control lifting and limit any crane slew to site air space except in agreed control procedures. Set lights on tall temporary structures;
- exclude public access without protective equipment and sign clearly.

These measures are based on total protection, considerate contracting, providing information and reducing nuisance. The Considerate Constructors Scheme has been set up as a voluntary compliance with the following principles:

- to ensure that work is carried out to limit exposure to the public;
- to keep sites and surrounding roads tidy;

- to control deliveries and the blockages they may cause in tight conditions;
- to keep neighbours well informed of what is going on and the measures which have been carried out to minimise the exposure to the public;
- to limit exposure to 'nuisance' activities.

Related environmental issues

Noise and dust

Noise and dust is a constant problem for building site neighbours and has given construction a bad name.

Dust is a problem because it means that extra cleaning is forced on others – some cleaning may be offered gratis and in particular there is a duty to keep the highway clear of mud at all times. Noise must be restricted to reasonable hours and should be for clearly delineated periods which are known. In public places, for example, road works, there is no excuse not to attenuate noise to a 60 dB level. Employees are often better protected than neighbours to such works. The creation of dust can be largely avoided by extraction or damping down.

Waste

Government regulations to encourage recycling of materials has had some effect on construction operations through the imposition of stiff land fill taxes to discourage the carting away of materials (also disruptive and noisy to neighbours and pollutive in its use of fossil fuels), and to encourage designs that reduce excavation, redistribute excavation materials on the site or recycle demolition materials. Other motivators are the desire for good publicity and several companies have indicated their achievements in reducing waste in environmental statements. A typical example is the construction of the Jubilee Line, when sand was delivered to the site by barge and essential excavated waste materials were also dumped to provide flood protection by the same method.

Many new schemes for recycling concrete and brick demolition materials have also become economically viable because of the land fill tax. Knock-on effects are the need for more intensive work on the site itself and the need for better sequencing to avoid increasing health and safety risks. Financial savings can be made against the costs of transport. Noise and dust must be a consideration in crushing plants.

Contaminated brown field development

More former heavy industrial sites are being opened for regeneration development schemes to cut down the amount of urban sprawl and to improve urban landscapes. This has resulted in the remediation measures, on or off site, to regenerate the soil and make it safe for development. Further methods involve dumping toxic materials on special sites with even stiffer taxes. Design solutions have been driven by economic solutions, but health and safety hazards to workers and the public have been increased by bringing workers closer to toxic deposits of uncertain composition. Carting away also puts a wider public at risk. Bio-regeneration is a popular, but more time consuming method and programmes need to be creative and well managed to incorporate the time periods required.

Emissions

There are two environmental aspects – the impact on others of more global emissions and the impact on site health of more local contact with toxic or carcinogenic emissions. Some emissions have a particular long-term effect on occupiers such as formaldehyde. Cleaner technology is another aim for the government and the level of pollutive emissions must be accounted for. Particular restrictions are placed on the reduction of 'green house' gases, CO_2, CFCs (from refrigerants) where substitutes are mandatory now, but where old equipment is still in use. All vehicles are taxed according to their emission levels. This has led to reduced deliveries and cart away and using local materials to reduce the embodied energy levels. It has also led to the introduction of cleaner sulphur free fuels.

This has the effect of improved healthy conditions on site, if clean fuel is used. Internal forklift trucks are often gas driven and tower cranes and cement mixers can be electric. Noisy and expensive brick or stone-cutting operations is another area where design technology could help.

Health issues

Clearly, the use of less hazardous and toxic materials such as oil or lead based paints or coatings is in itself environmentally friendly. Specific areas where environmental and health issues merge are the use of, or stripping out of, materials where fine fibres are given off. This includes the handling and disposal of asbestos, glass fibre, MDF and mineral fibres. Some are proven killers through diseases such as asbestosis and carcinogens and others are increasingly perceived as harmful, for example, Rockwool, MDF and fibreglass. The general principles of COSHH 1999 are designed to identify more hazardous materials and promote safe use. Environmentally, these materials will continue to be blacklisted by many clients. The main health responses are to isolate areas, to wet work down, work to reduce dust and to wear adequate and correctly designed masks and respirator. In the case of toxic fibres such as asbestos, the specialised de-pressuring of areas to stop the migration of dust and to provide dust barriers and washing down transition zones is critical and should be carried out by specialists and notified to the HSE. All building owners under The Management of Asbestos Regulations now have to identify the presence of asbestos and provide adequate monitoring and risk assessment and controls.

Other health issues are connected with the ergonomics of the workplace and these are controlled under the manual handling regulations. These cover all musco-skeletal conditions and they include repetitive strain disease as well as lifting, which causes back strain. Lost days at work is a major economic loss to construction companies, but it is still a major environmental, that is, quality of life problem with severe health and liability implications for incidences where work procedures have not been properly risk assessed.

A European planning requirement for major new construction developments is Environmental Impact Analysis. This is an analysis of the impact on the landscape and therefore a design issue, but also of the impact on emissions, rivers and other topographical features such as quarrying and contour features. Other issues are wild life habitats, noise, traffic impact and pollution. All of these need to be assessed as present or future health issues.

Emerging issues for project safety, health and the environment

Prevention of accidents

Research is commissioned by the HSE into causal factors by the examination of past accidents and into prevention factors by the use of focus groups contained a range of experienced professionals looking at particular areas of risk. Some recent causal research by the HSE [11] confirms the particularly strong connection of accidents with a lack of designer awareness of their impact on construction health and safety. They concluded that 46% of 100 accidents investigated, were due to permanent design issues and a further 36% were a lack of attention to design in temporary structures.

A prevention focus group study by the HSE (2003) [12] on falls from height (a major accident type in construction) estimates that 50% of accidents could be prevented by the application of risk control measures to just 2 areas out of 7 considered, which were design process (30%) and compliance to safety measures on site (20%). It was found that supervisors tended not to reprimand experienced workers about poor safety practices, either because they believed these people could look after themselves or because of lack of confidence. The study identified problem areas in design and compliance and then considered measures that could be most *influential* in correcting the problems at each level of the influence networks, including the design process. There is a gap in the communication process in that many accidents do happen to experienced workers because they are too confident of their ability and un-alert to the combinations of activities or circumstances that can increase the risk of a simple job exponentially.

This research concludes that regulatory and document access would improve the information needed to risk assess hazards which affect the construction process. This together with more training and direct access to the contractor at the design stage and the inclusion of health and safety items in the Bill of Quantities (BOQ) were considered to be key aspects in gaining a safer design. A similar influence network for compliance exposed the 'it won't happen to me' attitudes of experienced workers and their failure to assess the scale of the risk even though they were well aware of the hazards. Managers and supervisors leading by example and zero tolerance to the ignoring of safety measures were identified as a key prevention measure for non-compliance.

Safe working boils down to risk assessment each time which is hard in the case of familiar work. Chapter 8 also talks about the syndrome of increasing risk to a previous 'comfortable' level when new safety measures are introduced, for example, going faster.

Site conditions

Substandard welfare conditions on site are the result of limited resources being allocated. These have been accepted as the norm, but the downside is that work is less organised and that a 'new breed' of site worker may find these conditions affect their productivity. From a health point of view, it is a legal requirement to provide adequate washing facilities to promote personal hygiene and to provide adequate toilet facilities for the number and gender of all site workers. Drying out facilities for changing and wet weather are a legal requirement, though in practice this is often provided individually by separate contractors. All contractors need to make personal protective equipment available and to ensure they are maintained in good condition. Well-run site canteens and showers are also seen as a motivator. A clean site with facilities for personal hygiene is a safe and productive site.

Larger projects allocate greater space for offices and welfare facilities, but these may have to be creatively arranged if space is tight. Where the site is spread out, then repeat facilities may also save time so that workers do not have to go long distances for breaks or there are less, but longer breaks.

The CIOB Report[5] on site conditions conveyed a wish from both large and small contractors to clean up their act in order to continue to attract a competent and productive work force. This was considered to be a fundamental culture change in order to give better working conditions. They believe it should be management led, with more training, more policing and enforcement for minimum standards and the use of schemes like the Considerate Constructors to provide a better service to the public.

Accidents may happen because workers are tired and not taking proper breaks. The Working Time Regulations 1998[6] institutes a maximum average working week of 48 h on average and the requirement to institute regular breaks. However, there is a clause which allows workers to contract out of the Regulations and this is quite often used in construction projects, because of the complexity and uncertainty of construction processes. This may extend the working day at short notice and has potential for cumulative effect where workers away from home are only too keen to focus on their earnings. Future considerations may require a more disciplined use of shift work to avoid the wholesale ignoring of the Regulations.

Stress leading to depression and anxiety is a major target for the HSE. The Labour Force Survey in 1990 indicated that there were 183 000 people in England and Wales who indicated that they suffered from stress thought to be caused or made worse by their work. 105 000 workers actually thought it was caused by their work. It is also a 'top ten' workplace disease in America [13]. It is difficult to assess how much is caused by construction related organisations, but many employers are accepting that stress can be an adverse force and are now making available counselling to employees or encouraging support groups which seek ways to deal with stress better. Stress hardiness refers to the way in which individuals could cope with stress such as increased commitment (see Chapter 7) and project managers can take away negative and harmful connotations, by helping different individuals to gain commitment.

Management responsibility

In construction, it has been calculated that for every pound of insurable accident costs there are £11 of uninsurable costs HSE [14]. Uninsured costs include the costs of investigation, the cost of product and material damage, plant and building damage, tool and equipment damage, legal costs and compensation awards, cost of emergency supplies, clearing site, production delays, overtime working and temporary delays, diverted supervisors' time and management effort and fines. It is also true that insurers substantially raise premiums for unsuccessful organisations. Insurance companies themselves then have a beneficial role to play in encouraging improvements [15], by providing additional inspections. The same study [14] estimated that 8.5% of tender price was lost in accidents on construction sites. These sort of figures leave room for companies to put in place preventable measures and still make gains.

Corporate manslaughter is the proven connection of a negligent event at the operations level, with a controlling mind at the director or board level of the organisation. Prosecutable actions have to have clearly inhibited the ability of operations management to provide a safe provision. For example, short cuts in the treatment of hazardous materials which lead to the death of a person. In 1996 the first ever MD was imprisoned for an offence of exposing his employees and the public to the risk of asbestos for four months. No one was killed so this was not a corporate manslaughter charge. The MD of The Outdoor Activity company had been found guilty. The latter of four charges of corporate manslaughter when children died in a canoeing expedition planned and led by his organisation. He was jailed for 3 years [16]. As yet no large construction company MD has been successfully prosecuted for corporate manslaughter, but there is a desire in government to introduce corporate manslaughter legislation to make its prosecution easier. Another case when an access cage ran off the end of its supportive rails during bridge renovation work was unsuccessfully prosecuted due to the unproven link of director policy with operational negligence.

Environmental issues

The 'polluter pays' principle has not always been proven in the UK, but there is a move to introduce more green taxes to try and influence people's behaviour in reducing pollutive practice and encouraging recycling and the reduction of waste. The increasing landfill tax and so-called carbon taxes are having an effect on recycling economies on construction sites, by increasing the tax and fuel costs in taking material away from site. European countries need energy taxes as there are currently more incentives for energy producers to reduce prices for bulk buyers than there is to use alternative fuel sources.

Key issues in the environmental equation are reducing emissions that deplete the ozone layer such as CFCs in refrigerants, the net gain in the levels of CO_2 and certain other gases from combustion which causes global warming, the levels of SO_2 and NO_2 which create acid rain and the levels of atmospheric dust, both of which damage forests, streams and lakes and their wildlife. The huge levels of solid waste not recycled, creates expensive landfill problems that can create toxic leaching into gardens and water courses. Incineration also causes dust problems and pollution caused by transport which contributes to CO_2 and acid rain.

Significant improvements have been made environmentally in the UK and these indicate that emissions have fallen, for example, smaller particles known as PM10, CFCs (87%), carbon monoxide and dioxide, lead and nitrogen oxides (mainly due to unleaded petrol and catalytic converters) and volatile organic compounds (as used in paint) have reduced levels. UK rivers have reduced heavy metals, but increased nitrogen levels. Waste from mining and quarrying is a quarter lower. Sewerage plant are generally complying with the standards. On the poor side, air quality is regularly testing below standard, 10% of drinking water samples have higher chemical concentrations and noise complaints have risen. Global warming has increased slightly in the 1990s.

Although construction has moved in the right direction, there is still a continuing pressure [Rio and Kyoto Summits] to meet lower and lower standards and this requires a continuing effort to innovate. Decreased energy usage presents a challenge for designers in such areas as air leakage of buildings, better insulation, use of more local and less process energy materials, alternative forms of fuel and natural ventilation. The challenge for the construction process is to reduce waste, innovate wider recycling schemes, reduce landfill 'cart away' and develop less noisy and more energy efficient plant with alternative fuel sources. The influence of the client in the initial stages is a key to success in incentivising environmentally sensitive designers and contractors, and recognising the need in some cases to increase capital costs to decrease overall life cycle costs. The BRE Environmental Assessment Method (BREEAM) scheme and the Government supported a points scheme for housing has a better take up now, but is still a way behind other North European schemes mainly due to a poor customer uptake [17]. The main platform for legislation controlling new buildings design is The Buildings Regulations. Amendments reflect the increasing severity of measures such as reduced air leakage and μ values and tougher restrictions on emissions. The Environmental Protection Act 1992 identifies responsibilities for cleaning up pollution and contamination and these technologies are becoming more sophisticated and with the short supply of green field land and rising property prices, more brown field land is being treated and developed.

Integrated systems for quality, health, safety and the environment

Integrated management systems arise from the overlap and duplication which occurs between health and safety, quality and environmental management specified in the well-known BS 8800:1998 Health and Safety Management Systems, ISO 9000:1987 Quality Management and ISO 14001:1994 Environmental Systems. Griffith and Howarth [18] point out

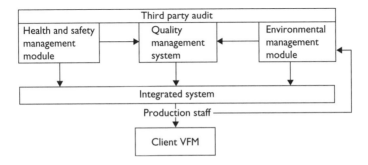

Figure 10.7 Integrated health, safety, environmental and quality system.

the need for a risk assessed *open* system (recognising external influences), which copes with 'unplanned events' and the appropriateness of the 'whole organisation' approach shared by each of the standard systems.

Holistic control in an integrated system (Figure 10.7) releases the potential for a better product for the client and better VFM, saving paperwork. The challenge is to keep the system simple and this is usually done by using a single system such as quality and adding on environmental and health and safety modules. To be effective the paperwork must be kept to a minimum and control remain in the hands of production staff accountable for outputs and not through a remote systems manager. The system should also be simplified for small projects.

A recent paper by HSC/CONIAC [19] has suggested that an integrated team approach which is forced upon PFI contracts is something that would benefit the introduction of a better health and safety culture. This method would have advantages for risk sharing and the early involvement of the supply chain in the health and safety plan. It is also suggested that, a solution which is good for health and safety at this stage can be value engineered so as to be good for cost saving, also moving away from the approach that health and safety costs money.

Conclusion

Health and safety regulation has multiplied in recent years with some encouragement from the European Union and continues to do so, as headline accident statistics refuse to reduce significantly in construction. Because of this health and safety training on construction sites is a substantial part of management training and most large contractors require evidence of training to a minimum competency for managers and workers. It is, however, easy to go through the motions of paperwork and inspections and not develop a rigorous culture of awareness by all the workforce, that is, all workers report poor practice and are committed to improving safety. This is difficult given the fragmented nature of the workforce and the many sub subcontractors in the supply chain make consistent communications difficult. The culture of induction is often substituted for specific training, but is not focused to the work area of the individual. It needs to be backed up with specific training, by each contractor represented.

The CDM Regulations have provided an integrated framework for the management of the construction project safety, by formally including the client and the designer with the traditional contractor role in health and safety management. This rightly recognises the process throughout the project life cycle from inception to completion. It has much potential in making sure that resources are available to ensure adequate tender prices for resourcing safety measures and risk assessment, but it is recognised by the HSE that changes are required, that do not kill initiatives. Problems are still encountered in implementing safer systems, if organisations do not understand, or apply RM co-operatively with others and enforce it in an integrated way between design, project team and facilities management. This is compounded by the problem of macho and complacent attitudes by workers that believe that 'it won't happen to me' and by supervisors and managers who turn a blind eye to shortcuts by specialist trades who have the experience to recognise the hazards, but have not entered the culture of *assessing the scale* of the risk.

Some would argue that SMEs want a more prescriptive system where generic standards become well known and can be used in a directive way to ensure compliance, others that self-regulation is a better way of owning the problem and developing universal awareness and accountability. As SMEs make up a major proportion of organisations on the site, there is a case for more prescription. Culture is still important, as prescriptive solutions suggest generic methods that do not take into account interruptions and changes, which create situations that often cause accidents. The culture in prescription is to meet minimum standards rather than to encourage continuous improvement.

Representation at worker level is poor in construction due to the fragmented nature of the work force and the limited trade union representation. The benefit of this is more ownership and integrated feedback by the project workforce in what is already perceived to be a dangerous environment. This may be less formally instituted by the use of a project-based health and safety committee, which would need to be given real power and route to feedback to project and employer management. In terms of encouraging self-supervision, more powers for worker inspection of the workplace need to be given.

The best practice movement is a way of engendering a sense of pride and peer pressure to reach better safety standards and to produce this culture of continuous improvement. This could work if clients recognise their role in insisting on certain standards of qualifying compliance, such as the compulsory use of CSCS cards which evidence competence, top management commitment to safe working improvements and an auditable safety management system which monitors and ensures good implementation by the use of site wide KPIs. If these requirements are not part of the selection process in competitive tendering, those that take the short cuts on health and safety will initially have an unfair financial advantage in a competitive market.

Closer connections have been made between safety, health, environmental and quality systems in the development of integrated management systems which reflect the numerous interfaces between these areas of management control. Here the client has a better chance for a VFM solution where quality is linked with health and safety and is seen as good business. This could be illustrated by more 'right first time' and less supervision. Many contractors have health and safety built into their QA systems.

Environmental management fits easily with health and safety at a strategic level and helps focus the long-term health aspects of health and safety. Some contractors are developing and advertising environmental policies. It also gives a wider perspective to issues such as COSHH assessments and considers impacts as well as consequences. Environmental responsibility is not so easy to enforce and depends upon a change of culture to one of sustainable construction and not short-term commercialism in context with the project environment.

Liability and negligence may be seen as good enforcement tools, but in practice they are seen as inevitable and as an insurable cost to the business. The HSE [14] study found that non-insurable costs were at least a factor of eight of insurable costs and this suggests that attention to management systems is critical to avoid waste in the system and even future loss in an environment where litigation may become more common place.

Appendix: guidelines to minimum practice requirements

Good practice design

It is quite clear that where design risks have been assessed then a safe régime may be established. This régime is also dependent on the production of a project HSF file which provides safe operating and maintenance instructions for the building user. Particular issues here are:

- knowledgeable use of plant and equipment provided and signage for authorised access;
- machinery settings to provide a healthy environment;
- awareness and safe usage of hazardous substances for cleaning, or maintenance or fuel;
- hazards associated with usage, proper lighting and storage;
- clear escape routes and direction, use and resetting of fire alarms and automatic doors;
- use, isolation and rescue from lifts;
- inspection and cleaning schedules, for example, legionnaires disease;
- safe component life cycles, for example, electrical hazards caused by worn components;
- use of machinery guards and protective equipment and training requirements.

There are a wide range of components and equipment used in buildings many of which are uniquely designed and poor usage and maintenance is a common form of accident or health hazard. Clearly the user has a strong accountability to maintain preventative maintenance and provide ongoing training and awareness for users and this can be helped by good information. The *Health and Safety File* compilation is the responsibility of the planning supervisor. It is not enough for the planning supervisor to merely collect and file information.

Many HSF are quite unwieldy and as such may not communicate well to a broad user audience. Training and induction programmes may help, but signage has a key role to play in instructing a public or occasional user. Usage should also be logical and research indicates that some control over environment, for example, heating levels, blinds or opening windows leads to more responsible and therefore safer use of the building. Many larger organisations have a help desk. Automatic lighting sensors and thermostats to save energy can create frustration without override or local control.

Health and safety planning

The pre-tender *Health and Safety Plan* (HSP) in the CDM Regulations was devised as a way of providing a 'level playing field' for competitive tendering. The plan includes risk assessments, which have been carried out at the design stage and indicates specific residual risks such as the use of substances hazardous to health and the presence of hazards inherent in the site itself such as asbestos locations, contaminated ground, underground cables and explosive substances. Designers have accountability to reduce risks to the construction process, but cannot double think the methods of the contractor. The emphasis is on providing full and proper information and eliminating design details which make components dangerous to fix, for example, heavy concrete blocks are not specified now due to the injurious effect of repetitive heavy lifting and twisting on a bricklayer's back.

Each tenderer is given health and safety information which is priced and the successful tenderer has to update the HSP before starting on site, indicating the risk assessments and methodology which is to be used for any hazards in the tender plan and all other hazards perceived to be relevant in the construction process itself. If heavy concrete blocks had to be used, then a method for using mechanical aids or employing more bricklayers to share lifting would be expected. Other issues concern the management and organisation of the health and safety function with named fire wardens, evacuation procedures and exclusion areas. The plan is on the basis of continuous development as subcontractors are appointed and submit their method statements, which need to be integrated into the plan as a whole. The aim of the construction plan is to

- eliminate or minimise construction risks and ensure safe method statements are being adhered to;
- provide safe systems of work which are planned and organised;

- ensure the appropriate and safe use of equipment;
- provide corporate or individual protection where there is residual risk including the public.

Reducing risk and corporate protection is of a higher order than personal protection and can be illustrated in the issue of ear muffs to those working on road works in the high street where the public would be better served by quieter machines, an exclusion zone or the use of acoustic barriers. The organisation and management of safe systems of work is often underestimated in the minutia of method statements.

A safe and healthy system of work

The PC is expected to organise and co-ordinate health and safety on site and they hold the ultimate responsibility for any accidents which occur due to inadequate planning, poor supervision, interfaces between different workplaces or within joint workplaces. They also need to risk manage issues which have an overall effect on the health and welfare of the site in general. Particular issues which may arise are:

- overall supervision of health and safety and welfare;
- co-ordination of risk assessments where they impact on others;
- the protection of the public by the proper protection of the boundaries;
- segregated pedestrian and vehicle access;
- induction training and awareness;
- visitor reporting and health and safety induction;
- site security to exclude trespassers or other unauthorised access;
- fire evacuation and signage;
- site rules for control of behaviour and creating exclusion zones, for example, wearing hard hats;
- general lifting operations, vertical transport and joint scaffold access;
- plant use and co-ordination;
- safe procedures for materials deliveries and storage;
- welfare facilities;
- access to PPE in good condition for visitors, staff and directly employed labour;
- noise and dust control and exclusion zones;
- integrated health measures and surveillance, for example, asbestos removal, contaminated ground and biological hazards;
- public relations to provide health and safety information to the surrounding community.

It is easy to pass on responsibility for health and safety to others. But the issue of co-ordination is a critical matter for the PC who has the overview and the authority to make reasonable decisions for the health and safety of the whole site. The project health and safety plan is the basis of this and should clearly designate responsibilities and remain a dynamic document.

On some large projects a logistics type package is devised to carry out some of the activities that are implied earlier, but it is clear that the successful management of health and safety remains the responsibility of the PC.

Site induction and rules

Most larger sites have now introduced a basic training for all new workers and sometimes visitors arriving at the site. It generally consists of information about the residual risks which have not been managed out and the protection which is offered, the need to report dangerous, hazardous incidents or behaviour thought to endanger others and the evacuation procedures in the case of fire or explosion or attack.

The problem is that they may be operated purely as a 'legal requirement' for the main contractor and/or are of a very general nature and so may be ignored. Effective induction should identify tangible benefits and be focused on risks that are real to the activities that are being carried out that day. They could also be followed up with other activities like tool box talks and tests, which reference issues on a regular and a systematic basis. Suggestion boxes would be used if people felt that their ideas were rewarded, or improvements were made as a result. This could help develop a culture of health and safety awareness and concern. Weekly rewards for best practice could also be given to produce a climate of continuous improvement.

Site rules need to be enforced to be effective, so that there are fines or other penalties for contravention which affect the bottom-line income of the individual or the company. Protective equipment needs to be available in good condition, free or to buy. Exclusion zones should be known and respected.

Evacuation of the site is an important part of induction, information on the alarm signals, escape routes, sounding the alarm, identifying fire wardens and safe assembly points must be clear. Signs should be graphic for those who cannot read and the maintenance of clear escape routes should be a part of the responsibility of the PC. There should also be an awareness of the effect of one contractor's operations on the other, for example, noise or dust so that activities are planned to cut down on nuisance and with a proper risk assessment adopted for the unique site conditions encountered.

Notes

1 Fewings and Laycock (1994) *Health and Safety in the Workplace*. University of Northumbria Built Environment Group. 26 safety related directives were adopted in EU countries in the period 1986–94 alone. pp. 8–10.
2 MHSW 1999 now incorporates special sections for young people and expectant mothers. It expects that where possible employers should use competent employees to carry out assessment of risk in order to give them ownership. Part 3 of the Fire Precautions (Workplace) Regulations have also been incorporated for escape measures.

3 Accident Advisory and Prevention Unit (HSE 1994) *Successful Health and Safety Management* (Revised Edition). The 1st edition was issued before the 'six pack' in 1992 and reflected much of the new culture of putting the onus on the employer to propose and develop new systems of work that was started in HSWA 1974.

4 TUC (1995) Commissioned National Opinion Poll to 1002 people, 98% feel they had the right to be represented by a trade union on health and safety (93% their pay).

5 CIOB (2003) Improving Site Conditions. Two workshops with major and SME contractors. March and July.

6 Working Time Regulations (1998). These regulations were amended in 2003 to include all non-mobile workers. Workers may opt out but all have the right to the 48 h limit, work breaks and annual leave.

References

1 Tye and Pearson (1974/5) Management Safety Manual. British Safety Council, London. 5 Star Health and Safety Management System, in HSE (1991) *Successful Health and Safety Management*.

2 85/374/EEC (1985) Product Liability Directive. OJ L210/29.

3 92/59/EEC (1992) Product Safety Directive. OJ L228/24.

4 APAU (1991) *Successful Health and Safety Management*. HSE Publications.

5 HSE (2004) http://www.hse.gov.uk/cdm

6 Loughborough University (2004) http://www.lboro.ac.uk/research/design4health

7 CIRIA R166 (1997) *CDM Regulations – Work Sector Guidance for Designers*. CIRIA.

8 HSE (1995) *Designing for H&S in Construction Guidance for Designers on the CDM. HSE Guidance Booklet C100*. HSE Books, Sudbury, UK.

9 Dalton A.J.P. (1998) *Safety, Health and Environmental Hazards at the Workplace*. Cassell, London.

10 APAU (1994) *Successful Health and Safety Management*. HSE Books. Revised Edition.

11 HSE (2003) *RR156 Causal Factors in Construction Accidents*. UMIST and Loughborough Universities, UK.

12 HSE (2003) *RR116 Falls from Height: Prevention and Risk Control Effectiveness*. http://www.hse.gov.uk/research/rrhtm/rr116.htm

13 Zenz C. (1994) *Occupational Medicine*. 3rd Edition. Baltimore. Mosby Books and Rom W.R. *Environmental and Occupational Medicine*. 2nd Edition. Little Brown and Co., London.

14 HSE (1993) *The Costs of Accidents at Work*. HS(G)96. HMSO, London.

15 Dalton A.J.P. (1998) *Safety, Health and Environmental Hazards in the Workplace*. Cassell, London. pp. 11 and 12.

16 Dalton A.J.P. (1998) *Safety, Health and Environmental Hazards in the Workplace*. Cassell, London, p. 36

17 ECI (2002) *Guide to Environmental Management in Construction*. European Construction Institute.

18 Griffith A. and Howarth T. (2000) *Construction Health and Safety Management*. Longman an imprint of Pearson Education, Edinburgh.

19 HSC/CONIAC (2003) *Integrated Teams – Managing the Process*. Work Paper HSC/M1/2. March.

Chapter 11

The PFI model

The PFI model has a real potential for radical integration of the project team and the development of productive long-term relationships. It is not a panacea for solving all the problems of the construction industry, but some of the principles for integrated project management which have been emerging from previous chapters such as early involvement of contractors, the committed use of RM and VM, integrated design and construction, life cycle costing and collaborative relationships are an integral part of this procurement. It is a flexible model and comes in several forms according to client need and might be expanded further. The purpose of this chapter is to introduce the principles and practice of the PFI as used by public bodies in the UK and to compare similar models such as the use of DBFO or build, own, operate, transfer (BOOT) that gives it wider applicability for private procurement. The chapter is used as a case study to assess the successes and failures of PFI and as a basis for identifying elements of a more progressive model which is a way forward for the construction industry. General issues of public procurement good practice will also be discussed without wholly applying to PFI forms of procurement. The objectives of this chapter are to

- introduce the PFI process and define its specialist vocabulary and rationale;
- discuss the principles for PFI and to assess its impact on the culture of contracting;
- evaluate some of the real and perceived problems that have beset the practice of PFI and look at the solutions which are being offered and developed as best practice;
- identify the differences in practice and application between the public and the private sectors;
- discuss the impact of expanding its use to the private sector in the form of BOOT, DBFO, prime contracting or other relevant forms and other future possibilities;
- consider the current issues and the future of PFI in construction projects.

Introduction

Public private partnership is a partnership which leverages private funding and the strengths of private entrepreneurship and management, for the maximum provision of public services in a climate of scarce public resources. PPP has been used selectively in the past by many governments globally, as a joint venture in public projects where private innovation and finance can reduce costs and enhance services. The PFI is a PPP special case where *all* the finance needed for the capital funding and its basic operation is supplied by the private sector in return for a service charge. The building or other asset is returned to the public body (the Authority) at the end of an agreed operating period. The advantage for the Authority is that substantial capital debt and assets are removed from the balance sheet and converted into a revenue expenditure underpinned by central government.

In the early 1990s, the need for major infrastructure renewal was becoming an increasing burden for the UK Exchequer and there was recognition that public purchasing was not producing the VFM that was expected in the private sector. The rationale for the introduction of the PFI in 1992 by the Conservative Government in the UK was to reduce the public borrowing requirement by the use of private funding and to reduce the risks of time and budget overruns. It was expected that VFM might be enhanced using private sector management skills and innovation and by transferring the specialist risks and liabilities involved to parties who are experienced in managing them. In 1997 'New' Labour Government agreed with the principles of PFI and adopted the policy and ironed out many of the structural problems that had delayed the process of signing up contracts with private sector consortiums. In June 2003, more than 500 contracts had been signed and £20b of private capital had been levered into public infrastructure projects such as schools, roads and prisons, which looked set to continue. Much central guidance has been developed by a series of Government agencies,[1] but it is the Treasury that has the veto on all larger PFI projects through its assessment panel. Other countries, such as South Africa, have followed the PFI model in the UK to procure certain public services. There is a large volume of material available on PFI publicly, which makes it a good case to analyse.

How does PFI work?

In the provision of built assets, PFI provides for 100% capital funding and building running costs by using a private joint venture (Special Purpose Vehicle (SPV)) for a concessionary period of 20–60 years, which may be renewed. During this concession, the Authority pay for the building or facility and its maintenance as a charge over the agreed period and do not normally have to show the building as an asset on their balance sheet. The capital repayment is included in the annual charge on a fixed or variable interest capital repayment basis, reducing as the capital reduces. The service charge is usually indexed with inflation to cover the cost of maintenance, updating and building operation as necessary. For instance in the case of a hospital, the maintenance of the buildings and provision of heating, lighting, building services and a help desk is usual, but much more may be supplied including the core service itself as in the prison service. The Authority will receive the asset back at the end of the initial capital repayment period, or renegotiate the service provision. If the asset is received back, it may be written off or valued by an actuary to go back on the balance sheet as an asset. Penalties are levelled by the Authority for any loss of facility use, or downgrade of the service quality. These are operated within tolerances. The charges are only payable from the point at which use of the facility or part of it is made available to the Authority, creating extended private funding requirements usually arranged by a funding partner in the SPV. The main principles are:

- Puts risk where it can be best managed and assesses it properly.
- VFM by harnessing private management efficiency and may also release land and assets for wider use.
- Affordability, which means it should be financially viable over the contract period.

To justify this method it is necessary to prove better VFM than traditional financing and a sensitivity analysis or simulation are required to test the assumptions that have been made in calculating the long-term NPV and the transferred risk. PFI allows the Authority to release the facilities management function and concentrate on its core business of ensuring VFM public services to its taxpayers and the transfer of substantial risk connected with facilities provision.

Authorities advertise in the European Communities Journal (OJEC) in accordance with European public procurement rules and need to show that fair competition has taken place in choosing a single preferred bidder with whom they negotiate the best deal. The provider, which is the SPV joint venture have vested interest in their own risk and value assessments for the capital and WLC of the building.

The structure of agreements

The structure of a SPV in larger PFIs is a joint venture company consisting of a funder, contractor and facilities company (Figure 11.1). The enterprise is financed by debt and equity capital with a 10–20% equity contribution from the members. Alternatively a company may choose to fund smaller

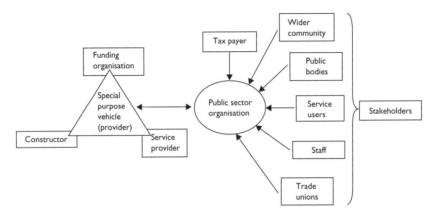

Figure 11.1 Structure of the SPV.

Source: Adapted from The Audit Commission (2000) Taking the Initiative: A framework for purchasing under PFI.

PFIs 100% off its own balance sheet. The SPV contracts directly with a public authority who commission the project on behalf of their stakeholders. The Authority will need to appoint a project leader – quite often the chief accountant, or a specialist member of their project staff, who will engage specialist advice as required.

Some research by the RICS [1] indicates that this model was used by all the case studies they investigated. The most common PFI projects (48%) in this research were in the NHS and Schools sector. In the NHS, the Trust chief executive is commonly the project sponsor. In the schools PFI, it is likely to be the Local Education Authority (LEA). The Case study 11.1 shows the organisational structure of a typical SPV and indicates the main cross relationships with the NHS Trust.

Case study 11.1 Redevelopment of major NHS teaching hospital, part one

The new hospital budget was agreed between the NHS Trust and the SPV as £100m construction costs. There was a three-year construction programme from 1999–2002. The SPV consisted of an Facilities Manager (FM) provider, a design build contractor and a financing partner. The SPV in Figure 11.2 consists of a joint venture between a contractor who is responsible for the construction and the facilities management, a medical services enterprise and sponsoring banks.

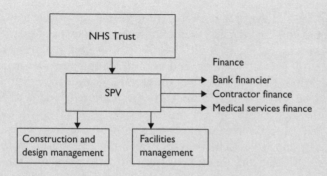

Figure 11.2 Case study SPV structure.

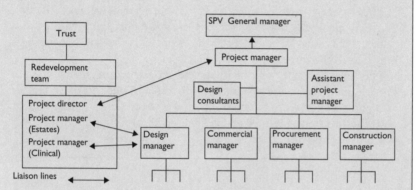

Figure 11.3 Project team structures and relationships.

The structure of the construction team varied from six to eight people in July 1997, which involved six key disciplines as the project developed towards its construction start date in 1999. These departments were populated to take on full responsibilities for the tendering and construction stages. An external design company worked in partnership on a design and build basis. Figure 11.3 shows the detail.

The structures above indicate the organisation of the Trust project team and the SPV project team. The Trust team has interaction points at both the user and the facilities level in order to determine the design requirements. There are regular design meetings and weekly user group meetings, that are attended by the design consultants and the design manager. The ultimate aim of these meetings is to determine scope and not detail design of the departments. Progress meetings are called by the Project Manager (PM) who will also approve the affordable scope.

The project manager is responsible for the delivery of the new build design and construction and maintaining the budget, programme and quality targets. He liaises at the level of Trust project director in order to establish project strategy and overall design and construction co-ordination. He is at the hub of the communication system and will be involved in the negotiations with the preferred bidder to reach financial close.

The SPV general manager liaises directly with the Trust Board at a strategic policy level. He is responsible for the liaison with the financiers, signing off financial close, the performance of the construction arm and on an ongoing basis throughout the operation phase of the hospital.

The services provider operates the facilities management function and may also be consulted as to the life cycle costs and maintenance issues at this stage. The FM function is provided by another business unit within the same contractor.

The procurement, commercial and construction managers are directly involved in the production process and need to liaise closely as a construction team which overlaps the detail design, procurement and construction stages. Package managers are appointed to cover specialist contractor control.

Competition

Competition for significant public contracts in Europe is governed by the Public Works Contracts Regulations [2] which require open and timely advertisement of the tender for forthcoming contracts >€5m. However due to the complexity of PFI projects there are provisions in the regulations for the use of two-stage competitive negotiated contracts. Where there is national sensitivity restricted procedures may be used. The procedure most used for PFI is the negotiated procedure, which tries to ensure an open competitive first phase followed by a negotiated phase to squeeze value from the original proposals (Figure 11.4).

A selection criterion to reach a short list has to be transparent to all bidders. Bidding costs in the *detailed* stages are very significant and the numbers of bidders being invited to tender on complex projects should not be more than six with one preferred bidder to move into detailed design development and negotiation. The competitiveness of projects is affected if there are few interested parties, because of the cost of tendering. Tenderers may need to be reassured of some remuneration for second stage bidding costs where they are short listed along with others. Compensation is now

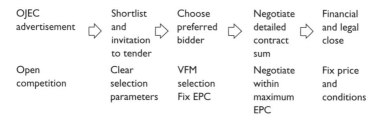

| OJEC advertisement | ⇨ | Shortlist and invitation to tender | ⇨ | Choose preferred bidder | ⇨ | Negotiate detailed contract sum | ⇨ | Financial and legal close |
| Open competition | | Clear selection parameters | | VFM selection Fix EPC | | Negotiate within maximum EPC | | Fix price and conditions |

Figure 11.4 Hierarchy of PFI selection.

being recommended where work is required in the detailed design stages of several bidders. The European Commission position on PFI has hardened indicating the need to ensure that competitive processes with the negotiated procedure are maintained.

Risk transfer

Risk transfer is a critical part of the equation in assessing VFM. Under PFI, many risks that are normally carried by the public sector are transferred to the provider. For instance, it is normal to transfer design risks and construction overrun risk and even planning delay risk. In order to deter-mine VFM, it is normal to calculate the cost of a conventional procurement method called a public sector comparator (PSC) and to add to it an estimate of the cost of risks that under PFI procurement are transferred to the private sector. A PSC is determined by the unit cost of similar type that has been procured under a conventional non-PFI route and adjusted to allow for the floor area and differences in other factors such as inflation, location and level of quality. To minimise the subjectivity of the value of the risk the Treasury has laid down certain rules in its calculation.

The principle of putting risks where they can *best* be managed is illustrated by the diagram and shows an optimum point where unreason-able risk transfer causes declining VFM because an unaccustomed risk liability is expensive to manage. For example the occupancy rate of prisons, if transferred as a risk to the SPV is not easy to manage, as the prison governor and the judiciary make the decision for prison allocation and safe occupancy levels. Without this knowledge a private risk assessment might err towards a low occupancy rate being priced in the contract, when in fact the prison may normally be full, creating economies of scale for services, which are not passed on to the Authority. Figure 11.5 shows this effect.

The main risks that are transferred are construction delay, cost overrun, design problems which are not to do with changing the scope, some planning risks and even some occupancy risks which are controllable. The

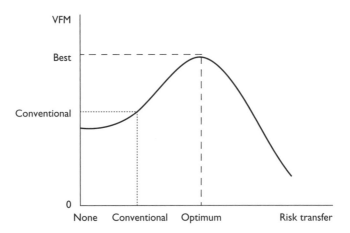

Figure 11.5 VFM and risk transfer.
Source: Treasury Taskforce Partnerships for Prosperity [23].

following case study indicates how a PFI for a large teaching hospital was set up and illustrates the relationship between the structural and risk transfer. The more risks transferred the better, the more that care is needed to check the VFM factor. Case study 11.2 provides some further information about how budgets were agreed on the same project.

Case study 11.2 Redevelopment of major NHS teaching hospital, part two

The budget was agreed between the NHS Trust and the SPV over a three-year inception programme from 1996–9 when financial close took place. The Trust provided the project team with a performance specification and clinical constraints from which a proposal outline design was developed a major architectural practice working for the contractor on a design and build basis. This design was developed with the design and build contractor, structural engineer and architect to change the footprint of the outline design so that a more efficient floor to wall ratio and circulation could be achieved. During negotiation detailed 1:200 design drawings were developed to ensure functionality and space planning and agreed to a given budget.

This budget was constrained by an acceptable annual unitary service charge, which covered the capital costs and yearly running costs. This allowed for adjustments between capital costs and WLC. Planning was obtained by the contractor for their developed 1:200 scheme design after sign off by the Trust. After planning approval, a detail design was developed with 1 : 50 plans, which were signed off by the client as compliant and these became the reference point for the structural, services and complete clinical functionality of the design. This was important so that detailed services design and performance specifications could be agreed with the Trust. In the case of components the contractor insisted on choice wherever possible, with a commitment to meet client performance requirements. Cost was controlled by the design and build contractor by elemental cost-planning checks with the architect.

In order to achieve sign off of the detail 1 : 50 design the architect attended the weekly clinician's project meetings to agree detail under the watchful eye of the design and build contractor's design manager. This allowed for innovations to be developed which would be acceptable to users and was supplemented by wider, less frequent user groups where specific areas of interest were being discussed. Each 1 : 50 floor plan was split into four zones to aid the more specialised inputs of individual departments and to finalise service requirements within the overall framework of the original 1 : 200 layouts. In order to co-ordinate, the hospital appointed an internal redevelopment project team. This team liaised with the contractor project manager and with the general manager of the SPV in order to ensure the continuity of relationships when the hospital was occupied and co-ordinate building operational procedures with the clinical, administration and support services employed by the Trust.

VFM considerations

There is considerable debate about the VFM balance and Figure 11.6 indicates the balance of extra costs such as the inflated cost of capital and the need to offer returns to equity holders.

A discounted cash flow calculation for the concession period generally estimates the impact cost of risks transferred and also the probability of those risks occurring. However good the arithmetic, there is bound to be some subjectivity in both of these estimates, especially as there is a very limited track record to run on. The temptation to 'make the figures work' is compounded by the fact that PFI credits granted from the Treasury may be the only way for some public authorities to gain credit, as other public borrowing credits are more scarce to allow for PFI quotas to be created. For the Treasury a PFI credit is a binding commitment to support the Authority for the service period. Treasury assessment procedures are therefore rigorous and provide a check on irresponsible or inexperienced commitments.

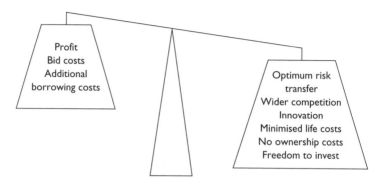

Figure 11.6 The VFM balance.

Source: Treasury Taskforce Partnerships for Prosperity [23].

Affordability

Affordability looks at the long-term effect of paying off the combined capital and service charge (Case study 11.3). Can the authority get the income over the concession period in order to pay the charge? This may be calculated by the use of discounted cash flow considering the NPV of cost and income as described in Chapter 3, which assesses the feasibility. In DBFO public infrastructure the outline business case will have valued the benefits as well as the financial income. Benefits must be presented to the public funding source as a convincing case of need. In private DBFO then a project will not go ahead unless the figures stack up and income has a reasonable chance of reaching expectation.

Case study 11.3 City of Bristol College, affordability and funding

Due to the holding of old buildings, expensive to run and situated in the suburbs awkward for student access, the College needed a more central modern campus. They looked around for a development partner who could offer a turnkey building development that could be maintained for an annual fee. The capital value of the new development offered was £13m. As a public authority, the College carried out a Public Sector Comparator (PSC) with a conventional approach to ascertain VFM and advertised in the EU to ensure broad competition. An OJEC notice was issued in June 1998 and the preferred bidder was chosen in October. Financial close was achieved in April 1999 on a 25-year lease with the developer responsible for FM of the building according to the college performance specification prepared mainly prior to the OJEC advertisement. The developer agreed to find a site and to carry out the planning design and finance risks for the project. At the end of the 25 years, the building would be transferred to the college for a nominal sum of £1. The agreement also allowed for an escape clause for the College to buy back on building completion for an agreed price. The developer added value by finding a site suitable to develop a hotel as well as the College in the City Centre and they completed construction according to the agreed performance specification. This defined the scale and scope of the accommodation, but gave flexibility to progress the design and to build in value savings. A small project management team was formed by the college headed up by the principal.

The College saved itself the job of buying land and controlling a major project and was able to spread the cost of the project over quarterly payments for the 25-year period to cover capital and running costs. VFM was passed on by the developer sharing opportunity savings on the land costs. The pre project period was slightly extended, but a quick construction stage was mutually acceptable to the developer. Credits were obtained from the Further Education Funding Council (FEFC) on the production of a properly assessed full business case assessing the risks and value in favourable comparison with a PSC.

The College learnt that it was important to allocate risks more closely in the negotiation period and it was able to take advantage of the current low interest rates to seal a long-term contract.

The length of the negotiation period has been a cause of escalating the contract price as it increases the resources required and the tender suffers from the inflation of building and facilities management prices and the negotiation phase becomes less competitive. Many NHS projects took more than two years of negotiation in the early days which effectively 'hung' the negotiations and frustrated tenderers. It is important to manage this phase effectively to encourage win–win agreements and to ensure client changes at this stage do not change the nature of the contract. One large PFI has been criticised by the NAO [3] and the PAC [4] for the extensive time that elapsed in the negotiation stage with a preferred bidder. This had the effect of escalating the bidder price by 21% before agreement was reached and could have been avoided by the better consideration of the problems of location and technical transition costs, which were still being sorted at this stage and impinging on the negotiations and giving the contractor a price advantage.

When is it right to use PFI?

This is not an easy question, but some general guidelines may be given. Authorities that are providing facilities with little room for innovation, shared use or who have very specialist use which requires limited public accesses are unlikely to prove VFM, for example, a laboratory building on a University campus. Public facilities which depend heavily on large and uncertain visitor numbers attract premium borrowing rates. These need very careful scrutiny and sensitivity analysis to look at worst case scenarios and backups. The Royal Armouries, Leeds [5] is a case in point where the SPV had to be bailed out by the substantial use of public money.

Authorities may enhance their business case figures by the release of unused assets such as buildings or land that have commercial value. They may also encourage joint private use of income generating facilities such as public use of school swimming pools at the evening or weekends and these are called *opportunity* benefits to be shared with the Authority or reduce the service charge. A large part of the success of PFI is responsible negotiation with a preferred bidder to reduce costs to the tax payer and to enhance affordability by innovative design and use, and economic running costs. Again independent guidelines are needed to protect long-term assets and offer sustainable services to the public that are at least as good as they have experienced previously. It is also important for providers to gain a reasonable profit margin to cover the additional risks they incur. Keeping this balance right throughout the project is critical for public trust in the system.

The culture of PFI

The culture of PFI is more open and flexible than conventional procurement and uses a performance specification expressed in outputs and not prescribed. It allows the provider to be much more innovative in its approach to the design and provision of the facility and it is important not to be strait jacketed by design rules or guidelines. Entrepreneurial approaches to design mean that alternative ways of providing space may be discussed in the spirit of VFM and the fundamental definition and inclusion of client values. The tendering culture is competitive in its initial stages as required by the EC procurement competition rules, but in the negotiated procedure it allows a final negotiated stage in order to bring discussions to the table with a preferred bidder who has passed the initial competitive stages. The culture is collaborative and seeks to build trust in order to optimise the project objectives within the budget which is available. The client brings their outline business case to the table and the contractor seeks to bring agreement to their design proposals by transparent pricing of alternatives in order to build up a detailed set of specifications, which are agreed as meeting the output specifications. With this culture, it should be possible to lever in extra value and to allow some flexibility for the client as user groups get involved.

Important to the success of the PFI contract is the formation of a project team and a leader. For the client this leader may well be the chief finance officer of the Authority in question, but the key issue is that they should have the team building and negotiating skills to provide an effective lead and response for the procurement process. Skills in negotiating and leadership and a knowledge of the whole PFI contractual process needs to be built up. It is also clear that the leader must have good abilities in building relationships, communications and dealing with conflict. They must also be able to set a timetable and oversee the production of a project brief and have authority to make the resources available. At completion, a proper evaluation needs to be organised and feedback arranged. Most of these things cannot be delegated to external advisors, who will provide specialist advice such as legal and financial planning as part of the team. The role and authority of the project leader needs to be formally established to deal with a cross section of department heads, advisors, users and other stakeholders so it needs to be empowered to make decisions.

The skills required in PFI

Research by the RICS [6] amongst surveyors has indicated that project managers are often not adequately skilled at driving PFI projects forward and that in the best PFI projects a partnership of skills between the public and private sectors is required. The Audit Office report [7] identified the need to thoroughly understand the project, be familiar with contract terms and how they work in PFI and have good relationship and communication skills. The RICS identified the mismatch between the technical skills of the consultants and the political and specialist knowledge of the client, but identified the need for more political, negotiating skills and client understanding. Overall it became clear that a new culture was emerging which for PFI to be successful needed to integrate the consultants and the client more.

Relationships in PFI contracts

In a recent survey of Authorities, (NAO [8]) the nurturing of good relationships is considered one of the most important issues for successful outcomes on PFI contracts. This relationship between the Authority and the contractors at the construction and operating stages is long term and the groundwork is laid at the negotiating stages with the preferred bidder. The key question is whether authorities manage their PFI relationships to secure a successful partnership. A happy contract is one where win–win solutions have been identified to the mutual benefit of both sides. This means that an authority gets good VFM and a contractor gets reasonable returns. It is also clear that poor communications can sour what was a good relationship and that building open communications is an ongoing process during construction and operating so that VFM is built on innovation and flexibility. The report found that 72% of Authorities and 80% of contractors believed that relationships were good and many believed that relationships were improving since the contract letting rather than getting worse. They considered that the key issues were:

- A good contractual framework with the correct allocation of risks.
- VFM mechanisms which encouraged rather than discouraged innovation.
- Benchmarking and monitoring the quality of service without being intrusive.
- Building in arrangements to deal with change.

The recommendations for the report were directed at sealing in the right skills for contract management by measures that encourage training and rewards that retain staff and pass on the experience gained in past PFI contracts by key staff. These experiences might well be shared between Authorities and need to be developed during the project by the Authorities having a greater commercial awareness.

Relationships need to be nurtured with the stakeholders of a project as their views may differ because of different perspectives. It is important to manage the priorities on the basis of the degree of influence (power) and the need to differentiate between consulting and informing. Communication systems can be used positively to break down barriers based on the belief that it will be worse than it is. The aim is to bring stakeholders on your side by confronting perceived problems (Case study 11.4).

Case study 11.4 Colfox school [9]

In this early, first school PFI project, the involvement of a comprehensive client project team helped to speed up decision making, which depended on agreement between the headmaster, Dorset County Council (DCC) and the school governors. The project team consisted of the deputy county treasurer, as project manager and included the school head and one school governor, officials of the local education authority and outside financial advisors. A smaller specialist core team carried out the negotiation process and reported periodically to the project team. The 4Ps was available for advice.

The use of all the school governors in the project decision-making process bought them on board and their enthusiasm and confidence was a help in convincing the funders that a viable operating phase was possible. The report also emphasised the success of open days in which potential bidders were introduced to the vision for a new school and the management imperatives for its operations and attracted 12 bidders to submit schemes. Eight of these were invited to detailed discussions with the team and four were invited to tender. It was believed that this improved the quality and VFM of the bids. A project room was also set up at County Hall as an accessible source of material to aid the bidding process.

Typical non-contractual relationships exist with the parents and the community and these were nurtured by occasional press and parents meetings on progress.

Key negotiation points with the contractor developed around the long-term agreement between the LEA and the School governing body for financial contributions, the compliance with the then Department for Employment and Education (DfEE) guidance for accommodation and contractual provisions such as The Local Government (Contracts) Act 1997 (LG(C)A 1997) to ensure proper standards of service.

The case study gives an insight into the extensive thought which went into the management and communication systems to bring all parties on board and to enhance relationships between the various parties to the contract and to bring other stakeholders.

Relationships with the SPV are likely to be between different levels such as the contractor project manager, the design team and the manager of the SPV. It is quite likely that they will not proceed as project manager after the main construction is complete and moves into the occupation stage.

(OGC [9]; used with permission)

Risk identification and allocation

The RICS categorises the typical risks in PFI as project cost and completion time, commissioning, future maintenance cost predictions, residual risk such as the value assigned to an asset at the end of a contract, functional and technical obsolescence, changed regulatory and legal standards and cost of project finance.

A survey carried out by Akintola *et al.* [10] indicated that contractors, clients and lenders viewed the priority of the risks quite differently. Table 11.1 shows the top three risks out of a list of 26 risks for each party.

The top 3 risks are covered in the top 10 of all three parties so it is interesting to see from the third column top risks that are not prioritised by the others, but may significantly determine the behaviour of that party

Table 11.1 PFI risk priorities

	Top 3 risks	Bottom 3 risks	High risks not considered high by either of the others
Contractor	Design, construction cost and risk of cost overrun	Euro legislation, credit and land purchase	Contractual
Client	Commissioning, performance and delay	Credit, bankers and debt risk	Operating and health and safety
Lender	Payment, volume and risk of cost overrun	Project life, land purchase and development	Credit risk

Source: Adapted from Akintola *et al.* [10] Ranking of PFI risks. Table 2.2.

individually. Some of the least prioritised risks may be because they are well known and easily managed by those who normally manage them and not because they have little impact if things go wrong. The way that the SPV prices the risk depends upon the reaction of the individual SPV parties.

The *allocation* of these risks should be with a view to reducing the cost of managing that risk and every project will be different. From a public body point of view the risks associated with procuring, financing, designing and constructing and maintaining the facility are considered to be a non-core business and they would like to see the private sector bear these risks. In many cases the client sees the risk associated with the service being offered within the facility, such as clinical health services, education or policing as a public responsibility and therefore retains the volume risk and functional obsolescence. The public body is also in a better position to predict as yet unplanned changes in regulatory and legal standards as a service provider. However, changes to the building regulations are not the forte of an unrelated Authority and could be argued to be better understood by the provider used to dealing with LA planning and building regulation departments. It is less usual to transfer technical obsolescence of this type, but for the reasons given earlier may well be cheaper to manage by a provider who can more easily mitigate the risk by building an allowance into the initial design. Shared risk is important where the impact caused by a failure is out of proportion to the impact cost of the risk itself. They are characterised by the need to collaborate closely, so that risks are mitigated by sharing information fully.

Political ideals may play a role in the level of services offered by the provider. Examples currently exist in the prison service, where custodial services have been handed over to a private contractor under what is called the design, construct, manage and finance (DCMF) form of PFI. In this case, a public interest is maintained through the oversight of a prison governor. Other examples exist in infrastructure or museum provision, where an agreed charge can be levied on the public over a concession period, which goes directly to the operator to offset all capital and operating costs. Logically, these risks can only be managed efficiently by the private sector where the experience of the provider is adequate, there is a partnership that allows minimal interference by the Authority and financial planning has been based on sound assumptions for volume and quality improvement. The Ashworth Young Offenders' Institution faltered because of a failure to recruit skilled staff locally at the rates of pay budgeted. On the other hand the Skye Bridge was delayed because it was recognised that re-negotiation of the bridge charges and concession period were needed to make the PFI viable.

The risk that the asset value at contract end is not as anticipated has not been tested by the completion of a PFI service period, but is closely associated with the maintenance of the asset during the concession period. Theoretically there is a much stronger incentive on the part of the provider to build in durability and low cost maintenance for the facility, as the additional life cycle costs are several times greater than the capital cost, even when comparing NPV. However, the continuation of use of the asset after the concession period also needs to be assessed and terms built in to write off the asset or to ensure its continuing value for the use for which it was built, or to ensure flexibility for functional change of use.

Of course, all the risks above may not work against the contract, because a higher cost in one may be balanced by a lower cost in another. This means that the sum of financial impact reduced by multiplying by a probability of occurrence for each of the risk factors is likely to be a reasonable presentation as long as the probabilities and impacts are realistic assessments. Some of the risks taken on by the provider may eventually return to the Authority and these risks should be properly priced.

Sector differences in PFI

Different public sectors have developed alternative provisions to reflect their particular needs Fewings (1999) compares some key criteria which indicate the differences between the higher education, health, prison, local government (schools) and infrastructure sectors. Interestingly there is very little standardisation between the different sectors. Table 11.2 gives a summary of the findings of the research in the areas of tender length, authority for go-ahead, risk allocation, affordability and VFM.

Table 11.2 Comparison of key criteria for sector case studies

Sector	OJEC to financial close	Approval and non-financial appraisal	Risk allocation and cost	Affordability	VFM	Improvements
Higher Education (hypothetical case study)	12 months minimum	Institutional with HEFCE sign off if requested – weighted criteria	Estimated cost impact and probability Covers design, construction, operations, finance costs and demand	Test against known financial income before and after procurement choice	NPV expressed over contract period compared with other options	NAO report suggests improvement of option appraisal and valuation of risk
Health (a) Dawlish (b) Bodmin	(a) 36 months (b) 13 months	NHSE Regional sign off. Criteria and weighting differed significantly	Standard risks are now identified centrally as checklist, but choice, probability and valuation of similar risk varied significantly	Compared new costs of preferred option with ring fenced incomes Sensitivity analysis (a) 88% occupancy (b) 90%	(a) 1.1%save, 30% > extending (b) 50% save + redistribution Full option costing using NPV before and after risk transfer	Reduce OJEC to FC from 3 years to 3 months

Local Authority (Colfox school)	16 months	DfEE sign off and full council approval Consultation with the community and Governors	Risk allocation proposal and value from bidders compared against LEA risk matrix at short list stage	Initial feasibility study test against PSC capital and revenue grant	2% savings Use PSC to compare initial bids and estimated risk transfer	Suggested: bring in financier earlier, more detailed PSC and risk valuation before shortlist bids
Prisons (a) Bridgend (b) Fazakerley	17 months 17 months	HMPS Political imperative for expanded provision	Option appraisal including HMPS valuation of availability and service risk compared with public sector cost of risk	Compared each option with police cell option	10% saving Compare with artificial PSC and estimated risk transfer	Nine months OJEC to start early identification of non-transferable risks Financial viability

A further comparison indicates the difference of procurement methods:

- *Prisons*   DCMF. Sign off by HM Prison Service (HMPS).
- *Health*   DBFO. Sign off by National Health Service Executive (NHSE).
- *Schools*   DBFO. Sign off by Department for Education and Employment (DfEE).
- *Roads and bridges*   BOOT. Sign off by Department of Transport (DoT).
- *Higher Education*   There is a variety of models possible but generally DBFO. Guidance by Higher Education Funding Council (HEFCE).

The efficiency of the tendering process from OJEC to financial close varies from 12 months to 36 months and there is an indication of expected improvements to reduce this to 9 months in some cases. In most cases approval in these case studies is shown to be by the relevant government department. The affordability does not always use a PSC. The risk analysis is normally against a pre-arranged matrix, but no indication is given of the way in which the risk is valued. VFM is assessed generally by doing an NPV calculation for an option appraisal. The value of saving varies from 1.1% to 10%.

Issues affecting the operation of PFI projects

The NAO [11] survey of performance of PFIs in the UK compared with the traditional public procurement has claimed to be much improved from 30% to 78% finishing on time and from 28% to 78% finishing on budget. The claims for budget compatibility have still to be tested in the full term of the contract. Particular concerns are also expressed if the building use becomes obsolete or needs to be substantially adapted within the term of the contract and what flexibility there is to make changes with the provider. Members of the SPV in their turn may wish to sell on their interest and Authorities need assurances about the quality of their new partners. Authorities are generally given powers to terminate contracts, which are substantially in breach of the terms.

There are a growing number of reports that have dealt with various aspects of PFI identifying problems that have arisen such as the extended tender period, the sharing of savings gained by rescheduling finance, standardised accounting treatment in distinguishing between capital and revenue, the development of standard contracts, control on design quality, VFM and affordability, risk transfer values, cost of tendering and EU competition rules. In the main they point areas for improvement rather than terminal problems.

At the centre of the arguments, the concerns lead back to properly assessing VFM. Whilst some believe that the paying out of profit to private providers and the less favourable borrowing terms cancel out benefits for almost any project, most believe that the proper assessment and the credible use of tools in that assessment will make some projects suitable and some not for this form of procurement. This facilitates against the uninformed use of PFI by public bodies as a 'cure all' and begs the question as to whether government funding force fits PFI by making other funding difficult to access for public authorities. This in itself is not necessarily bad, as long as the projects remain affordable over the long term and the policy has other benefits. However, VFM needs to be properly tested for each project against the PSC and other political or extra project benefits have to be properly assessed if the project is to proceed. There is also a feeling that wholesale long-term ring fencing of funds for future charges cuts down the flexibility for changes in policy and that the risk of obsolescence should properly be valued in the functional context and charge periods properly matched to the risk. For example, if the Millennium Dome had been built under a PFI procurement then a 25-year ongoing charge for a building redundant after one year would have to be totally unacceptable, but PFI may have been more acceptable for the Channel Tunnel as its eventual return to public ownership is desirable given the number of times the project has been 'bailed out'.

Track record

The other fundamental problem is the lack of track record for PFI and the huge commitment made for public funds on untested risk analysis in a wide range of sectors. However, this has led to one of the merits of PFI as extensive public assessment is an audited requirement and private assessment is good business, so in partnership the risk is significantly better planned and managed than standard procurement. Conventional procurement has often led to poor maintenance regimes and the premature demise or expensive renovation of public buildings when insufficient maintenance or poor quality buildings have been accepted to reduce capital borrowing. This puts risk factors in, on the opposite side of the balance sheet as a saving and a promoter of integrated best practice.

For the project manager, PFI and DBFO is a pecuniary responsibility and thorough assessment must be seen to be done. In addition, later undermining of the business case should not occur through poor negotiations in the preferred tenderer period, where the pressure to proceed or to heavily compensate tender costs in the case of pulling out, may unbalance the negotiation leverage for a good deal on the one hand. On the other hand, a fair deal for the provider is critical in order to engender the partnership that is so important to VFM in long-term relationships.

Funding

Funders generally build in step-in-rights to protect their interests in the case of non-performance and may, 'in extremis', replace contractors who are under performing. These rights are part of the standard contracts in most sectors. Interest levels for the repayment of long-term debt have been lowered competitively to gain business in this comparatively new sector as risks have been better understood and mitigated against funders to improve the security of their investment. Funders also take an interest in the finalisation of terms in the negotiation stage.

Refinancing the SPV debt has potential for savings and was first controversially carried out on Fazakerley Prison by the contractor, which one report [12] quotes as trebling the contractor's profit from 13% at the letting of the contract to 39% after refinancing. Savings may be gained from reducing the interest repayments, lengthening the term of payment, or by combining the debt to get better terms. This led to new OGC guidance to build in contractual provisions for sharing any refinancing windfalls which may be gained on the original deal with the Authority. Refinancing may also lead to transferred liabilities for the Authority in the case of termination of the contract or selling on of interest and it is now usual for contractors to require consent for any refinancing that may affect third parties.

For instance, the Calderdale Hospital [13] gained a windfall saving of £4m for the taxpayer when it was refinanced after the more risky construction stage and a deal between the contractor and their financiers was struck to reduce the interest rate levels. Thirty per cent was voluntarily shared with the NHS Trust in accordance with the OGC code [14] for historic ontracts and recommendation for a 50% sharing in standard contract conditions.

Operating issues

Fazakerley Prison was one of the early contracts let in 1995 [15] and many lessons were learned. The issue of insurer of last resort was raised as it is often difficult to insure for some of the risks that arise in a PFI project. Where this has occurred through no particular fault of the contractor then the Authority will provide insurance for these risks on their behalf. The review also points forward to halving the procurement period of 18 months on this contract to 9 months on Lowdham Grange, which was the next prison contract to be let.

Ashfield Prison near Bristol was the scene of further investigation [16] into the custodial service which is provided under the DCMF PFIs for the prison service. Here, the Prison Service had to put in its own management team for five months in 2002 as there was a concern for security when the custodial service were unable to recruit staff to meet indicative staffing levels as shown in the contract. New measures for joint training of public and private prison personnel are in place to enhance exchange of information and improve best practice across the board. Further NAO recommendations [17] suggest that, continuous improvements may be made to PFI in general in making examples of good practice collected by CABE and OGC available, to stimulate further innovation in future projects. Other problems in performance were identified by PAC [18] as the need to reduce major changes and additions in the early part of the operation of the facility. Where necessary, they recommend benchmarking them and rigorously testing the PFI service contractor costs by having conditions which allow transparency of pricing or competition from other contractors. To meet criticisms of excessive profits from PFI contractors, benchmarking their margins was also suggested.

VFM

Other concerns have been expressed with the VFM by the NAO and PAC. For example, the Dartford and Gravesham Hospital [19] PFI had a poorly researched PSC and it was recommended that the PSC should be properly priced and risks from traditional contracts should be calculated accurately from several previous contracts whilst maintaining a competitive climate and by ensuring sufficient bidders. Likewise in another contract [20], it was felt that the long negotiation period of 21 months with a preferred bidder had created non-competitive escalation in costs of 9%. Although this was considered not to be out of scale for the time period, the period itself could have been reduced if fundamental matters such as the suitability of possible project sites had been more thoroughly researched prior to going to tender.

Legal issues

Another issue with PFI contracts is 'ultra vires' which refers to the validity and the ability of the public body to commit the sums to the PFI project over the long term. This has affected the confidence of lenders and inflated the cost of finance. Two pieces of legislation have been passed here to under gird the credibility of Local Authorities and NHS Trusts by providing Government guarantees as well as tightening up the use of monies offered as PFI credits.

Design quality

Design quality has often been criticised in PFI by its opponents and its users. Building design quality has been defined as the extent to which the asset has a high standard of space, light and sensory comfort, whilst retaining all functional requirements and relating to the surroundings

Treasury [21]. The government, however, is anxious not to strait jacket the process and to allow some innovation into the system to achieve results.

The guidance note on PFI indicates the following key principles:

- Functionality must be agreed by the sponsor to satisfy the existing use, but also allow for expected future change, growth and adaptability.
- Provision for service enhancement refers to an airy, clean and well-lit environment in which customers and staff feel valued and respected.
- Particular architectural requirements should be given a strong steer at the bidding stage so that bidders may make comparable proposals. For example, the Berlin Embassy PFI was put out to an architectural competition prior to the bidding proposal. The winning design was then made available to the bidders so that they could take account of the key features in their own detailed design.
- A consideration of the social and environmental impact of the building.

Design responsibility is handed over to the SPV on the basis that an output specification provides greater freedom for innovation. There is also a need to give the contractor an incentive to provide high quality design without affecting their ability to win the contract.

Supplier risks

The supplier risks are well outlined in CIC [22] that indicate the importance of asking the question 'Is this project for me?' from a business strategy and financial point of view. This is because of the significant involvement which is required and for a major partner of the SPV there is also a large financial commitment. It is also an extended commitment and if the terms are not carefully negotiated and the partnership is not right, a haemorrhaging of resources is also possible with poor financial returns over a long period.

The wider supplier risks that are cited as one of the main rationales for PFI are indicated as bidder risks, design, construction, commissioning and operating risks, the consequences of delay or defective quality, operating costs and indexation, demand risk, residual value risk, maintaining constant quality risk and finance risk. The latter risks are unaccustomed risks in conventional contracting and arise out of a longer involvement. Once the contracting phase is over, however, the most risky components are seen to be past by the lenders and it is possible to favourably refinance on the basis of the successful achievement of the risky phase.

The compensation for taking on more risk is that the remuneration should be greater and returns of 7–15% should be expected. Risks under priced or overlooked can offset the compensation and realistic pricing of new risks comes from experience. It may also be possible to sell on an equity share in the PFI, if it is found that the conditions are not suitable to the trading profile of a future business strategy. As yet, the market in buying and selling PFI equity is not firmly established and significant 'write offs' may need to be made.

The other more significant cost compared with traditional projects is the tender bidding cost. These are relatively low in the first round, but increase significantly in the short list and even more during the negotiation phase to financial close, because of the degree of financial planning and design detail which is required to convince the client that they have VFM. The use of an output specification by the client means that quite a bit of design is done to determine the functional form before a price can be worked up. More work is also carried out on forming the SPV, bringing a funder on board with the right financing deal and determining the basis of the business relationship. The formula for a competitive bid is also more complex with a balance of non-financial factors such as design, team partnerships and innovation arising out of VM. Opportunities may also arise out of asset offers by the Authority (e.g. excess land), or the widening use of the premises for private as well as public use. Each of these takes time and costs money for consultancy advice.

The financing of a PFI project may be either asset based or on balance sheet involving the commitment of assets already held by suppliers on their balance sheets and unless there are many, there is the lowest limit to the amount of funding to be raised. Alternatively project debt or equity financing can be taken out against the cash flow of the project. In practice on a project of any size some of each is taken out with an 80–90% debt funding being the normal range. Some equity commitment from the contractors will also be required. The influence of the lender on that amount of debt stretches to contractual requirements for step-in-rights to retrieve debt in the case of project difficulties and the presentation of a viable business case and risks contingency plan. In the early days the validity (vires) of certain quasi government organisations (e.g. NHS trusts and Local Authorities) to place and underwrite such long payments was questioned. This has been solved latterly by the passing of the Local Government Contracts Act and a similar legislation for the NHS Trust.

Construction performance in PFI projects

Overall performance needs to take account of efficiency gains which may be given by contracting out facilities management services and by transferring risk. Blackwell [23] quotes a Treasury [24] example as follows, based on their average experience of 20% savings on contracting out facilities management type services and average cost overruns of 24% on construction costs, which under PFI the risk would be borne by the private sector.

There is a Government estimation of 30% :70% breakdown of construction costs to costs in use

24% improvement in construction efficiency on 30% of the contract	7% overall saving
20% saving on letting out facilities management on 70% of the contract	<u>14%</u> overall saving
	21% efficiency gain
3% interest premium for private finance	−3%
Total saving (21% − 3%)	<u>18%</u> improvement

These figures are very general and different types of projects may not produce the savings claimed. They also do not take into account the need to distribute more profit to the private sector for taking on more risk, neither the cost of the bidding nor the due diligence required by funding banks, so efficiency can be expected to be substantially less. It is however, an indication of the potential of PFI. Savings however may be enhanced by the release of dormant or under-used assets for private sector use.

There is evidence that PFI has been making savings by looking at the before and after situation:

- An NAO report [25] called Modernising Construction confirms that 73% of public projects were running over budget and 70% of them were running over time.
- In comparison, a later report for the NAO [26] gives comparative figures of 22% and 24% for 37 PFI projects.

These figures indicate a significant improvement on the construction phase performance, especially as, in all cases, the budget overruns represented changes or additions to the initially stated specification by the Authority. The overrun was more than two months in only 8% (three) of the projects, this compared with an average overrun of 13% across all projects found by Agile [27].

Private DBFO or BOOT

The use of DBFO or BOOT contracts in the private sector has potential for savings, if risk can be transferred to the SPV and managed better and clients can concentrate on their core business. There is no reason why significant additional risk cannot be transferred in the area of design, construction and maintenance if there is a clear assessment of this risk by the client and the DBFO provider. The client will also get a significant release of capital funds for other investments and business activities. Again the important issue is that the integration of facilities management and building design by the same provider will lead to reduced WLC, because the provider has a vested interest. The cost of borrowing is unlikely to be significantly different. Benefits should also accord from the partnership itself and work done to make this mutually beneficial will seal the productivity of the negotiating construction and operating stages. The performance of the public sector indicates that some sectors are more suitable than others.

The private client however, is required to make a long-term commitment based on their current outlooks and PFI may not provide the flexibility they need. There is also a need to convince the private funders of the collateral of their investment as there needs to be ring fencing of funds over a long period of time.

The more general DBFO model should build in the flexibility that is required to ensure response to the market and this may require shorter concession periods and insurance bonds for the client to provide funder assurances. This will not be a long way different from the funding regimes on more conventional procurement. It is possible that the building may become obsolete and the provider will have to pay out clauses in the contract.

Conclusions

The development of PFI has meant that a large proportion of public work in the UK is now carried out using PFI procurement. The long-term relationships which are necessary encourage partnership and the long-term commitment encourages business planning and RM at the beginning of the project, where the problems can be ironed out with less conflict. The VFM exercises have led to greater discussion between the user and the provider. Integration between the various parties of the SPV and the client is much stronger than when funders, contractors, clients and consultants are all operating under conventional procurement. The project manager has the potential to play a leading role to drive the project, particularly where they have a good understanding of the client business.

There are still concerns about the justification of PFI projects that are induced by the lack of alternatives to other finance. It is possible in these cases that public institutions are motivated to make up cases that are affordable because they attract the funding, but are not proven VFM. Collapses in these types of schemes have been illustrated where over optimistic incomes have been forecast. Some go as far as to argue that PFI as a concept is less efficient and argue that we are just delaying the extra costs. Others are also concerned that handing over design responsibility on land mark developments and pressure on the planning authorities, may lead to compromised building and urban design quality. SPVs often operate in name only with contractors keen to be paid off at the end of the construction phase, nullifying the commitment to long-term success predicted earlier.

Evidence collected so far by the NAO and others suggests that capital cost savings and occupational efficiencies are being achieved. These same reports stress that conclusions about the full life cycle costs cannot be proven until projects' concessions are complete as other factors such as inflation, obsolescence and handover costs are uncertain. It is also clear that certain projects are more suitable for PFI and these include schemes that are more easily accessible to the general public so that multiple uses can be made of capacity to generate more income. The release of land and other assets that can only be redeveloped because of the PFI scheme is also another way of reducing capital and service charges.

The use of PFI type arrangements such as DBFO and BOOT are occasionally used by the private or quasi-public sector, but have yet to see a real uptake outside the public realm.

Note

1 The Private Finance Panel 1992–7, The Treasury Taskforce 1997–2000, partnerships UK was formed as a 50% equity with the Government in 2000 to do the same for central government and Government agencies and the OGC has become heavily involved in policy guidance and training materials for PFI since 2000, 4Ps gives support to local authority PFI projects and other large agencies like the NHS have their own support units. In addition, PFI projects have been scrutinized regularly by the National Audit Office (NAO), the Audit Commission and Accounts committee (PAC). Verdicts guidance are published in the public real.

References

1 RICS Project Management Forum (2003) PFI and the Skills of the Project Manager. Project Management Research Papers on CD. July.
2 These derive from The Public Works Procurement Directive EEC 1992.
3 NAO (2003) Government Communications Headquarters: New Accommodation Programme. HC955. 16 July.
4 PAC (2004) Government Communications Headquarters: New Accommodation Programme. HC65. 15 June.
5 NAO Guidance Notes (2001) The Renegotiation of the PFI Type Deal for the Royal Armouries Museum in Leeds HC103. 18 January.
6 RICS Project Management Forum (2003) PFI and the Skills of a Manager. RICS. July.
7 NAO (2001) Successful Partnerships in PFI Projects. NAO.
8 NAO (2001) Managing the relationship to secure a successful partnership in PFI projects. HC375. 26 November. HMSO, London. This survey was based on 121 PFI projects with views from contractors and sponsoring clients.
9 OGC (1998) PFI Material Colfox School Case Study. www.ogc.gocv.uk (accessed 18 August 2004).

10 Akintola A., Taylor C. and Fitzgerald E. (1998) Risk Analysis and Management of Private Finance Projects. *Engineering, Construction and Architectural Management*, 5(1): 9–21.

11 NAO (2003) PFI Construction Performance. HC371. 3 February. HMSO, London.

12 CIOB (2003) 'PFI value for money two'. *International News*. 25 August. www.ciobInternational.org.com (accessed 28 July 2004).

13 NAO (2000) The Refinancing of the Fazakerley PFI Prison Contract. HOC Session HC584. 29 June.

14 OGC (2002) *Voluntary Code of Conduct*. Speech by Peter Gershon, Chief Executive of Office of Government Commerce, at the UK Government Conference on Public–Private Partnerships in London on 16 October 2002.

15 OGC (1996) Report on the procurement of custodial services for the DCMF Prisons at Bridgend and Fazakerley. PFI Material. OGC.

16 PAC (2004) The Operational Performance of PFI Prisons. 10 February.

17 NAO (2003) PFI Construction Performance. HC371. 5 February.

18 PAC (2003) PFI Construction Performance. HC567. 15 October.

19 NAO (1999) The PFI Report for the new Dartford and Gravesham Hospital. HC423. 19 May.

20 NAO (2003) Government Communications Headquarters: New Accommodation Programme. HC955. 16 July.

21 Great Britain Treasury (2003) *PFI: Construction Performance*. Report HC371 for the House of Commons. 3 February. Comptroller and Auditor General. HMSO, London.

22 Ive G. and Edkins A. (1998) *CIC Constructors Guide to PFI*. Thomas Telford Publication, London.

23 Blackwell M. (2000) *PFI/PPP and Property*. Chandos Publishing, Oxford.

24 Treasury Committee (1996) The Private Finance Initiative. The Government's Response to the 6 Report. HMSO, London.

25 NAO (1999) Modernising Construction HC87. NAO.

26 Mott MacDonald (2002) Review of Large Public Procurement in the UK. July. HMSO, London. The PFI projects included central government, including NHS projects, but not including local authority projects. The prices compared were those agreed at contract signing.

27 Agile Construction (1999) Benchmarking the Government Client.

Chapter 12

Supply chain management in construction

Martyn Jones

The objectives of this chapter are:

- To outline the historical development of supply chain management in other sectors and why it has been important to business.
- To analyse the key features of SCM and its impact on the way business is conducted.
- To consider how it can be applied to construction projects and supply systems and the rationale for its use.
- To outline the reasons why its implementation in construction has been comparatively slow and the main factors inhibiting its take up.

Introduction

The diversity of clients, buildings, sites, materials and components means that construction projects have a wide range of supply systems supplying labour, skills, materials components and sub-assemblies. These supply systems have developed to service the range of technologies and processes associated with the different phases of the project and the different parts of the building under construction, as shown in Figure 12.1.

These supply systems provide incomplete sets of resources and services, parts of which are assembled to make up supply chains to individual elements or work packages. For each of these elements the most appropriate supply chains have to be selected, organised, aligned and managed in a way that best fits with the client's requirements and the overall nature, aims and objectives of the project.

As well as being shaped by the technological requirements of each element, these supply chains are further conditioned and shaped by the complex linkages between individuals, firms and projects and the procurement strategies and forms of contract adopted by the main project participants, but particularly by the client.

Encouraged by substantial improvements in performance, resulting form the successful implementation of first and second generation partnering, as

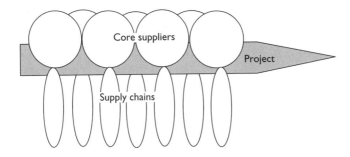

Figure 12.1 Interaction of project supply systems.

discussed in Chapter 6 and on p. 330, and the adoption of SCM by other sectors of the economy, parts of construction started moving towards the adoption of SCM in the late 1990s. This is essentially being led by more informed private-sector clients who adopted first generation partnering during the late 1980s and early 1990s, and then second generation partnering in the late 1990s. During the 1990s considerable effort went into defining and developing partnering to meet the different needs of the clients in various sectors of the construction industry. Unlike partnering, SCM is still very much perceived as a concept 'borrowed' from other sectors and which is in a very early stage of development in construction. It is also important to recognise that SCM is a complex innovation, which has proved to be difficult to implement and sustain even in sectors such as the automotive industry in which it originated. It is described as a fifth generation innovation [1].

Given that it is a multi-factor process built around close and long-term intra- and inter-organisational relationships, it demands a strategic and long-term approach. As it is such a strategic innovation, it necessitates continuous learning and commitment from top management. As it is so challenging, it is dependent upon links with, and support from, the external environment.

Given the complexity of this type of innovation, this chapter reviews the concept of SCM before addressing its relevance to construction projects. In order to fully understand this increasing interest in, and use of, SCM in construction, it is first necessary to understand the main factors that have been driving and shaping its development, not only in construction, but in a number of sectors of the economy. The two most important factors include the new environment within which business is increasingly being conducted and the new thinking in relation to management approaches – in the case of the latter, not only within organisations, but also externally in relationships with other organisations in supply chains.

Changes in the business environment

Many writers, from a wide range of perspectives, have been arguing for some years that the wider environment is changing significantly and that business is moving to a new model or paradigm. From the late 1970s, new methods of production and ways of doing business began to emerge. The key ideas associated with the new emerging paradigm include:

- A significant shift towards empowerment to encourage continuous improvement through learning and innovation.
- A change from dedicated mass-production systems towards more flexible systems that can accommodate a wider range of products, smaller batches and more frequent design changes – hence the 'economies of scale' are increasingly being replaced by 'economies of scope' and greater mass customisation.
- A customer-focused approach with the customer brought into the centre of the business in order to align the organisation with the evolving needs and expectations of customers.
- A move towards the greater integration of processes and systems within companies and between suppliers and customers, in order to align resources more closely to market and customer requirements.
- Developing the appropriate organisational structure and culture to support and sustain learning and innovation.

It is the emergence of these new possibilities and challenges associated with the new technological-economic paradigm that is driving organisations – including an increasing number in construction – to undertake a fundamental reassessment of the way in which they achieve their objectives through innovation, not only within their organisation but also in their networks of customers and suppliers. Another significant feature of the new paradigm is the growth of outsourcing and the increasing reliance on suppliers and subcontractors to build and maintain competitive advantage. Outsourcing refers to the activity of purchasing goods or services from external sources, as opposed to internal sourcing (either by internal production or by purchasing from a subsidiary of the organisation). This growing reliance on suppliers means that, for instance, a typical company's products or services are obtained from other companies in their supply chains. This means that unless the quality, cost, design and delivery of inputs from their suppliers are appropriate the final goods or service is unlikely to meet the needs of their end customers.

The main motives for the increase in outsourcing include cost reduction (as external sources can enjoy greater economies of scale), access to specialist expertise and greater concentration on an organisation's core competence, by discarding peripheral operations. The potential disadvantages of outsourcing include reduced control over the operations involved, fragmentation and concerns over delivery and quality. SCM has emerged as an approach to obtain the benefits of outsourcing but also address its difficulties, to introduce greater co-ordination and integration and to respond as effectively as possible to the needs of customers and changes in market and wider environments.

Outsourcing in construction

Outsourcing in construction grew with the increasing complexity of the design and specification of construction products and processes during the twentieth century. More finely subdivided engineering specialisms began to emerge to support the work of the architect. Also general contractors were no longer able to undertake all the work or provide the substantial capital investment needed in the emerging specialisms as many of them required, for example, expensive plant and equipment. In addition, subcontracting has been giving general and main contractors the flexibility to deal with fluctuations in overall demand for construction products and services as they respond to the diverse needs of clients and their wide range of construction needs. It has also been argued that the greater use of subcontracting by general contractors, during the period between the early 1950s to the early 1970s was a means of circumventing difficulties such as the poor relations that had developed with their workers [2]. Subcontracting either pushed the labour relations problem on to another firm or removed the role of trade unions altogether, for example, by employing workers on a labour-only self-employment basis. A further factor was the introduction of Selective Employment Tax in the 1960s, which stimulated an acceleration in the use of labour-only workers. All of these developments contributed to the fragmentation of the industry, the marginalisation of subcontractors and thus increased the need for more effective co-ordination and integration of construction's supply chains.

Origins of SCM

The term SCM began to be used in the early 1980s to refer to the management of material flows across functional boundaries within organisations. This innovation, and others such as Just-in-Time (JIT) and TQM, often provided substantial internal improvements by breaking down barriers between departments and focusing on efficiency in managing core processes. With the increased reliance on suppliers as a result of increased outsourcing, SCM began to be extended beyond the boundaries of a single business unit to comprise all those organisations and business units that have to interact in order to deliver a product or service to the end customer [3].

The theoretical roots of SCM

Similar to partnering, SCM derives from two roots of practically oriented management theory: operations management and partnership philosophies. Within operations management a typical definition of a 'supply chain' as proposed by Aitken [4]:

> A network of connected and interdependent organisations mutually and co-operatively working together to control, manage and improve the flow of materials and information from suppliers to end users a system whose constituent parts include material suppliers, production facilities, distribution and customers linked together via feed forward of materials and feedback flow of information.

Figure 12.2 illustrates the origin of SCM in the context of operations management. These early internal and increasingly external improvement programmes led to the broader concept of SCM. This includes:

- purchasing and supply management;
- physical distribution;
- logistics;
- materials management.

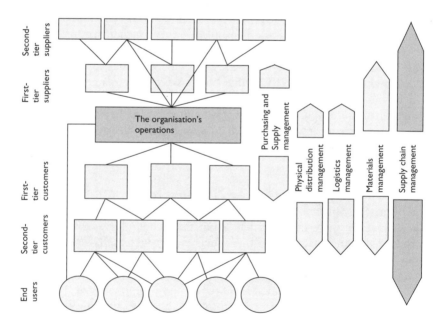

Figure 12.2 Supply chain for consumer goods manufacturer.

Source: Derived from Slack *et al.* [5] and Hill [6].

The given figure identifies the key management functions involved in the management of the supply chain for consumer goods manufacture. As can be seen, *purchasing and supply management* is concerned with the links with first-tier suppliers. More than merely buying, the aim is to obtain supplies according to the five 'rights': the right price, the right quantity, the right schedule, the right quality and from the right source. It is now widely accepted that these features contribute significantly to the value offered by organisations in their own products and services to their customers. Establishing deep relationships with a few suppliers or even a single supplier, the so-called 'single sourcing', has resulted in improvements not available in traditional approaches to purchasing.

Physical distribution management (PDM) refers to managing the flows to first-tier customers. It deals with issues such as the number and location of depots, the type of transport and the scheduling of flows. The skills and investment required in this activity mean it is frequently outsourced to specialists. Logistics is usually seen as an extension of PDM in considering the flow of products to consumers. While PDM considers the best way to deliver to the next tier, logistics seeks to optimise the whole chain. For example, to take joint decisions on packaging so that it not only protects the product during transit but also carries useful information and serves as a unit for display for the retailer. *Materials management* considers the flows of products both within and outside the organisation including many sites spread throughout different regions and countries in order to offer high efficiency and flexibility. It involves issues such as purchasing, location of plant and warehouses, stock control and design of transport systems.

SCM has emerged as a broad concept covering flows within and between organisations. It focuses on integrating all these functions, their processes and interfaces. This emergence of SCM has been driven not only by internal pressures to reduce costs and add value, improve efficiency and satisfy customers, but by a number of external factors which characterise the new paradigm such as globalisation, shifts in the nature of competition, systems development and ICT.

Working more closely with suppliers requires high levels of information sharing, co-operation, increased openness and transparency, which highlights the importance of the partnership philosophies root. This has led to an increasing adoption of partnership approaches and inter-organisational collaboration to achieve significant mutual benefits involving sharing resources, information, learning and other assets Mowery [7]. Effectiveness in adopting partnership approaches is also linked to creating the appropriate internal organisational and cultural changes.

The partnership philosophies emphasise the nature of the relationships between the organisations involved in the supply chain. The central concept is about building mutual competitive advantage or 'win–win' thinking through better buyer–seller relationships. At one end of the spectrum is the 'arms–length' or 'hands-off' contract for goods or services where a price is agreed for a completely specified, or a standard off-the-shelf, product or service. Apart from agreeing the product or service and the price, the buyer and the supplier need know nothing of each other's processes and operations. In contrast to this approach, at the other end of the spectrum is a more involved and explicitly interdependent model based on a common purpose which is mutually beneficial leading to a sharing of profit and risk. This needs to take place in an atmosphere of trust based on sharing of information and knowledge in order to understand issues and problems as they emerge, and devise appropriate solutions. This often requires going well beyond the commitment of customers and suppliers associated with traditional approaches to procurement and relationships based on a formal contract.

Definition of supply chain management

As SCM is a relatively new concept, it is still in the process of being clearly defined. Christopher [8] defines a supply chain as,

> the network of organisations that are involved through upstream and downstream linkages, in the different processes and activities that produce value in the form of products and services in the hand of the ultimate customer.

There are numerous definitions of the management of supply chains. A definition provided by Christopher is,

> the management of upstream and downstream relationships with suppliers and customers to deliver superior customer value at less cost to the supply chain as a whole.

Most definitions link SCM with the integration of systems and processes within and between organisations, which include the upstream suppliers and downstream customers. Accordingly, SCM can be seen as a set of practices aimed at managing and co-ordinating the whole supply chain from raw materials suppliers to the end consumer. It is also viewed as the co-ordination of manufacturing, logistics and material management functions across the organisations [9].

SCM is therefore, closely associated with improvement programmes that have been broadened to include methods of reducing waste and adding value across the entire supply chain [10, 11]. The main objective is to develop greater collaboration and synergy throughout the whole network of suppliers through a better integration of both upstream and downstream processes [11, 12]. This significant emphasis on co-ordination and integration is strongly dependent on the development of more effective and longer-term relationships between buyers and suppliers with increased trust and commitment [13, 14]. It is about adopting a more holistic approach in order to optimise the overall activities of companies working together to build greater mutual competitive advantage and greater customer focus. Figure 12.3 shows a common way of visualising a supply chain.

However, this view is increasingly seen as an oversimplification of the context within which SCM is to be applied. In reality, most companies have several, if not scores of customers and suppliers. Often, several companies compete for the same customers and have common suppliers. A more realistic picture is more complex, with a multitude or spider web of relationships between customers and suppliers as shown in Figure 12.4.

The complexity of this innovation suggests that its implementation requires a number of actions which have to be considered concurrently rather than sequentially. It implies a process involving some degree of experimentation, learning and feedback mechanisms. It also needs to be contingent upon the organisations involved and their present relationships and wider environment. Hence, the implementation of SCM should not be seen as a linear process and the following features should not be seen as a sequential list of actions for its successful implementation.

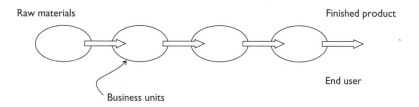

Figure 12.3 A common view of a supply chain.

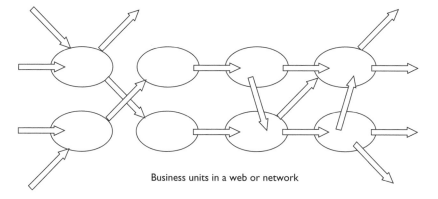

Figure 12.4 Supplier networks.

Agreeing a common purpose to ensure mutual competitive advantage

As with partnering, effective SCM requires a significantly higher level of joint strategy development, where the members of supply chain collectively agree a common purpose and jointly set strategic goals that are mutually beneficial. It is this concept of mutual competitive advantage or 'win–win' thinking which is at the heart of successful SCM. Most failures of SCM are mainly attributable to members of supply chains continuing to being in competition with each other, behaving opportunistically and not sharing their strategic thinking or agreeing a common goal and strategy. Establishing a common purpose based on mutual advantage needs to reflect the degree of preparedness of the organisations involved and the anticipated scope of the relationship. In the first instance this should focus on the development of contractual trust. A common purpose should be based on

- reciprocity by which one is contractually obliged and subsequently morally obliged to give something in return for something received;
- fair rates of exchange between costs and benefits;
- distributive justice through which all parties receive benefits that are proportional to their contribution and commitment.

Ensuring mutual benefit is perhaps the biggest challenge in removing hidden agendas and shifting from the often adversarial nature of buyer–supplier relationships to long-term and trusting relationships. In this approach, competitive advantage is seen as no longer residing only with a company's own innate capabilities, but also with the effectiveness of the relationships and linkages that the firm can forge with other organisations in the supply chain. The new dynamics of these supply chain relationships is built upon the fundamental idea of sharing both profit and risk. In this context it is no longer appropriate for suppliers and customers to view themselves as independent entities competing with each other.

Developing more collaborative inter-organisational relationships

Evidence shows that, in the appropriate circumstances, long-term agreements can improve mutual understanding, build closer relationships and reduce conflict and transaction costs. Instead of pursuing short-term contracts characterised by frequent bidding and switching costs and the costs of pursuing claims and resolving conflicts, purchasers and suppliers can direct their energy and efforts toward value-adding activities. Long-term purchase agreements are seen as a prerequisite to achieving ongoing and closer co-operation between purchaser and supplier leading to, for example, early or even ongoing involvement of the supplier in understanding markets and the design of new products, services and processes.

SCM, which is dependent upon 'win–win' thinking permeating the culture of the supply chain, can help overcome the often deeply ingrained adversarial attitudes in traditional business relationships. This involves the two-way exchange of information and the building of closer, long-term relationships so that a supplier may have a guarantee of the business for several years, rather than having to re-bid for each order. In return the supplier is required to hold or reduce prices, improve quality and delivery, introduce more efficient processes to drive out waste and add more value. The greater trust developed between customer and supplier can lead to the acceptance of open-book accounting, which can also lead to more transparency in relation to true costs and open communication between the members of the supply chain. Often, the buying company will also commit itself to helping the supplier to improve its competencies through supplier development activities and the sharing of key resources.

Selecting the right partners

The move to close, long-term and more collaborative customer–supplier rela-tionships has increased the importance of choosing the most appropriate partners. It requires both customers and suppliers to have an overall under-standing of the nature of their businesses and their links with supply chains. In this complex setting, companies need to realise that some relationships between customers and suppliers will be more important than others in terms of their own competitive advantage. Increasingly companies are also looking at a wider set of competencies to guide their selection of their closest partners. Vollman et al. [13] have identified three different types of competencies which should influence the selection of suppliers and their subsequent development:

- *Distinctive* competencies that provide the organisation with a unique competitive advantage.
- *Essential* competencies that are vital to the effectiveness of the organisation at the operational level.
- *Plain* competencies that have no direct effect on the product or service delivered.

Reducing and shaping of supplier bases

Developing closer customer–supplier relationships requires considerable time and resources. This explains the need to reduce the number of suppli-ers significantly (often from several thousands to less than a hundred) in the supply chain. This smaller number of suppliers are then often organised into a hierarchy of tiers, as shown in Figure 12.5, which reflects the importance

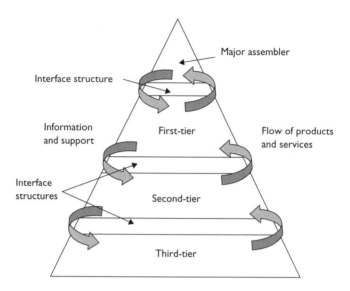

Figure 12.5 A pyramidal supply system.
Source: Jones and Saad [3].

and competencies of the suppliers. This rationalisation of the supplier and customer bases allows organisations to focus their efforts on their most significant customers and suppliers.

This hierarchy, which often takes the form of a pyramid, demonstrates the tiering commonly found in the automotive industry. In this example, there are multiple layers or tiers roughly delineated by the size of the firms and their roles in the supply chain. At the apex of the pyramid sits the final assembler who is supplied with sub-assemblies by first tier suppliers. In turn, these first tier companies are supplied by a larger number of second tier suppliers. These second tier suppliers have their own subcontractors who provide them with specialist process abilities. In some instances there may even be fourth and fifth tier suppliers [15].

Managing the interfaces between the organisations involved

Managing the interfaces between customers and suppliers, based on closer inter-organisational relationships, is a critical element of SCM. The effective management of this relationship requires that the organisations have the appropriate internal cultures and organisations, as summarised in Figure 12.6. This facilitates the sharing of information and learning in order to achieve a greater co-ordination and integration of processes across the interfaces.

An advanced form of interface structure between a customer and a supplier is where organisational boundaries become blurred. This interface

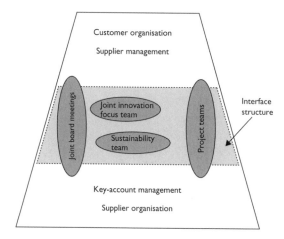

Figure 12.6 An advanced interface structure.

Source: Derived from Christopher and Jüttner [12] in Jones and Saad [3].

structure allows for both joint strategy development and day-to-day problem solving and operational activities. Joint teams can also be formed to co-operate on specific issues such as innovation, R&D, market environment and end user satisfaction.

Shifting from inter-firm competition to network competition

In SCM, organisations seek to make the supply chain as a whole more competitive through the value it adds and the costs it reduces overall. Real competition is not between organisations but rather between supply chains. Reducing costs and increasing value, are both important parts of gaining competitive advantage. The traditional Fordist approach sought to achieve cost reductions or profit improvement at the expense of other participants in the supply chain. Increasingly, however, organisations effectively embracing SCM, are realising that performance improvement should not be bought at the expense of reduced profits, but from the reduction of waste. Also, simply transferring costs upstream or downstream does not make the supply chain any more competitive as ultimately all costs will make their way to the end customer. The shift from classic inter-firm competition to supply chain competition is illustrated in Figure 12.7.

Performance improvement requirements and management

Continuous improvement and performance measurements are key elements of SCM. Robust measures of performance and targets for improvement are also

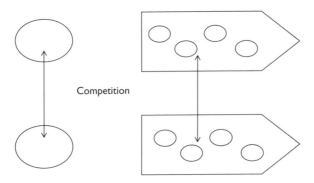

Figure 12.7 The shift to supply chain competition.

Source: Jones and Saad [3].

necessary to prevent the danger of complacency developing in long-term relationships. Specifically, organisations need measurement systems to identify

- supplier performance and improvement opportunities;
- performance trends;
- appropriate suppliers;
- areas for supplier development and the type and scale of resources needed;
- the overall effectiveness of supply chain management in improving performance and adding value.

Empowerment through supplier development is another key factor of SCM and its ability to unlock additional value and competitive advantage from suppliers. Supplier (or customer) development is where a partner in a relationship modifies or influences the behaviour of the other partner, with a view to increasing mutual benefit. Cross-functional teams from the organisations work closely to share learning, solve problems and seek improvements in their internal and interface processes.

Governance

It is important to understand power in the context of SCM. In supply chains it can be seen as the ability of one individual or organisation to influence the behaviour of another individual or organisation, in order to achieve their desired situation and outcome. To have power is relative not absolute, since it is contingent upon the context and relationships within the supply chain. There are a number of types of power in supply chains:

- The power to reward, which is related to the extent that one individual, group or organisation can reward another in the supply chain.

- The power to coerce, which is determined by the ability of an individual, group or organisation to sanction or punish others for unwanted behaviour.
- The possession of legitimate power, which corresponds to the extent to which an individual, group or firm feels that it is right for another individual, group or organisation to take action, thus exercising power.
- The expert power which an individual, group or firm perceives another to have, in relation to key knowledge or specialised technical skills.
- The referent power, which corresponds to the extent to which others in an organisation or supply chain wish to identify with a single individual, group or organisation on the basis of, for example, their leadership style, position or approaches in dealing with a difficult issue.
- The power arising from an individual or group ability to cope with uncertainty and ambiguity. In creating certainty for others in the supply chain individuals, groups and organisations become more powerful.
- The power arising from indispensability which an individual, group or organisation has, if it cannot be substituted in the supply chain.
- the power from relationships and centrality which is conferred by being well-networked into the whole supply chain.

The more types of power that an individual or group of organisations possesses, the greater their influence in the supply chain. For example, a supplier which is indispensable from the perspective of the customer, will be more powerful and will have potentially more influence than a supplier undifferentiated from its competitors. Key customers who buy large volumes of products or services on a regular basis are also more likely to have greater power over their supply chains than those who are small or only occasional customers.

It is important to understand, how the exercise of power in supply chains can determine whether the relationships between suppliers and customers are either loosely or tightly coupled. For example, if supply relationships are tightly coupled they may impinge upon the operation of another firm in the network and paralyse its strategic actions. On the other hand, if relationships are too loosely coupled it can affect the management of the interfaces and reduce the overall effectiveness of the supply chain.

This brief review of SCM suggests that its successful implementation is associated with the following key actions:

- Choosing the best suppliers and integrating them into a rationalised supply base.
- Developing long-term, close, stable, win–win and trusting relationships.
- Breaking down the barriers between internal departments, business processes and between companies to improve the management of interfaces.

- Co-ordinating the supply chain to manage fluctuations in demand and provide more predictability.
- Sharing information, learning and resources to develop the capacities and competencies of the whole supply chain.

Potential benefits

A review of literature in a number of sectors of the economy provides evidence of the significant benefits to be gained from the synergies developed from the successful implementation of SCM. Sako *et al.* [16] suggest that, for instance, closer collaboration in European vehicles manufacturing has helped speed up the rate of performance of improvements in order to meet global competition. The success of the Japanese car manufacturers in introducing new models faster and with fewer labour hours has been attributed to the early, and more proactive involvement, of their suppliers in product design [17]. Shorter lead times, reduced total cycle time and smoother and more responsive flow of materials and products are also achieved through an effective management of interfaces of the entire supply chain [15, 18]. Towil [19] argues that time compression has had a major impact on the accuracy of demand forecasting, the time taken to detect defects, the time to bring new products to markets, and the amount of work in progress. He also identifies a number of other benefits including: elimination (removing a process); compression (removing time within a process); integration (changing interfaces between processes) and concurrency (operating processes in parallel).

This greater collaboration between organisations within the supply chain can also result in increased internal and external customer focus [20]. The collaborative SCM approach, based on frequent and direct communications between suppliers and customers, can help in defining and agreeing the customer requirements and the means by which they can be satisfied. These new types of relationships are increasingly perceived as a means to better utilise resources better throughout the whole supply chain [21].

There are examples of where SCM is delivering significant performance improvements and increased competitive advantage [22]. It can also be an important element in innovation in products, processes and organisation [23] as information can be more readily shared and knowledge identified, captured and disseminated throughout the organisations in the supply chain [24]. This sharing of information and knowledge has led to joint learning resulting in significant improvements in a number of aspects of the supply chain. This shared learning can also lead to better problem identification and joint solving, and allows the more ready application of techniques such as VM, analysis and engineering. It can also improve the predictability of changes in the external environment as a result of more effective communication and greater synergy. Greater compatibility and integration of processes and systems achieved through collaboration and partnership within the supply chain are more likely to lead to increased flexibility and more responsiveness to changes in the external environment. All these potential benefits are linked to the development of trust through greater mutual understanding, transparency in transactions and commitment [25].

Increased trust

The concept of trust is very complex, not clearly defined and interpreted differently by the literature. Most of the literature relates trust to predictability, risk, vulnerability, co-operation, personal traits and confidence. Trust is, on the whole, associated with the confidence that a partner will produce a mutual beneficial behaviour. This view challenges the prevailing view of economic theory that organisations and individuals are self-interested and will pursue their interests opportunistically and often with guile. Trust is based on the belief that a partner is reliable and will fulfil the perceived obligations of the relationship and hence lead to a reduction of the risk that they will not perform an action detrimental to the relationship. Trust is also defined in terms of good intentions and confidence. There is a clear connection between predictability of behaviour and trust which are seen as means to manage and reduce uncertainty [3]. However, it must be recognised that risk can be increased in the early stages of building closer relationships as partners increase their vulnerability as a precondition to the formation of trusting behaviour. Trust is therefore, related to a degree of expectations which, if not fulfilled, may lead to disappointment. This implies the need to recognise and accept that risk exists.

When trust is low, other control mechanisms need to be employed. Typically, legalistic remedies (e.g. accrediting organisations, insurance, bonds and guarantees) are used either to compensate for the lack of trust in exchange or to create the conditions under which trust might be restored. However, these kinds of mechanisms can be counterproductive and lead to even higher levels of mistrust.

In addition, these remedies to high levels of mistrust are costly and impede performance. When trust is present within and between organisations, the cost of transactions can be lower as fewer controls are needed to measure, monitor control performance. Forecasts of future events are more realistic, as are budget projections. Co-operation improves along the value chain, productivity improves, profitability is enhanced and the sharing of learning and innovation is encouraged.

It is also important to recognise that trust must go beyond predictability if it is to produce effective and lasting relationships and partnerships based not only on contractual and competence trust but also in the longer term on good will trust [26]. Contractual trust emerges when each partner will

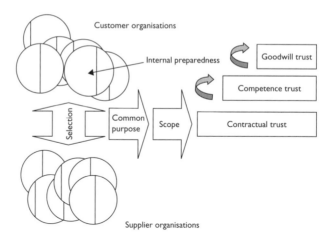

Figure 12.8 Development of trust.
Source: Jones and Saad [3].

follow written or oral contracts. Competence trust refers to the ability of the partner to complete tasks to a given standard whilst goodwill trust corresponds to the situation where a partner driven by mutual benefits does more than expected and goes beyond predictability and the agreements of a contract. As illustrated in Figure 12.8, trust is cumulative and takes time to build. Contractual trust needs to be established prior to competence trust which in turn will lead to the development of goodwill trust. The latter can continue as the partnership and the relationships develop.

What distinguishes 'goodwill trust' from 'contractual trust' is the expectation in the former case that partners are committed to take initiatives (or exercise discretion), to exploit new opportunities over and above what was explicitly promised. The key difference here is that partners are not only looking after their own interests but are also seeking to offer their partner a competitive advantage. If both partners do the same then the combined efforts of both customer and supplier will lead to a mutual competitive advantage, which will help in selling their collective product or service to the end user. However, it is unrealistic to describe the development of trust as a linear process, as the three types of trust are interlinked. In general, contractual trust is often the starting point for the development of trust. Competence trust cannot develop without contractual trust and goodwill trust cannot exist if the former two are not present. It is a long process and takes time and resources to develop.

Although the literature identifies trust at the individual, organisational and collective, level, there is consensus that the common denominator or rather the starting point is indeed individual or *interpersonal* trust. Interpersonal

traits are considered as forming the foundation of trust [27, 28]. Interpersonal trust is associated with the level of expectation that the word,
promise, verbal or written statement of another individual or group can be
relied upon [29]. However, the propensity of groups and individuals to trust
others is often determined by different developmental experiences, personality
types and cultural backgrounds [30].

Ongoing interaction allows individuals to know and understand each
other and develop higher levels of trust [31]. This is why Kanter and
Myers [32] claim that personal relationships and personal connections are
crucial in forming effective relationships and greater trust between groups
or organisations. These interpersonal and informal ties can increase over
time and lead to greater trust in managing partnerships. In such situations,
where trust emerges as a consequence of effective and informal relationships rather than a pre-requisite to the relationship, informal contracts are
used rather than adversarial, detailed and formal contracts. It is clear that
trust between organisations is progressively and incrementally built as
organisations and individuals repeatedly interact [33]. Repeated personal
interactions across firms can encourage higher levels of courtesy and consideration whilst the prospect of ostracism among peers can discourage
opportunism. Mutual expectation of repeated trading over the long run is
major incentive for co-operation and effective partnerships [34–36].

There is however a risk associated with predicting behaviour and
sustaining trust merely through informal relationships. A wider level of
trust needs to be achieved within and between the organisations in order to
ensure an acceptable degree of predictability and sustainability in the
relationship. Emphasis needs, therefore, to be placed on organisational
traits, learning and culture which can be conducive to an organisations'
identity, image and reputation upon which trust can be based. This can be
embedded within the relationship through a clear common purpose, mutual
objectives, openness, transparency and the sharing of information and
knowledge. Hence, *internal preparedness* based on a culture of trust can be
seen as a vital pre-requisite for effective and sustainable inter-organisational
relationships.

The degree of internal preparedness, the selection of customers and
suppliers and the degree of common purpose, should help determine the
type and nature of trust to be achieved in the relationship. In the case of
contractual trust, this can be based on a contract which sets out the
expected behaviour of each party. Such contractual trust should lead to the
elimination of disputes, conflict and adversarial relationships. Improved
relationships coupled with more appropriate internal preparedness should
lead to more sharing of information, greater joint problem solving, joint
learning and innovation and competence trust. Feedback on the behaviour
of the individuals and the organisations involved will influence the scope of
the common purpose. Negative feedback may result in reverting back to

contractual trust or even opportunistic behaviour, while positive outcomes would take the organisations towards the next step on the journey: goodwill trust. Achieving goodwill trust is a long process which evolves slowly and is based on a continuous improvement approach.

The concerns associated with SCM

As in all innovations, SCM has some significant weaknesses and inhibitors to its successful adoption and implementation. It is a long-term, complex and dynamic process and its implementation requires a thorough understanding of the concept [24, 37, 38]. It is also seen as closely dependent upon the ability to create, manage and reshape relationships between individuals and organisations within the supply chain [11, 39, 40]. The implementation of SCM requires new intra- and inter-organisational arrangements and culture all of which require considerable commitment and resources and take time to develop. Time is required for learning, achieving mutual understanding, agreeing a common purpose, selecting the appropriate customers and suppliers, building trust, negotiating objectives and co-ordinating activities. Difficulties which can emerge include multiple and often hidden goals, power imbalances, different cultures and procedures, incompatible collaborative capability, the tension between autonomy and accountability, over dependence, complacency leading to a reduction of long-term competitiveness and continuing lack of openness and opportunistic behaviour [41, 42]. It needs to be recognised that over tight networks can reduce agility and responsiveness to new market conditions and opportunities and lead to the risk of contravening legislation aimed at promoting competition. It can also result in having to continually pass inappropriate levels of value to the customer and missing out on technological and organisational innovations in other supply networks.

There is an emerging recognition that a major difficulty associated with SCM is what constitutes effective relationships and how their effectiveness can be assessed. SCM is based on soft or less tangible aspects which current metrics of performance are not sufficiently adequate to measure. The main tools currently include the Business Excellence model (Chapter 6), Competitive Positioning Matrices and the Partnership Sourcing model. A number of aspects of the Business Excellence approach relate specifically to customer–supplier relationships.

The barriers to be overcome in implementing SCM

A number of barriers to the successful implementation of SCM are attributable mainly to insufficient internal preparedness by the participating organisations. These include:

- lack of commitment to SCM;
- insufficient understanding of the concept of SCM;

- lack of strategic leadership;
- lack of understanding of the new model of network competition;
- inappropriate organisational structure;
- lack of leverage by any of the members of the supply chain to affect change and modify behaviour;
- unwillingness to adopt win–win thinking;
- insufficient allocation of time and resources to build internal and external relationships;
- lack of common purpose and transparent and mutually beneficial goals;
- resistance to the sharing of information, procedures and processes;
- inappropriate distribution of risk;
- inappropriate exercise of power;
- lack of commitment to innovation and learning.

As can be seen from the review discussed, SCM demonstrates the key features of a complex innovation. It is a multi-factor process which involves different functions, stakeholders and variables and a whole sequence of events. It is a complex, dynamic and long process which involves individuals and groups from within and between organisations. SCM has shifted the emphasis from internal structure to external linkages and processes, and is dependent on the interaction between the organisation and its external environment, with strong feedback linkages and collective learning. Its success is associated with the challenging and difficult development of a new culture based on long-term and closer intra- and inter-organisational relationships, shared learning, greater transparency and trust. Intra-organisational preparedness is an important prerequisite for the effective implementation of an SCM strategy.

Given that the main objective of SCM is to increase mutual competitive advantage through improved relationships, integrated processes and increased customer focus, it may well be highly relevant to construction with its adversarial relationships, fragmented processes and lack of internal and external customer focus. However, contingency theory suggests that there may well be specific issues related to implementing SCM in the context and culture of construction.

SCM in construction

The emergence of SCM in construction

Construction has a long history of introducing changes aimed at improving its performance. Partnering, for example, is a recent key change in the shift from fragmentation and short-term adversarial relationships to greater integration and longer-term inter-organisational relationships. This has led to a growing interest in introducing SCM in construction to further integrate processes, manage interfaces between organisations in projects and supply systems, reduce uncertainty and increase overall effectiveness in a greater part of the supply chain.

As discussed in previous chapters, over the past fifty years or so there has been a succession of reports into the state of the UK construction industry and there have been many calls for action to improve its performance and competitiveness. An analysis of these reports indicates that the problems facing construction can be categorised into three broad areas: poor relationships, insufficient understanding of processes and a lack of customer focus, as shown in Figure 12.9.

A number of problems in construction's supply chains downstream of the main or general contractor Jones and Saad are outlined in Table 12.1 [43].

SCM, with its strong emphasis on improving relationships, a process-oriented approach and increasing customer focus, is an appropriate strategy for improvement in construction. Consequently, there is now a growing interest in SCM as an innovation to address the problems impeding construction's performance, and to tackle the issues of uncertainty and interdependence. This is being led by more informed private-sector clients who adopted partnering in the early 1990s in their attempt to both increase the degree of collaboration between their preferred consultants and contractors and to extend this approach downstream to include key specialist and trade subcontractors and suppliers. Some main contractors are also playing a key role in implementation of SCM throughout a greater integration of both upstream and downstream participants and processes.

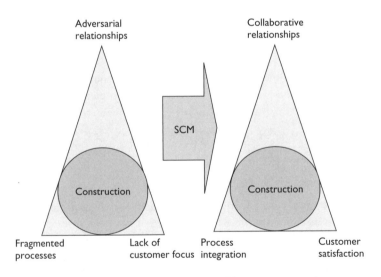

Figure 12.9 The role of SCM in addressing the key problems in construction.
Source: Jones and Saad [3].

Table 12.1 The main problems of construction [3]

Relationships	Processes	Customer focus
1 Lack of trust leading to conflict	1 Fragmented nature of the design process	1 Fragmented nature of the construction process and poorly integrated value chain
2 Onerous contract conditions and unfair loading of risk	2 Inadequate design period	
3 Lack of understanding of the risk involved and their consequences	3 Late, poor and incomplete design information lacking specialist contractors input	2 Insufficient focus on internal and external customer requirements
4 Unfair selection procedures	4 Poor overall planning with inadequate lead in time	3 Ambiguous tender packages
5 Unfair payment procedures		4 Insufficient understanding of specialist contractors requirements
6 Perceived poor status of specialist contractors	5 Fluctuations in demand for the products and services of the specialists	
7 Failure to communicate with specialist contractors and to view them as equal project partners	6 Failure to involve specialist contractors early enough in the process	5 Unclear statements of requirements and ambiguous project information and tender packages

Encouraged by improvements brought about through partnering, these pioneers of more collaborative approaches have continued to develop closer relationships with their partners and to further integrate project and supply chain processes. A number of these clients and their consultant and main contractor partners are also beginning to extend the adoption of longer term and more collaborative relationships downstream of the main contractor, to include key specialist and trade subcontractors and materials and component suppliers. Using their leverage in their supply chains, these frequent users of construction services have been able to successfully make the transition from project-specific partnering, through strategic partnering, and onto SCM as shown in Figure 12.10.

With the encouragement provided from a number of reports into the inadequacies of construction procurement in the public sector, HM Treasury has been advocating partnering and SCM has ways of delivering VFM. Some public-sector clients have responded by attempting to build the new purchaser–supplier relationships associated with SCM into their procurement of construction products and services.

Another example with more ambitious objectives and greater scope is provided by Defence Estates (DE), an agency of the UK's Ministry of Defence, who are adopting prime contracting which includes many of the key elements of partnering, TQM and SCM. Its aim is to promote collaboration through leadership, facilitation, training and incentives and replace short-term, contractually driven, project-by-project, adversarial relationships with

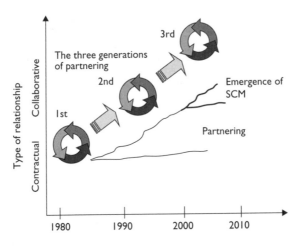

Figure 12.10 The emergence of SCM.
Source: Jones and O'Brien [44].

long term, multiple-project relationships, based on trust and co-operation. It includes the restructuring and integration of project processes and supply networks with fewer strategic supplier partners. These new relationships incorporate continuous improvement targets to reduce costs, to enhance quality and to focus on the WLC and functional performance of buildings [45].

Some clients and their advisors are adopting the tiering structure outlined earlier in this chapter. In this approach it is the responsibility of the customer tier to organize, communicate and nurture the level below. Thus, the assembler takes responsibility for the welfare of the first-tier suppliers, the first-tier that of the second-tier firms and so on down the pyramid as shown in Figure 12.5. Given the leading role played by clients, they are at the apex of the pyramid with cost and design consultants and the main contractor as their first-tier suppliers. The second-tier comprises the specialist and trade subcontractors, with their material and component suppliers forming the third-tier.

A variation on this structure is where the main contractor occupies the apex of the pyramid. This is rather like the prime contracting approach and cluster structure being adopted by Defence Estates. The cluster structure, which is illustrated in Figure 12.11, is an idea developed by the Reading Construction Forum and piloted by DE in the 'Building Down Barriers' project. In this approach a prime contractor (PC) is appointed to work with key supply partners, known as Cluster Leaders, who set the general direction of designing and delivering a significant element of the building – the ground works and substructure, the superstructure, the services and so on.

This allows the PC to improve in an integrated fashion the process for designing and delivering the overall building as well as the materials and the components that go into its main elements. It means the client has a single

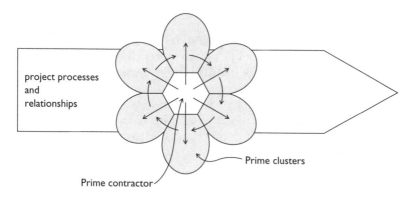

Figure 12.11 Structuring of supply systems using clusters.

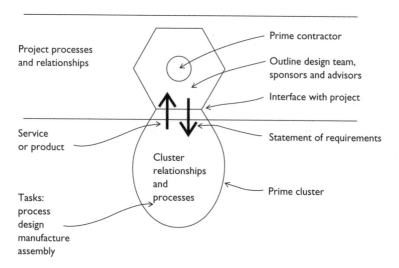

Figure 12.12 Interface between the project and a cluster.

contractual relationship with the construction team. The contract is held by the prime contractor, who is usually a main contractor, although the role can be undertaken by other members of the construction team. The team is stable and consists of designers, the prime contractor and suppliers who have a long-term contract with the client for a series of projects. This means the team's performance can be improved from project to project and communication across the interface structures developed as shown in Figure 12.12.

There is growing evidence of the effectiveness of SCM in certain parts of the industry. It appears to be most effective in the case of fairly standard buildings for regular and frequent clients of the industry as shown in Case study 12.1.

Contractor Kajima UK Engineering completed a $9250\,m^2$, £11.5m Asda superstore in Swansea under a JCT81 contract. The project, is based on Asda's model store, which is a standard design modified to suit each site. The building has a steel frame structure with a standing-seam built-up roof, composite cladding and brickwork panel walls. The floor layout is based around a central open-plan sales area. A two-storey office and restaurant block is located in front of the sales area, with a warehouse and plant area at the rear.

Key features

The Swansea project made a cost saving of 5.75% on a similar Asda store built in Gateshead 18 months earlier. This gives the store a benchmark score of 65% on the government's KPI for construction costs, which means it is 15% better than the industry average (50%) for all projects.

The store was completed in 15 weeks, 39% faster than the Gateshead store. This gives it an 85% score on the indicator for construction time – 35% better than the industry average.

Kajima used modular units to speed up construction time. Efficiency was improved by performance monitoring. Site processes were analysed to assess how much time was spent on productive and non-productive activities.

Asda had post-project workshops with design team members and subcontractors to discuss what should be carried forward from this project.

The contribution of modularisation to cutting build time

The retailer's store development manager says

> Asda was looking to make a quantum leap in the construction process... The project aim was to bring down construction time without compromising safety or increasing costs, while producing a replicable build process that could be carried forward to the next project. Asda has a standard store model that is adapted to suit each site. It also uses a set of partnering contractors and consultants to achieve year-on-year savings of 20% for store construction time and 10% in store costs.

Cutting construction time means that more activities have to be carried out at the same time. Kajima proposed several changes to

Asda's standard construction model, including using modularisation to speed up the build time and to take activities off the critical path. At the front of the store, on either side of the main entrance, are the double-storey office and restaurant. Kajima modularised these by bringing them to site as 28 fully fitted-out modules, taking them off the critical path. However, installing the units proved to be more difficult than the team had anticipated. Kajima's contracts manager says, 'It was an experiment...Ultimately, if you are looking to bring down construction time, you have to start taking work off site. The lessons learned from this will be applied to future projects.'

Modular construction was also used in other parts of the building. Kajima clad the first 3 m of the external façade using modularised, brick-faced panels rather than building brick walls in situ. As the wall was constructed in February, using wall panels took weather dependency out of the operation.

The panels also meant the mess associated with a wet trade such as brick-laying was avoided, and there was no need to have scaffolding around the outside of the building, freeing up space on site. Kajima also used modularisation for the mechanical, refrigeration and electrical plantrooms. These were supplied as packaged units, fully assembled and commissioned, and installed in a separate area at the rear of the building. By mounting the units separate to the main store, Kajima ensured that they, too, remained clear of the critical path.

How monitoring improved productivity

One area targeted for improvement by Asda was site productivity. Even before Kajima had been awarded the contract, it arranged a three day workshop with its selected subcontractors and Asda's team. This was to develop the process through which the project objectives would be achieved. These objectives included building the store in less than 20 weeks, and for 10% less than a similar store built a year earlier. Before starting on site, Kajima gave all its subcontractors an induction to explain how the teamwork culture would work and to allow them to commit to delivering the objectives. The contract manager describes this as 'creating the atmosphere to allow operatives to challenge the obvious, to try to think outside their particular box and to think of others'.

Asda used Calibre, the Building Research Establishment's site performance monitoring system, to measure who was doing what, how much time they were taking, how much their activity contributed to advancing the project and how much time was spent on non-productive activities. 'We wanted to know where we could do better', says Asda's general manager of research and development. All site operatives were given an identifying number to wear on their vest, allowing the two BRE observers to record who was doing what task, and the time it took to do them. To ensure communication between the trades, Kajima insisted on meeting in a 'huddle' every morning and evening. The idea was borrowed from Asda's management team, and was seen as a chance to sort out problems.

At Swansea, the results of the previous day's Calibre monitoring were reviewed by the contractors and subcontractors at the morning huddle. The aim was not to look at what went wrong but at how the process could be improved the next day. For example, Calibre showed that the productivity of the lining contractors had decreased. At the huddle, the problem was found to have been caused by ductwork stored on site, which was restricting the liners' access. Calibre showed that the Swansea team spent 8% more time on productive activities than the industry maximum from previously monitored sites. The lower figures at the start of the project were caused by the high number of managers on site relative to the number of operatives. Kajima's contracts manager, thinks the high productivity figure can be attributed to a change in the attitude of everyone working on the project – something that was targeted before the project was on site.

(Adapted from Pearson [46])

There are considerable difficulties in applying SCM in construction. These include short termism, lack of trust and adversarial relationships, the transient nature of construction projects and the significant number of irregular clients. The lack of initial preparedness of the organisations involved is also a significant barrier.

Given its traditional short term, contractual relationships and fragmentation, adopting a more collaborative approach is not straightforward. The integration of organisations with different cultures, power and knowledge bases can be problematic in construction given its multiple and often hidden goals; power imbalances; differences in professional language, culture and procedures; incompatible collaborative capability; and the tension between autonomy and accountability. Addressing these differences provides the basis of the benefits to be derived from SCM, but also explain the reluctance and slowness of many construction organisations to meet the challenges and complexities of working more closely together. The main barriers which need to be addressed in implementing SCM in construction are outlined in Table 12.2.

Table 12.2 The main barriers to the implementation of SCM in construction Saad et al. [1]

- The 80% of the industry's clients with their small or infrequent construction programmes – this means that only 20% of construction's clients having the commitment, knowledge and the necessary leverage to adopt network competition.
- The transient and short-term nature of construction projects, processes, teams and relationships.
- Present procurement strategies with their emphasis on contracts and price competition.
- Deeply embedded adversarial relationships and opportunistic behaviour.
- Fragmented processes.
- Lack of possible partners with the appropriate collaborative capability.
- Multiple and hidden goals.
- Major power imbalances.
- Lack of contractual and competence trust.
- Insufficient resources and time to build relationships, integrate processes and manage logistics within a one-off project environment.
- Differences in professional language, culture and procedures.
- Lack of experience of innovations such as JIT and TQM.

Conclusion

Effective SCM requires stable and long-term relationships between organisations in supply chains. Clearly this basic requirement exists in the case of the regular construction client such as British Airports Authority (BAA), ASDA Walmart and Defence Estates where effective long-term relationships have been developed. These significant and frequent customers of construction services are denoting that elements of SCM including closer long-term relationships are leading to greater transparency in transactions, trust and commitment which are seen as being central to the development of mutual competitive advantage.

However, this is very difficult for infrequent clients of construction and their one-off projects. Such clients have little opportunity or indeed motivation to stabilise and improve their construction suppliers. Centrally co-ordinating and integrating their process with project processes is difficult where the client is unwilling, or unable, to exercise this degree of leadership and co-ordination. In these circumstances, leadership and management of supply chains will need to be undertaken by other project participants such as main contractors. It can be argued that these barriers to implementing SCM in construction also show that it offers a highly relevant approach to improving its performance.

If appropriately implemented, SCM can offer a way forward for improving relationships, integrating processes and increasing customer focus. The benefits to all participants can include, improvements in quality and delivery, more repeat work suppliers, reductions in the overheads associated with obtaining work, increasing profitability and the acquisition of new specialist knowledge and skills in significant sectors of the construction market.

SCM can help construction manage the relationships and processes between the different participants by providing a more inclusive environment where construction organisations such as specialist and trade subcontractors and suppliers can be given a more participative role in order to ensure a greater co-ordination of an increasing part of the process. Both, the upstream processes to the client and the end user and downstream processes involving the specialist and trade subcontractors and their suppliers, need to be more effectively and fully involved as shown in the BAA Case study 12.2.

Case study 12.2 BAA Pavement team framework agreement

> The pavement team agreement is a partnering agreement set up over five years between AMEC, the client BAA Plc and the design consultants to deliver aircraft pavement aprons and ancillary works. It is worth a total of £130m and it has included more than 50 contracts in the five-year time period. This includes conceptual and feasibility studies, but the majority of the value is for construction works. The supply chain management consists of four primary suppliers subject to the agreement. The framework agreement is not formal and contracts are placed for each project.
>
> The project teams are collocated so that staff are seconded and the team is led by a general manager from BAA utilising BAA client skills in project management. The benefits of keeping the best teams together have met the objectives to reduce cost by 30% over 5 years and to reduce the programme time They reduced programme time by one third between the first and second jobs. The team also has a particular interest in meeting some other headline improvements which include a reduced accident frequency rate compared with the construction industry average and an enhanced productivity rate measured by an improvement in turnover per head of its employees.
>
> Recently the second-tier suppliers have accepted management roles in the team. Current challenges are to integrate the third-tier supply chain into the partnering agreement and to use skills to further blur management roles in the integrated team, which includes client and specialist managers.
>
> (Adapted from Constructing Excellence [47])

Extending a comprehensive SCM approach to the whole of the construction market will be difficult and challenging. A greater part of the industry needs to be more adept at segmenting the market, identifying and developing critical supply chain assets, being aware of market conditions and aligning all project and supply chain processes to meet the needs of the external customer.

There is strong evidence that there is a growing awareness of the concept of SCM amongst construction practitioners. However, it must be recognised that it is yet too early to undertake a comprehensive evaluation of its implementation in and its impact on construction. Also, it is clear that the present complexity, fragmentation, interdependency and uncertainty which characterise the construction market and process will influence the way in which SCM and other innovations are adopted and implemented.

In addition, construction is moving to the adoption of SCM without having benefited from earlier innovations, such as JIT and TQM. It is only, relatively recently, with the emergence of partnering, that the industry has started moving towards more collaborative relationships and integrated processes that can be interpreted as laying the foundations for SCM. Even where partnering has been adopted, it is still largely misunderstood and has not yet led to a widespread change of culture which remains essentially adversarial with arms-length relationships and a significant use of price-competitive procurement approaches and rigid contracts. In addition, partnering is mainly being adopted upstream and essentially between clients, consultants and main contractors and has yet to be extended to many of the suppliers downstream of the main contractor.

There are significant difficulties which need to be addressed if SCM is to be effectively implemented. These include the lack of preparedness of construction organisations in adopting SCM; the limited understanding of the concept and the pre-requisites associated with its implementation; the unwillingness to rationalise supplier and customer bases; the difficulty in establishing a clear common purpose, exchanging information and sharing learning. This can be interpreted as a further indication of their lack of awareness and reluctance to embrace a new culture associated with SCM relationships. Although learning is perceived as increasingly important in supporting innovation in construction, the types of learning being under-taken do not always match the competencies and the cultural changes needed for such a complex, multi-factor and dynamic innovation [3].

References

1 Saad M., Jones M. and James P. (2002) 'A review of the progress towards the adoption of supply chain management (SCM) relationships in construction'. *European Journal of Purchasing and Supply Management*, 8: 173–83.

2 Ball M. (1988) *Rebuilding Construction: Economic Change in the British Construction Industry*. Routledge, London.

3 Jones M. and Saad M. (2003) *Managing Innovation in Construction*. Thomas Telford Publishing, London.

4 Aitken J. (1998) 'Supply Chain Integration within the Context of Supplier Associations, Cranfield University, PhD Thesis'. In M. Christopher (ed.), *Logistics and Supply Chain Management*. 2nd Edition. Pearson Education, London, p. 19.

5 Slack N., Chambers S. and Johnson R. (2001) *Operations Management*. 3rd Edition. Financial Times Prentice Hall, Engelwood Cliffs, NJ.

6 Hill T. (2000) *Operations Management – Strategic Context and Managerial Analysis*. MacMillan Business, Oxford.

7 Mowery, David C. (ed.) (1988) *International Collaborative Ventures in US Manufacturing*. Ballinger, Cambridge, MA.

8 Christopher M. (1998) *Logistics and Supply Chain Management – Strategies for Reducing Cost and Improving Service*. Financial Times Prentice Hall, Engelwood Cliffs, NJ.

9 Harland C.M. (1996) 'Supply Chain Management: Relations, Chains and Networks'. *British Journal of Management*, 7(special issue): 563–80.

10 Tan K.C. (2000) 'A frame work of supply chain management literature'. *European Journal of Purchasing and Supply Management*, 7: 39–48.

11 New S. and Ramsay J. (1997) 'A critical appraisal of aspects of the lean approach'. *European Journal of Purchasing and Supply Management*, 3(2): 93–102.

12 Christopher M. and Jüttner U. (2000) 'Developing strategic partnerships in the supply chain: a practitioner perspective'. *European Journal of Purchasing and Supply Management*, 6: 117–27.

13 Vollman T., Cordon C. and Raabe H. (1997) *Supply Chain Management: Mastering Management*. Financial Times/Pitman Publishing, London. pp. 316–22.

14 Kosela L. (1999) 'Management of production construction: a theoretical view'. *Proceedings of the Seventh Annual Conference of the International Group for Lean Construction IGLC-7*. 26–28 July, Berkeley, CA. pp. 241–52.

15 Hines P. (1994) *Creating World-Class Suppliers: Unlocking Mutual Competitive Advantage*. Pitman Publishing, London.

16 Sako M., Lamming R. and Helper R.S. (1994) 'Good News–Bad News'. *European Journal of Purchasing and Supply Management*, 1(4): 237–48.

17 Lamming R. (1993) *Beyond Partnership: Strategies for Innovation and Lean Supply*. Prentice Hall, New York.

18 Womack J.P. and Jones D.T. (1996) *Lean Thinking: Banish Waste and Create Wealth in Your Corporation*. Simon and Schuster, New York.

19 Towil D.R. (1996) 'Time compression and supply chain management – a guided tour'. *Supply Chain Management*, 1(1): 15–27.

20 Lipparini A. and Sobrero M. (1994) 'The glue and the pieces: entrepreneurship and innovation in small-firms' networks'. *Journal of Business Venturing*, 9: 125–40.

21 Dubois A. and Gadde L. (2000) 'Supply strategy and network effects – purchasing behaviour in the construction industry'. *Supply Chain Management in Construction – Special Issue, European Journal of Purchasing and Supply Management*, 6: 207–15.

22 Burgess R. (1998) 'Avoiding supply chain management failure: lessons from business process re-engineering'. *International Journal of Logistics Management*, 9: 15–23.

23 Holti R. (1997) 'Adapting supply chain for construction'. Workshop Report, CPN727, Construction Productivity Network, CIRIA.

24 Edum-Fotwe F.T., Thorpe A. and McCaffer R. (2001) 'Information procurement practices of key actors in construction supply chains'. *European Journal of Purchasing and Supply Management*, 7: 155–64.

25 Ali F., Smith G. and Saker J. (1997) 'Developing buyer–supplier relationships in the automobile industry, a study of Jaguar and Nippondenson'. *European Journal of Purchasing and Supply Management*, 3(1): 33–42.

26 Sako M. (1992) *Prices, Quality and Trust: Inter-firm Relations in Britain and Japan*. Cambridge University Press, Cambridge.

27 Dasgupta P. (1988) 'Trust as a commodity'. In D. Gambetta (ed.), *Trust: Making and Breaking and Co-operative Relations*. Sage, London.

28 Farris G., Senner E. and Butterfield D. (1973) 'Trust culture and organisational behaviour'. *Industrial Relations*, 12(2) May: 144–57.

29 Rotter J.B. (1967) 'A new scale for the measurement of interpersonal trust'. *Journal of Personality*, 35(4).

30 Hofstede G. (1980) *Cultures' Consequences, International differences in work related values*. Sage, Newbury Park, London.

31 Shapiro D., Sheppard B.H. and Cheraskin L. (1992) 'Business on a handshake'. *Negotiation Journal*, 8.

32 Kanter R.M. and Myers P. (1989) 'Inter-organisational bonds and intra-organisational behaviour; how alliances and partnerships change the organisations forming them'. *Paper presented at the First Annual Meeting of the Society for the Advancement of Socio-Economics*, Cambridge, MA.

33 Good D. (1988) 'Individuals, interpersonal relationships and trust'. In D. Gambetta (ed.), *Trust: Making and Breaking Co-operative Relations*. Blackwell Publishing, Oxford. pp. 31–48.

34 Axelrod (1984) *The Evolution of Co-operation*. Penguin, London.

35 Kreps D.M. (1990) 'Corporate culture and economic theory'. In J.E.A.A.K.A. Shepsle (ed.), *Perspectives on Positive Political Economy*. Cambridge University, New York. pp. 90–143.

36 Telser L.G. (1987) A theory of efficient co-operation and competition. Cambridge University Press, Cambridge, New York and Melbourne.

37 Akintoye A., McIntosh G. and Fitzgerald E. (2000) 'A survey of supply chain collaboration and management in the UK construction industry'. *Supply Chain Management in Construction – Special Issue. European Journal of Purchasing and Supply Management*, 6: 159–68.

38 Whipple J.M. and Frankel R. (2000) 'Strategic alliance success factors'. *The Journal of Supply Chain Management*, 36(3): 21–8.

39 Spekman R.E., Kamauff J.W. Jr and Myhr N. (1998) 'An empirical investigation into supply chain management: a perspective on partnerships'. *International Journal of Physical Distribution and Logistics Management*, 28(8): 630–50.

40 Harland C.M., Lamming R.C. and Cousins P.D. (1999) 'Developing the concept of supply strategy'. *International Journal of Operations and Production Management*, 19(7) 650–73.

41 Huxham C. (ed.) (1996) *Creating Collaborative Advantage*. Sage Publications, London.

42 Cox A. and Townsend M. (1998) *Strategic Procurement in Construction: Towards Better Practice in the Management of Construction Supply Chains*. Thomas Telford Publishing, London.

43 Jones M. and Saad M. (1998) *Unlocking Specialist Potential: A More Participative Role For Specialist Contractors*. Thomas Telford Publishing, London.

44 Jones M. and O'Brien V. (2003) *Best Practice Partnering in Social Housing Development*. Thomas Telford Publishing, London.

45 Holti R., Nicolini D. and Smalley M. (1999) *Prime Contracting Handbook of Supply Chain Management Sections 1 and 2*. Tavistock Institute, London.

46 Andy Pearson (1999) *Building*. The Benchmark series, Builder Group. 9 July.

47 Adapted from Constructing Excellence. http://www.constructingexcellence.org.uk. Project Number 64 (accessed 29 September 2004).

Chapter 13

Quality and customer care

A project manager's ultimate concern is to provide customer satisfaction. The primary link to the customer is through the project manager who also needs to understand the needs of the end user. For the contractor, repeat business from a knowledgeable, satisfied client is good business, because it allows a relationship to be set up and a better understanding of the product to be gained. The contractors are the key provider of customer care.

The objectives of this chapter are:

- Evaluation of the Egan targets for quality and customer care.
- Handy's principles of customer care as a model for working with construction customers and adapt it to make it relevant in a project environment, which is a virtual organisation where many organisations are working together.
- The principles of quality planning to provide a custom-built plan that specifically meet a client's needs.
- The implementation and management of quality and care principles and their impact on traditional contractual relationships.
- The use of relational marketing in developing sustainable customer relations.

The Egan threshold

Egan [1] has identified a focus on the customer as a key driver in the improvement of the construction industry. The report believes that in the best companies,

> the customer drives everything. These companies provide exactly what the end customer needs, when the customer needs it and at a price that reflects the product's value to the customer. Activities which do not add value from the customers viewpoint are classified as waste.

The report claims that recent studies show that up to 30% of construction is rework. This provides a challenge for the fragmented supply chain in the construction industry, particularly in the separation of the design and assembly functions, to improve outputs. There is a need for the project manager to identify the customer in customer care. In many cases, it is necessary to look beyond the client to the stakeholders who have a continuing interest in the finish project.

Quality is another driver and is closely associated with customer care and can be defined as fit for purpose and the elimination of defects. In the context of customer care, however, it is also connected with best value and the cutting out of waste. Egan [1] recognises quality as a more integrated package, which is,

> not only zero defects, but right first time, delivery on time and to budget, innovating for the benefit of the client and stripping out waste...it also means after-sales care and reduced costs in use. Quality means the total package – exceeding customer expectations and providing real service.

The Latham Review [2] also recommended improving quality by the adoption of quality/value procedures in tendering selection to replace widespread lowest price practice. The review looked at ways of achieving this whilst not contravening the competitive approach required for public organisations. The Construction Client Forum (CCF) [3] went on in similar vein and supported a call for right first time and zero defects. Again seeing that, eliminating waste was an important part of increasing productivity.

One of the main Egan targets is to reduce defects by 20% year on year and to challenge the concept that it is not possible to have zero defects. The Constructing Excellence [4] benchmarks of 8–10 is a few defects and none that substantially effect the client's business. When you look at the chart for this, only 58% of the benchmark population achieved this scale of performance in 2002. The M4i demonstration projects, which were committed to run with Egan principles were scoring significantly better with 86% achieving this level of quality. Of course, the definition of these projects should put them in the top 10% of the population, but it would indicate some verification of the target.

On a broader base, the Atkins Report [5] for the EU proposes that the level of defects should be reduced together with the cost of 'non quality' and that the level of specifications should gradually be raised on buildings and infrastructure projects. They also recommend that Quality Assurance (QA) systems should be more appropriate for construction companies and that construction firms should move from QA to full quality management. It is unclear as to whether this is taking place in such a wide market place.

In this wider definition it is necessary to move away from the model of inspection to a model of quality planning which is focused on clients' needs and quality improvement. This will not work in a climate of selecting designers and contractors on the basis of lowest cost. Clients it seems are prepared to look at best value approaches as long as it is clear that the supply side are making a joint effort to affect changes in practice. To this end the CCF [3], who represented 80% of construction investment in the UK, produced a 'Pact' document in response to the launch of the Construction Best Practice Programme emerging from the work of the CIB. This pact set out areas of good practice to the construction industry to

- Present objective and appropriate advice.
- Introducing a right first time culture, finishing on time and to budget.
- Eliminating waste streamline processes and work towards continuous improvement.
- Work towards component standardisation.
- Use competent work force, kept up to date.
- Improve management of supply chains.
- Keep abreast of changing technology, innovation and investing in R&D.

As it is a pact the clients also agreed to set clearly defined objectives, benchmark their own performance, communicate better, promote teamwork and relationships based on trust, share anticipated savings gained through innovation, appraise WLC, influence statutory legislation favourably, support training, educate decision makers, not exploit purchasing power, apportion risk fairly and improve management techniques.

The European Construction Institute (ECI) has a similar system called Achieving Competitiveness Through Innovation and Value Enhancement (ACTIVE) [6] and this puts forward an eight point system for the improvement of the industry.

Handy's principles of customer care [7]

According to Brown [8], 'customer care is an attitude of mind' and the way it is carried out is foundational to the type of relationships, which have developed between the supplier and the customer. Customer care aims to close the gap between what the customer expects and what the customer actually got.

Charles Handy [7] has three principles, which he believes will transform ability to attract customers:

- 'Customers are forever' which reinforces the need to provide full customer satisfaction so that they return again and again.
- 'Customers are everywhere' encourages us to develop a culture so that the whole workforce has an acute sense of who the customer is, so that they feel accountable for the standard of service they give. Looking to the internal user of your output as your customer attains this. It moves

away from concentrating customer satisfaction in the hands of the marketing department.

- 'Customers come first' assumes that we listen to what the customer wants and then design the product to suit. This is on the basis that all businesses need customers. He argues it may drive people away if we change the language before we follow the principle.

Principle 1   Customers are forever

Attracting and keeping customers can be an illusive process if you are not sure who your customer is. A project has a number of stakeholders. The key issue in customer care is to provide a good quality of work on time and to budget. Benefits will be gained from a good track record and the possibility for tendering future work by remaining on tender lists, or preferably negotiating further work.

If the contractor is required to take on greater risk such as design and planning applications then, there is a public element to the project as to how it impacts on the community. Construction has to recognise the issues, which effect the core business of the client commissioning the construction project and not just concentrate on the efficiency of the project. In the commissioning of the Skye Bridge the joint venture had a responsibility to themselves for ensuring that they had a viable income to cover their expenses and the payments to their shareholders, but also a core responsibility to the community and the environment. They would therefore have to consider the quality of the design and the effect on the lives of the people. Cost is important, but it cannot override these public responsibilities. This meant that when they were pushed on the planning appeal more money had to be spent on a redesign.

Principle 2   The internal customer

This principle involves everyone in customer care and it introduces the concept of the internal customer. The internal customer is another member of the project team. This involves asking questions such as 'What is the standard required?' 'Am I supplying it on time?' and 'How can I contribute to the process of adding value to the service?'

The internal customer concept has several implications:

- There is accountability of each person in the project organisation for the quality of their output which means there is no where to hide.
- There is a knowledge by all of the overall requirements of the process and the final client.
- Everybody is looking for ways of improving what they do, believing that it makes a difference to the big picture.
- Everybody is an ambassador for his or her organisation.

To achieve this, 'ownership' changes in culture have to be made and comparatively junior personnel have to be given exposure and trusted with clients at their level of operation. A new respect for other members of the project team with accountability has to be set up.

One housing developer arranges three progress visits for their customers and this puts site managers directly in the front line with the buyers. Their internal customer is the marketing department and they require reliable promises for completion dates, clean and preferably segregated access to their show homes, with reduced noise and dust for those who are already moving in.

As an ambassador for their own organization, they are showing new owners around their partially completed houses and the way they do this affects how the customer thinks about the organisation. Their communication skills, the site conditions and the structural quality will make a strong impression on the customer. This in turn may change the site manager's attitude to a safe and clean environment and the promises they make about interim progress, because they are now directly accountable to the customer. Customers also notice the mistakes that are made in progress like the blocking up of a door in the wrong place.

Principle 3   Customers come first

Putting customers first means listening to what the customer wants and then trying to enhance the value that is added. They will respond to this in different ways depending upon their experience. Customers come first in construction, not by giving away contractor profits, but by:

- Including the client on the team and making sure that the user and the maintenance staff are involved in the early stages of the design and giving time for adaptation.
- Giving time to add value to the original specifications and iterate the design.
- Using feedback from other projects and from current user problems and concerns, which may be relevant for design and occupation of the new building. Lessons learnt should be incorporated into the design.
- Implementing outstanding defects and adjustments in deference to the occupier's priorities. An aftercare and training programme may be appropriate.
- Obtaining feedback from the customers to inform future work.

Putting the customer first is often given more credibility in some of the newer forms of procurement. PFI procurement allows by definition a negotiation period when a preferred bidder sits down with the client to agree what can be achieved within the available budget framework and how value may be enhanced. This offers the chance to build up relationships and to more formally value and manage the risks. The occupation stages are more integrated with the continuing facilities management role. In design and build, there is an opportunity to have a single point of contact and closer relationships between the contractor and the client which helps an understanding of the client's business. In strategic partnering, there is the opportunity to use a no blame contract, to develop collaboration with more 'transparent' pricing and to hold out incentives for sharing the benefits of innovation and ongoing improvements from project to project.

The Strategic Forum for Construction [9] has developed a maturity assessment grid, which recognises different stages of client involvement and with increasing levels of customer satisfaction. These stages are listed by the degree of their integration with the client and are differentiated by the sequence of the four main components of development.

- Historical (traditional)   Need > develop > procure and implement.
- Transitional   Need > procure > develop and implement.
- Aspirational   Procure > need > develop and implement.

These models can be mapped with different procurement systems, which essentially bring in a working relationship of a project team with the client earlier and, earlier as is indicated by the procure activity. The last one allows the greatest amount of integration because of the more fundamental integration of client and project team clarifying the need together. The Strategic Forum [10] claims there are significant financial benefits to integration as well as having more comfortable relationships with the whole supply chain and they have developed an 'integration toolkit'.

The smaller contractor is likely to operate within the competitive tendering field and new ways of integrating with the client will be over laid on model one. However many small contractors are used to providing a turnkey package for the client and have relied on close collaboration and repeat business for many years. It is still important to be able to benchmark performance so that progress may be measured.

The Client Pact [3] mentioned earlier has several concerns for improvement. Some of these aspirations are being worked out and it is now compulsory for many contractors that workers on site have a CSCS card before they can start. To encourage progress, clients are promising to share savings resulting from innovative methods, to choose contractors on the basis of WLC and to not unfairly exploit their purchasing power by squeezing the competition, but to look to form lasting relationships and to educate their decision-makers in 'clientship'.

Models of quality

In the construction industry there is more than £1b worth of basic waste per year. This is due to design faults, inadequate quality of materials and components, lack of management of the process and workmanship problems. It may also be due to the culture of the organisation.

The Chartered Institute of Suppliers' (CIPS) mission statement as communicated to the Latham Review in [11] suggested that the key objective was

> to improve the quality of the UK construction industry and reduce the average costs by at least 30% within 5 years and to introduce continuous improvement programmes.

Chapter 5 considered the importance of planning for construction quality in it and is now important to take this a stage further and look at the systems for improvement. Quality management of projects depends on the continuous improvement philosophy of the organisation. This section discusses the merits of the various quality regimes commonly used and their relationship to productivity and client satisfaction for the project.

The Atkins Report [12] suggests that there are five tools to be used to control quality in construction:

- Product standards and design codes.
- Technical compliance with regulations, standards and specifications.
- Liabilities and guarantees covering risk of defects.
- Registration and qualification of contractors and consultants.
- QA and quality management systems.

We shall deal with the last category as the others are mainly self-explanatory and are not the main concern of this section.

Quality management

In the past, quality has been achieved by inspection of the finished output (quality control). This is considered wasteful due to abortive costs and on one off projects mistakes are less easy to retrieve so should be fewer. Clients are demanding a higher standard of process management, design and manufacturing quality that will reduce defects.

There are two approaches which promote the 'right first time' philosophy:

- Quality assurance (QA).
- Total quality management (TQM).

These approaches both depend upon an early planning approach to quality in order to save on abortive costs. Quality systems needs to be set up each time a new project is to be started in order to ensure that the project quality proposed by the supplier is equal to the level expected by the client for the project.

The QA procedures are well established and require a production process to be properly defined so that checks can be made at key points in the process to ensure compliance with procedure. In theory, the cost of the quality checks is less than the saving in abortive costs of not getting it right first time. The quality system must be sound, otherwise QA will simply 'seal in' the faults. This may be worsened if there is only a periodic review of the system to make improvements. Another problem may be that few people feel ownership of the system, so senior management only gets limited feedback for improvement that is, from their viewpoint. Another problem is that a quality system has no senior management leadership – it has been deserted. For the client, third party QA is important evidence of potential, but does not guarantee achievement.

Hellard [13] argues that there is a need to co-ordinate the different QA systems operated by the different suppliers to the project and to overlay a project quality system. This may not so easily fit into the ISO 9000 series of third party accreditation which has been set up for manufacturing systems (9001 covers design, manufacture and installation, 9002 just manufacture and installation and 9003 covers inspection and testing) and thus assumes a repeatable production process with the quality of outcome depending on one organisation. This is clearly not the case in construction projects. His claims however that the process of self-assessment and client assessment of quality is the most effective line of attack, but requires a senior commitment to drive through changes in attitude.

Taguchi's model identifies loss as costs incurred and profit lost. Loss should be minimised by quality management. He identifies offline (Pre-production) and online (production) quality management. Offline management covers:

- Systems design (reflecting appropriate technology).
- Parameter design – where it is easier to design a product insensitive to manufacturing variances (e.g. brickwork) than it is to control those variances (e.g. rainscreen curtain walling which must be 100% impervious).
- Tolerance design – the degree of variance permitted in assembly.

Online (production) quality management, should aim to minimise losses due to variations and to reduce control where possible.

Quality function deployment

Design quality can be established by the quality function deployment (QFD) method. It starts by listening to the customer and assessing in detail the customers' needs and including them in detail in the design specification as relevant technical requirements. You continue by translating customer demands 'the roof will be flat and not leak' into quality requirements 'the roof will use a high performance single skin polymer membrane, which will be installed by trained operatives and tested every 10 years'. Note that the re-testing is an important part of the quality deal in the light of current technology.

Production quality is best served by preventing mistakes and minimising loss. A number of techniques can be used to anticipate problems such as statistical process control (SPC) and QA to ensure that a process is more watertight.

Culture

The corporate philosophies, traditions, values and commitment create the culture of an organisation and provide a basis for the mission statement. Theory Z is a term coined by William Ouchi, which describes an ideal culture, based mainly on the successful methods and approach used by the large Japanese companies. This culture is less bureaucratic and hierarchical and encourages responsibility for quality and its improvement at all levels of the organisation by encouraging a consensus in decision making, reached by agreement with peers and subordinates. He argues that self-direction and mutual trust lead to high levels of performance and job satisfaction. This may also lead to less paperwork and less supervision.

Applying this cultural approach to quality improvement is the basis of TQM, which seeks to involve the whole workforce thinking about quality inputs. It advocates not only 'right first time', but continuously improving the process by encouraging feedback and review.

TQM and customer focus

The basic principle of TQM is that the cost of preventing mistakes is less than the cost of correcting them and the cost of the loss potential of future sales. Amongst other things, it is spreading the marketing function to the whole workforce.

It has been well established in the industrial sector and recently construction organisations are recognising the importance of quality as an essential factor of competitiveness and progress. It is also an aid to the development of the firm.

The rationale for a quality system is that it saves money for all parties because:

- The search for quality makes it possible to reduce production costs by less abortive work due to defects.
- Quality is part of the brand image of the firm and client has more confidence in suppliers and there is less disruption due to latent defects.
- More self-regulation leads to less independent supervision and control and good quality guarantees are available at low cost.

It is acknowledged that this is an evolutionary process that works in conjunction with the other 'tools' and should be adapted to suit the smaller organisation. It provides a holistic response to the Latham [2], Atkins [5] and Egan [1] which also have a lot to say about integrating structures to remain internationally competitive.

One customer-focused approach is to provide cast-iron guarantees on the performance specification of the product. For example, Bekaert fencing [14] offer the client a guaranteed level of security for 15 years to the real estate, that is surrounded by the fence. Here, they are selling a service (security) for a given sum and not a product (fencing). Any worries the client has about the durability and continuing efficacy of a fence is taken away from them and the effort has been invested in building a relationship with Bekaert who install and maintain the fence in a suitable state with compensation for any breaches of security. This assured through their contractor certification scheme and an insurance called Beksure.

TQM is difficult to define, but its objective according to Hellard [13] is to set out a framework for action to fully satisfy customer requirements. From a survey of the literature, including Deming, Crosby, Duran and Ishikawa he also derives common principles of

- management leadership to develop a quality culture;
- improvement of processes (not just better motivation);
- a wide-ranging education programme for management and workforce;
- defect prevention rather than inspection;
- using of data and statistics tools for benchmarking;
- developing a team approach between departments and between all authority levels;
- continuous improvement.

The TQM approach, by definition, is customer focused and must identify customer needs and expectations first and set standards consistent with customer requirements. This will provide the basis for establishing quality standards and enabling and empowering middle management and the workforce to achieve it.

In construction projects, TQM is not seen as an easy approach and it is not surprising that it is resisted. Some are not convinced that they can make the most money this way, as profits are often supplemented by contractual claims. It is also clear that there are several complications that make it harder to adapt principles, that were first devised in the context of manufacturing, to the one off nature of projects. The entrenched 'them and us' attitudes fostered by the segregation of the client and the supply chain are also unhelpful.

Turvey [15] in a study of a major private contractor looked at their programme to make the company more customer-focused and identified various 'key inhibitors' to the adoption of such a policy due to the nature of the industry itself. These included the project-based nature of the building industry, which means that the focus tends to be on the building and not on the customer. The sheer number of different organisations involved who may have never worked together before, all trying to make a profit in whatever way possible, made selling the idea difficult. There was also a tendency to revert to old confrontational attitudes when a significant problem occurred. The company, however, remained optimistic and identified enablers that were remarkably similar to the principles that have been outlined for the development of a TQM policy. These were establishing customer requirements at the outset and operating an open book policy with the client, including a measure of co-location of project staff and developing partnerships with suppliers and training programmes to change the culture.

Hellard [13] mentions several more issues of difficulty in the construction project culture and puts his biggest emphasis on working together in partnership with the client and all the contractors. In this respect, the client becomes part of the 'senior management' commitment that enables a TQM approach to proceed on their project. This will initially involve a financial commitment, which should pay dividends later by adding value to the project.

Hellard [13] and Turvey [15] mention the importance of meeting the needs of your internal customer in achieving TQM. The internal customer is the next person in the work flow chain. For example, this makes the estimator the customer of the quantity surveyor and design team as they put the tender documents together. It also makes the painter the customer of the plasterer and the Mechanical and Electrical (M&E) contractor the customer of the services consultants and main contractor. Customer focus turns the traditional attitude of selling contracting or consulting services on its head and represents a major change in culture. Obligations through the work flow process such as reliable, timely information in an understandable format, should serve to cut out waste and help make the internal client's input more effective. The efficiency of the workflow will definitely benefit the external client also.

The sponsoring client represents the external project customer, but in a holistic project management approach the end user of the facility is important. The end user is the client of the project manager responsible for the operation of the various services and building systems.

Hellard [13] has a further analysis of the principles for TQM at three of the life cycle stages of project inception, project design and construction stages.

Implementing a construction project quality plan

Performance concerns the assurance that quality standards have been met and equipment works. Merna [16] indicates a hierarchy of quality policy, quality procedures and work instructions and stress the importance of allocating responsibilities to the management team and leadership. The author also suggests the periodic review of quality actions. Benchmarking service and product is possible in order to ensure that a standard has been reached and can be improved.

Customer satisfaction is centred round their confidence in the system and by the effective operation of the system and they will interfere less where there is evidence of effective controls. Zero defects should apply to the performance of the building and equipment and some expectation of minor problems can be tolerated where business productivity is not at stake as suggested in the Constructing Excellence benchmarks. The client experiences much more frustration when the project overruns time and budget expectations. More information on how this works may be found in Chapter 4.

Relationship marketing

Construction services are traditionally bought through hiring consultant time or by arranging a competitive tender, but how are they sold? Getting on to a tender list or in a position to negotiate is a full time job for a marketing department, but it is critically about building up relationships between the buyer and the seller. The price is dependent on competitive market forces or, as suggested previously, may be adjusted strategically to enhance value and WLC. In both cases, firms are compelled to build up relationships in order to win work. For the client and project manager, it is important to recognise the need to work with the supplier. Previous chapters have discussed procurement alternatives for building in best value into the price by, for example, having a two-stage tender. The business relationship is a means to choose the best supplier.

Smyth [17] has differentiated between 'selling to people' and 'selling through relationships'. The latter implies getting to know the client beforehand in order to optimise what you can offer. A client needs good reasons for considering your product and this may be achieved by a number of issues including reputation, value, quality, ability to deliver and attention and differentiation to show understanding of the client's needs. Smyth's four foundations for relationship selling are vision, expectations, authority and trust. This allows a two-dimensional approach of getting more for the money and/or enhancing quality for the same amount. Success bodes well for ongoing contractual relationships and in developing strategic partnership.

In construction, the vision and expectations are often generated in isolation between the key designer and the client and the contractor comes along with a higher price or a reduced specification. Where there is no mechanism for negotiation, trust is lost by the client. This points to the need for an earlier involvement by the contractor especially where there is less certainty and more choice of options. Smyth [17] emphasises the parallel activity of clinching the sale through the stepwise process of encourage, exhort, enlighten, engage and empower to bring the client on board. These five steps interact with the foundation principles in relationship selling with an ultimate aim by both parties to do repeat business so that, the trust relationship might be further developed. Case study 13.1 illustrates how a contractor has developed his business to focus on the client.

Case study 13.1 Client accounts

It is not unusual for contractors to develop special relationships with clients. A medium-sized contractor originally offering a service in the South West divided their services into business units which represented retail, leisure and general construction and within this client accounts were created. Organisationally, although the company kept a head office in the South, they were now servicing client buildings nationally as the focus of each unit was to build up their relationship with particular clients in the niche market and follow their orders across geographical boundaries. The staff in each unit were trained to get to know the market and the particular client and an accounts manager rather than a contract manager was responsible for allocating teams and resources who were trained to understand their needs. Clients ask for teams to stay intact and move from contract to contract, which impacted on the expectations for the staff to be mobile and changed the staff profile. This impacted on the management of the supply chain and the equal willingness for key staff to be mobile to maintain the shallow learning curve.

The client base for the specialist markets also shrunk substantially to match the capacity of the contractor to develop ongoing relationships and their potential to secure their best profit margins in a business they knew intimately. The synergy from better professional teamwork was also released.

Smyth points out the relationship between more emphasis on relationship marketing as described earlier and the more integrated approach to procurement as indicated in Figure 13.1. Interestingly there is a greater transactional cost and more complaint with relationship marketing, which means that savings must be offered in other places and longer term.

One argument against this is the risk to the supplier of a client running out of work after substantial investment in working with the client. Influential larger clients may also squeeze smaller contracting partners so that they operate on reduced margins and at the same time have created dependency

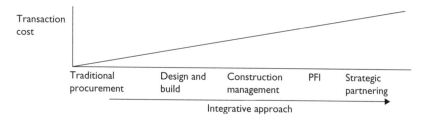

Figure 13.1 Integration of procurement versus transaction costs.
Source: Adapted from Smyth [17].

by specialising the product and demanding more exclusive supply. Relational marketing should therefore remain in the control of the supplier.

The contractor appointment is based on the procurement route chosen (Figure 13.1) and should be made at the earliest stage possible. The CIOB Code of Practice suggest that early involvement of the contractor may usefully contribute to

- resolution of buildability issues at design stage;
- choice of the most efficient materials to be used;
- practical costing, about costing issues to support VM;
- giving an opportunity to involve specialist contractors in more fundamental design issues;
- giving the contractor a better chance to work with the client and understand the client's requirements;
- inputs on health and safety issues that influence the design.

These things are relatively limited in traditional procurement and yet if incorporated by more integrative procurement methods it can produce valuable inputs that are likely to save money.

Conclusion

Customer satisfaction is an important part of the process of project management. The Constructing Excellence model divides it into essentials that are the basics of doing what you said you were going to do and differentiators, which are the things that make you stand out in the crowd. Starting from a low base in this model is not an option, as clients become more confident in their requirements and use benchmarking as a way of comparing performance on a wide variety of parameters. This means they will be looking for excellent service at a reasonable price. The culture of quality improvement will also inflate expectations as time goes by, so waiting to improve is not an option either. Clients operate their own client charter system [18], which requires them to keep up to reciprocal standards of 'clientship', which enables contractors to assess clients when choosing who can provide a sustainable profitability.

Future pathways could include the longer-term commitment for a contractor to a service rather than a product so that clients may be guaranteed a certain level of performance over a period of time and not just buy a product. This will come at the price of a proper guarantee scheme, but not necessarily by year-on-year payments.

The use of client accounts is more associated with partnering, but reducing the client base with the intention to agree a programme of repeat work with a small number of trusted clients is the chosen route for many larger contractors who have become fed up with the unpredictable and unfulfilling low margins of the competitive tendering market. Relationship marketing has played a role here in introducing the concept of 'buying in' to and engaging the clients' vision and adopting its culture in best meeting its needs. It has an element of selection for the contractor in the type of and breadth of clients any one contractor works with.

Performance measurement to facilitate benchmarking is required in order to satisfy the QA procedure and to prove quality and value improvement. However, TQM is a step further in generating a culture of improvement that pervades throughout the project at all levels of the supply chain to ensure that everyone is on board to work for quality, making the need for a third-party quality accreditation system less critical. This is a model which is not yet mature in the industry, but has claimed success for other industries such as retail, car and steel. It also requires a much more integrated project team for its ideal to be released.

References

1 Egan J. (1998) *Rethinking Construction: The Report of the Construction Task Force*. DETR July.
2 Latham M. (1994) *Constructing the Team*. Final Report of the Government Industry Review of Procurement and Contractual Arrangements in the UK Construction Industry. HMSO, London.
3 Construction Clients Forum (1998) *Constructing Improvements: The Clients Proposals for a Pact with the Industry*. CCF, London.
4 Constructing Excellence (1992) 'KPIzone, all construction KPI'. In *Strategic Forum for Construction (Egan)* (2002), Accelerating Construction. DTI, London.
5 Atkins W.S. (1994) Secteur – *Strategic Study on the Construction Sector*. Final Report. European Union.
6 European Construction Institute (2003) *Active Implementation Handbook*. A guide on how to implement the ACTIVE implementation process in your company, http://www.eci.online.org
7 Handy C. (1991) Inside Organisations: 21 Ideas for Managers. BBC publications, London.

8 Brown (1989) *Quoted in ICSA Study Text (1994)*. Professional Stage 1 Management Practice. BPP Publishing, London.

9 Strategic Forum for Construction (2004) http://www.strategicforum.org.uk/sfctoolkit2/help/maturity_model.html (accessed 31 August 2004).

10 Strategic Forum for Construction (2003) http://www.strategicforum.org.uk/sfctoolkit2/home/home.html

11 The Chartered Institute of Suppliers (2003) Final report. Productivity and Costs. As reported in appendix vi of the Latham Review 1994. p. 128.

12 Atkins W.S. (1994) Secteur – *Strategic Study on the Construction Sector*. Final, Brussels.

13 Hellard R.B. (1993) *Total Quality in Construction Projects: Achieving Profitability with Customer Satisfaction*. Thomas Telford Publishing, London.

14 Constructing Excellence (2004) Best Practice Case Study. http://www.constructingexcellence.org.uk/resourcecentre/publications/document.jsp?documentID = 116220 (accessed 1 December 2004).

15 Turvey Jane (1999) Customer Focus in the Construction Industry. Unpublished thesis report. University of the West of England.

16 Merna T. (2002) 'Project quality and management'. In Smith N.J. (ed.), *Engineering Project Management*. 2nd Edition. Blackwell Publishing, Oxford.

17 Smyth H. (2000) *Marketing and Selling Construction Services*. Blackwell Science. Oxford.

18 Construction Clients Confederation (2002) http://www.clientsuccess.org.uk (accessed August 2004).

Chapter 14

Project close and systems improvement

with Martyn Jones

The objectives of this chapter are to:

- briefly consider the coverage and management of commissioning;
- look at the requirements for successful handover and close out;
- consider the process of post project appraisal and how to assess success;
- discuss systems improvement measures that are relevant to construction;
- consider the role of lean construction.

The final stage of the project life cycle is to commission, handover and review the project. It is common for the importance of these stages to be overlooked and the benefits of reviewing the project not to be evaluated in the rush of getting on to the next project. The APM Body of Knowledge identifies a handover stage which involves the completion of the project to the satisfaction of the sponsor. It also involves introducing the user to the operation of the facility and compiling documentation that provides information on the safe, efficient and effective use of the building.

Post project evaluation includes a benefits assessment which compares performance with the original client objectives. Was the project a success? Why/why not? The end stage is one of the opportunities to look at systems improvement, but this should be an ongoing process through out the project in the spirit of a continuous improvement. It can be a fully integrated evaluation at points throughout the project.

Finishing a project off

The final stage of the project includes commissioning and handover to the client. Commissioning ensures that the building functions according to specification and that should be tested so that, all the different services and the building fabric interfaces work together.

It is a critical operation that is planned at the beginning of the project so that adequate time for full, unobstructed access to the whole building, or

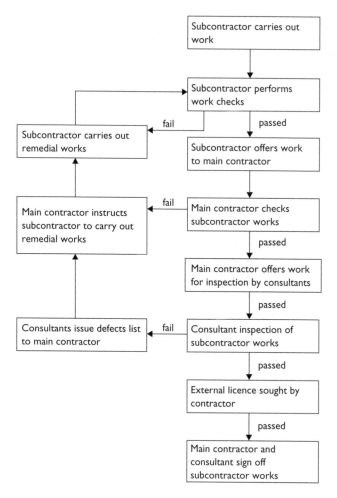

Figure 14.1 Construction inspection and commissioning.
Source: Adapted from the CIOB [1]. Used with permission.

distinct service zones is achieved. Figure 14.1 indicates the commission process. A strategy for commissioning to suit requirements is incorporated in the master plan and the construction plan from the beginning may be phased. Commissioning plays an important role in the latter part of construction and beyond into fitting out and occupation and final commissioning controls the methodology of the main programme, as it is the last activity before handover.

The major stages of *engineering* commissioning are connected with approvals by the client and the statutory authorities. The fabric needs several 'OKs' connected with the building regulations such as a fire certificate, an energy efficiency compliance and a structural stability. It also needs compliance for the connection of services and for insurance purposes. Tests could be applied for water leakage, drainage, sprinkler supply, electricity and gas supplies and should demonstrate safe and healthy operation. In addition, engineering services which supply and extract air and heat have to be checked for conformity to standards in use and across seasonal variations. Health and safety issues, apart from fire, will include contamination of the site, asbestos clearance and sensible and safe access for maintenance and repairs during operation. Under the CDM Regulations this comes as a designer responsibility. It is also co-ordinated in a document called the health and safety file (HSF). Building envelope leakage tests as required in some building regulations need to be carried out prior to handover. They may generate some significant remedial works that need to be carried out before services can properly be commissioned.

The *client commissioning* is distinguished from engineering commissioning as the stage after practical completion and handover when the facilities are prepared for occupation. The CIOB Code of Practice distinguishes three parts to the task of client accommodation works, operational commissioning and migration of the workforce. In considering a new office building, this could be the setting out of furniture and second fix IT installation, the adjustment of air conditioning, the moving in and the familiarisation of office staff, their files and personal equipment and the adjustment of heating and cooling to meet the live load of the building. The client is likely to consider this as a separate project with an in-house project manager to liaise with the main project manager.

The *handover* is the direct interface with the client, the user and the maintenance team to ensure that full information about operation and use including health and safety procedures is passed on. Documentation includes as built drawings, a range of component specifications, manuals for use and the HSF. It also needs to record warranties, contact numbers, spares availability and product codes. It should include training in the use of equipment, familiarisation with maintenance cycles and fire practices for users. The latter role is reduced where there is minimum fitting out managed by the project team such as shell and core buildings handed over to developers. Good practice is likely to involve direct contacts between the user and the managing contractor as would be the case in the selling of houses.

Contractors may also wish to get feedback from their customers, who will have a unique perspective as to how effective the processes are and whether their expectations have been met (Case study 14.1). A feedback

Case study 14.1 Stage visits on housing

One housing developer has three progress visits for the buyer connected to stage completions. The arrangement clearly states cut off points for making certain decisions, such as the choice of kitchen or wall colours, but goes further in letting the buyer see the quality of the work. This is a problem where there is poor control of quality or progress on the site, but it expresses confidence by the developer that there is nothing to hide and seeks to build up relationships. Practical completion is marked by the production of an insurer's final certificate and in return the customer agrees to a short financial and legal close period following this. Drawn out post-occupational defects of any substance result in compensation under the industry insurance scheme that is put right by the developer. In turn, the developer keeps their insurance contributions down by eliminating significant defects. A pro forma indicating the checks carried out is handed to the customer making the quality process more transparent.

The argument against this practice is that many domestic clients are likely to be one-off for a unique project in a unique location and that customers do not understand building practices. However, indirect marketing is gained by recommendations made to others. This is used in the replacement windows and kitchen markets, but less so by housing developers. A clean site is often associated with an efficient and healthy site which is also of benefit in building up a reputation, where clients are on site or near the site.

scale needs to identify specific areas of improvement, and perceptions and expectations need to be unravelled. Research indicates that customers can tolerate reasonable waiting periods for putting things right, but are less tolerant of promises that are not kept.

Documentation needs to be comprehensive, but also needs to be easy to access. It often takes up considerable amounts of time to gather the documents, especially if it is left to the end of the project, when items are less accessible. Key documents are:

- As built drawings for the fabric and structure indicating exact services routes and access points.
- Manuals for the safe and effective use of the equipment and the upkeep of the fabric.
- Test certificates and warranties that give underlying confidence in quality and functioning efficiency.

- Information that gives knowledge for future projects such as the as-built drawings, contacts and spares availability. Spares may also be handed over.
- Information contained in the HSF about hazardous materials and their safe use and dismantling.

Project systems need to be closed down so that personnel may be reassigned and responsibilities redistributed or phased out. A continuing interest in settling the final account becomes the main concern of the quantity surveyor. Senior project managers may wish to reassign finishing off activities and commissioning in the hands of their deputies. Final project reviews need to be carried out before key personnel disappear so that lessons may be learnt and passed on.

Project reviews

Traditionally projects have been reviewed according to their ability to deliver on time, budget and quality targets. These are generally termed the project objectives and refer to a closed system unaffected by external factors. Clearly construction projects are open systems and consider the impact of external factors which are beyond the control of the project team and also the impact of the project on the environment and community.

In a client's business context, the main purpose is to objectively evaluate the successes/failures of the project, to ensure that benefits are optimised in operation and improvements are carried forward. The client may well be helped by the project team in this review, but the review is initiated by the client. In a project context there may not be direct parallels between this project and the next, or the next project team may work differently. Therefore, it is important to distinguish more generic lessons and to make the report accessible and helpful to a new team. Buttrick's [2] key areas for evaluation refer to issues that may be raised in both types of review.

Business objectives	Were they met? Did they change?
Benefits	A restatement of the tangible benefits, how they will be measured in the future and by whom.
Outstanding issues	These are listed, together within the necessary action, accountability and deliverables.
Project efficiency	Actual costs and resources used are compared with the plan Indicators such as cost/m^2 or cost/bed may be compared with a benchmark.
Lessons learned	Identify areas for improvement as well as noting improvements to plan.
Close out	Note contributions, pass on responsibilities and celebrate success.

The issues ensure that outstanding accountabilities are handed over, that the project has been carried out in an appropriate manner, as well as providing accurate and objective information for future projects.

The main improvements, which can be passed on, are going to relate to process. Generic learning points may be used by the project manager to apply to any future project and specific ones to projects of the same nature, client and team. This is one of the main arguments for long term client project team partnerships.

The idea of continuous project evaluation has been developed by Timms [3]. In this model, there is a regular meeting to review the main benchmarks for the project in a spirit of continuous improvement. In the construction best practice programme benchmarking measurements are made on a continuing basis to provide control and to keep on target. These benchmarks need to be chosen for their relevance to the particular project so that what is compared is considered to be a critical indicator of success. They also need to measure parameters which are measurable, can be fed back quickly and clearly point to corrective action. Timms [3] advocates that the whole team should be involved in the evaluation meetings and that this should lead to integrated construction and design action.

Post project appraisal is more related to the review of client objectives and continuous evaluation to the control of the project. CIOB [1] endorses post completion review as a basis for the assessment of fitness of purpose and satisfaction of requirements, but they also suggest that certain lessons can be carried forward by measuring performance by assessing the strengths and weaknesses of the project and to encourage appropriate action in the future.

Systems improvement

Continuous improvement

Continuous improvement or 'kaizen' consists of measuring key quality and other process indicators in all areas, and taking actions, normally small and incremental to improve them. It focuses on *processes* and should be pursued in all areas. Maylor [4] distinguishes improvement between 'learning by doing' as those elements which can be learned from previous experience and 'learning before doing', using results from external inputs when we break new ground or experiment or develop other people's ideas and may have an element of training or research and development.

Continuous improvement is a fundamental element of TQM. It is seen as a very challenging approach because it is based on developing a learning culture with the aim of improving continuously through an endless search for excellence. It is a bottom-up approach in that workers are the recognised experts because they have the detailed knowledge of how the work is done, thus they are the best equipped to improve the process. This view contrasts with the Tayloristic approach in which the experts were the engineers, managers and supervisors and an operator was a skilled 'pair of hands'.

The top management must initiate and support and co-ordinate continuous improvement, as it is a democratic process. The approach is built around teams who are responsible for the individual operations they perform and for improving the process. The role of the supervisors and managers changes to that of team coaches, making sure that the teams have the resources they need to fulfil their missions to improve performance. In construction, project teams are in the learning before doing category in the sense that the projects may be working to unique designs, unique locations and with unique clients, but are 'learning by doing' category, in the sense that a lot of construction applies traditional solutions and familiar technology to provide a tried and tested solution.

Constructing Excellence [5] is a UK programme to promote construction excellence by creating continuous improvement through the exchange of best practice. It does this by organising events, usually locally, to exchange information on best practice in workshops and seminars and by the use of demonstration projects. Demonstrations embrace particular aspects of sustainability, an integrated team approach and promoting respect for people. The demonstrations also look at business improvement in the organisation as a whole. In addition there is a structure by which to drive change more widely by raising the awareness of the benefits of cultural changes as seen in leading edge companies. This leads to the three objectives of improving performance, improving industry image and engagement to take action. The Constructing Excellence website is an example of a knowledge portal to support best practice, but as a sharing resource it is also updated by feedback from best practice.

There are several tools that have been developed in order to achieve better productivity. These are continuous improvement, lean construction and benchmarking. Each of these presents challenging changes in culture that requires a sustained effort. In the context of productivity long lasting and not short-term improvements are needed. Improvements will be lasting if they continue to respect and motivate the workforce whose co-operation is needed at all points.

Case study 14.2 indicates the important role that the client has in facilitating change in the construction industry. Many of the improvements given earlier need at least tacit understanding by the client for implementation. Construction companies then also need to work with the client to ensure maximum utility.

Case study 14.2 Fusion [6]

Fusion stands for six values which are considered critical in developing a new culture of openness. These are Fairness, Unity, Seamless, Initiative, Openness and No blame. It is a system that has been used by a large pharmaceutical client to carry out refurbishment projects of the size £12–20m. They developed the system to turn around the traditional procurement system from client need > develop > procure > implement to

Client need > procure > develop > implement.

This allows all partners to come on board early, to take part in the development process and contribute to systems improvement by giving them equal status. The system starts with a 'stock take' which identifies strengths and weaknesses in each organisation for the development of each of the seven values and commits that organisation to be a contributor to the FUSION culture. For the members of the construction team this means a commitment to non-confrontation methods and an open books approach to accounts. The projects where this has been applied have ended up more innovative, more flexible and have met the time cost and quality targets very easily. The essence of the culture has been driven by the client who has only engaged partners who are prepared to use this approach. This has meant that single project partnering has had more major savings than usual.

Lean construction

Lean construction is a business improvement approach, back to our theme of cutting back on wasteful procedures and increasing value. The short-term view of squeezing the profit out of contractors is not a sustainable approach. Lean thinking as explained by Womack *et al.* [7] is a system to 'change the world' based on simple but challenging ideas to help convert waste into value. Its three main pillars are:

- management of processes and integrated logistics flow;
- management of relationships with employees, teams and suppliers;
- management of the process of change.

Waste is classified as wasted time, materials, labour, capacity and management effort. A proper analysis of the process must take place so that existing prejudices are not compounded and real problems remain uncovered. This leads to a new culture of co-operation and integration of the different processes in the business, with better planning and more ambitious objectives. Lean principles were first applied to the Japanese manufacturing industry and resulted in startling improvements in production methods by better supply chain management, reduction of defects by getting things right first time and by introducing better delivery methods like 'JIT', which reduced multiple handling. These were obvious areas of waste that were identified by looking at better ways of 'doing production'. The 'kaizen' principle must continue by looking at further ways of fitting people to the right job as suggested in the second of the three pillars.

Other ways show up as a more informed assessment is made. According to Towil [8] there is dramatic evidence of real savings in time spans by using what he calls business systems engineering. He recommends that a 'quick hit' should be made by focusing on the basis of the 'Pareto' principle (20% of the problems causes 80% of the waste). The solutions must be implemented and monitored properly otherwise regression to the old level is likely.

Partnering is a key aspect in systems improvement as it presents the opportunity for innovation, extended relationships over a series of projects and allows a culture of improvement to develop within a fixed team. It is not however the only method by which improvements may be made as can be seen in Case study 14.3, which looks at the profile of a main contractor that has developed its own culture and has given its committed business support working with a lot of different clients.

Case study 14.3 Integrating business improvements [9]

> The corporate Group employs 3000 staff in 40 offices around the UK. It has grown over the past 10 years through expansion and acquisition.
>
> > If you look within the sector you are always going to follow. You need to look outside to lead.
> > (Business Improvement Director)
>
> It is useful to look at the profile of this company who have developed systems improvement over a period of more than 10 years. This profile represents a continuous improvement process that has been integrated to build upon each of the 11 initiatives that have been incorporated.
>
> Prompt project completion was an initiative to meet client concerns and finish projects on time and to gain the financial benefits for the company of timely completion.
>
> Process re-engineering was a mapping exercise to capture the flow of activities and to improve the process where there were problems. It began to look at increasing VFM for the client by introducing

initiatives such as supply chain management and customer service. Task forces were set up with a director to champion the cause and to put forward proposals that might be adopted by the whole group.

This resulted in a need to develop a partnering culture and they used workshops in order to develop ideas and to streamline their supply chain, which they were able to sell to their clients.

Customer service involved bringing in an outside consultant to tune up their customer approach. This resulted in coaching for 600 members of staff so that customer care could be extended in depth across its projects and culture.

By now, they were ready to look at efficiency again and consider ways in which lean construction might apply to them and to introduce efficiencies to reduce waste, including integrating their management processes. Innovative ideas by individual staff, to help reach Egan targets, were rewarded. A series of further workshops were facilitated to help move into the new ideas and to fly the flag for culture change.

This has lead to a major investment in IT to introduce more collaborative software across the group and this has been backed up by extensive training.

The key issues in making progress from a traditional approach contractor to an integrated supply, customer focused one has been a sustained management commitment which has involved the staff at all levels, through the workshops and the task force. These workshops also help build the culture widely. The development of initiatives has been done by task forces represented by all parts of the group and company performance has been benchmarked in these areas. As a result, the Group has substantially increased its turnover and profit margins in 5 years. It has also won a number of industry awards and increased repeat business to 85%.

The Company have been able to make a change of culture because it has involved its staff and made it a 'top down' commitment and a 'bottom up' change to meet the challenges of new client expectations, whilst becoming more competitive.

(Used with permission)

Benchmarking outside the industry

Benchmarking is an approach which aims to create and sustain excellence by rigorously examining and comparing business and technical processes with best practice. The Royal Academy of Engineering [10] suggest that in spite of evidence of best practice in the construction industry there is a long tail of ineptitude against which improvements can be made by adapting best practice in other industries.

The more successful companies have always compared themselves with their competitors and other companies in an informal way. In the last decade benchmarking has become more formal. Cordon [11] sets out his steps benchmarking with a *specific* company. These can apply for a company outside the industry, but caution is needed in the choice of comparable metrics:

- select the process to improve;
- select the team;
- determine the 'as-is', how the process is currently undertaken;
- develop a set of key measures or metrics;
- choose an organisation to benchmark with – increasingly, firms are seeing the advantages to be gained from benchmarking with a company from outside their industry;
- obtain agreement for collaboration which is often not as difficult as it sounds because it should provide useful information for the benchmarked company;
- identify and understand the processes in the benchmarked company through visits, interviews and questionnaires;
- analyse the results to identify the differences in processes and establish best practice;
- develop an action plan to implement the new processes;
- disseminate the results to both participating organisations.

Cordon [11] claims that one of the main advantages of benchmarking is that it can help to overcome the natural inclination to believe that major changes and improvements in processes are not possible. It also helps to ensure that improvement targets are set sufficiently high. A further advantage is that it contributes to the creation of an outward-looking and learning culture in the company.

But again, there are problems, although the risks associated with failure are low. Some of the problems associated with this type of benchmarking include:

- what to measure and the metrics to focus on;
- the natural reluctance to fully embrace an idea from somewhere else;
- the tendency to over-emphasise the uniqueness of the organisation and the market environment.

Conclusion

The commissioning stages of a project should be planned first so that sufficient time is allocated throughout the project for checking functionality and putting defects right. The traditional culture has looked to others to provide a checking service which comes in the form of a list of snags or defects, for which there is an intensive programme of remedial works. Snagging and re-testing in front of a third party is wasteful as work is being carried out twice. Pre-inspections should be carried out in the spirit of 'right first time' for finished work areas. It is also clear that a known test should pass first time as a pre-test should be done to identify any malfunctioning, which is then already corrected. Continuous control of the quality standards is required to ensure that defects are not compounded by successive trades.

The process of *benefits management* allows the client to review the achievement of their objectives in different aspects like budget, programme, relationships and team building, health and safety, life cycle running costs and user reactions. This may take the form of an audit trail that takes each goal and looks at the outcomes from the new project. The project manager is unlikely to be involved in this process unless they have responsibility for a programme of client projects. Lessons learnt may guide management of the new building and feed forward into other projects and guide the valuation and allocation of risk, if facilities managers and users are consulted in ongoing developments. A client needs to assess the overall VFM, which may need an assessment of economic profit as well as financial profit derived from the project. For a new factory building this takes into account productivity gains made from more efficient layouts, better staff motivation and more integration.

Systems improvement is connected to the culture of continuous improvement or 'kaizen' and is proactive rather than reactive to iron out past problems and to reduce system waste still further the next time around. It does not assume blame, but looks to learn from mistakes. It will also depend on using a system of measurement so that the improvement is based on known performance. The measurement system is hard work and needs to be selective to suit the key parameters of the project in hand, otherwise it increases the paper chase without the benefit of identifying causes of poor performance. It also needs to feed forward information during the same project if possible. This might be achieved by having a series of evaluation meetings [3] at key stages of the project so that lessons learnt may be used to adjust systems immediately or passed on to down the supply chain. The no-blame culture means that mistakes are admitted, conflict is resolved as it arises and problems are treated as challenges. More traditional post project evaluation has often been neglected because the project team breaks up before the end of the project and it is difficult to communicate lessons between projects unless the project team follows on.

Integration is being promoted as a way of setting up project teams that are often termed 'virtual organisations'. The rationalisation for such teams is to cut out the waste caused by poor interaction between organisations and systems. At the heart of such a system is a determination to build a common culture and expectation together with an integrated communication system that allows an instantaneous distribution of information through a dedicated website or similar. The IT Communications systems are expensive as a large number of organisations in the supply chain need to commit to standardising protocols and security firewalls may cause problems. Information overload requires that distribution is focused to relevant parties only. The common culture is equally challenging as team objectives override individual ones. This requires the added principle of trust and openness between contractors, consultants and clients. This theme is taken up in the final chapter.

References

1 CIOB (2002) *Code of Practice for Project Management for Construction and Development*. 3rd Edition. Blackwell Publishing, Oxford.
2 Buttrick R. (1997) *The Project Workout*. Financial Times/Pitman Publishing, London.
3 Timms S. (2003) Project Evaluation. Unpublished dissertation. University of the West of England.
4 Maylor H. (2003) *Project Management*. 4th Edition. Pearson Education, Edinburgh.
5 Constructing Excellence (2004) Best Practice Knowledge. http://www.constructingexcellence.org.uk/bpknowledge/default.jsp?level=0 (accessed 27 September 2004).
6 Fusion (2004) Team Stocktake. http://www.fusion-approach.com (accessed 27 November 2004).
7 Womack J.P., Jones D.J. and Roos D. (1990) *The Machine that Changed the World*. Rawson Associates, New York.
8 Towil D. (1998) Business System Engineering: a way forward for improving construction productivity. Construction Manager. April 1998. pp. 18–20. CIOB.
9 Construction Best Practice Programme (2001) Best Practice Profile. http://www.constructingexcellence.org.uk/pdf/profiles/mansell.pdf (accessed December 2004).
10 George B.V. (1996) *A Statement on the Construction Industry*. Royal Academy of Engineering, London.
11 Cordon C. (1997) 'Ways to improve the company', in T. Dickson and G. Bickerstaffe (eds), *Mastering Management*, Financial Times/Pitman Publishing, London.

Chapter 15

Towards a more integrated approach

> To form into one whole; to make entire; to complete; to renew; to restore; to perfect. That conquest rounded and integrated the glorious empire.
>
> (Thomas De Quincey 1795–1842)

The final chapter of this book is an attempt to pull together the varied strands of project management which have been portrayed and integrate them as a whole. Project management is often portrayed as a set of tools and it is important to see that different actions need to be related in the *whole* process of delivering a successful project to a client. There is a need to develop an integrated team, because there are so many fragmentary processes in the construction industry.

There are five key themes that have emerged throughout the book that are described in terms of the project manger's job:

- meeting a client's requirements;
- creating a sustainable, integrated strategy to optimise risk and value;
- organising flexible structures for continuous improvement in the whole process;
- engineering an integrated project team and an effective leadership;
- planning and controlling performance.

Each of the chapters has considered at least one of these themes and they are important to the overall delivery of the project. The strand of these themes can be bought together by looking at the umbrella theme of integration.

The principle of integration is the bringing together of disparate parties into a holistic approach to the delivery of a project. The first half, 'inter', has the sense of mutual, reciprocal, between and among. The second half has the opposite intonation, conjuring up pictures of friction, being too close together and rubbing up people the wrong way, sounding a note of caution for the principle relating to words such as interfere, interject and interrupt. Integration in the project context is not a toolkit, but a whole philosophy for bringing people together in more productive ways to bring more value to the project.

The opposite to integration, differentiation [1], is not a bad thing and has come about as a cyclical phase in construction when projects have been subject to specialisation and project activities have been divided up to cope with the increasing complexity of projects, because of the range and sophistication of technologies expected and the wider division of labour. Walker [2] describes the need for interdependence where lots of specialist contributors from independent organisations have interfaces where their work overlaps and depends upon the work of others. This occurs through each stage of the life cycle:

- during project definition (client and project team members);
- during design (design team members);
- during construction (downstream supply chain).

He gives an ascending hierarchy of pooled, sequential and reciprocal interdependency. The hierarchy is defined by the degree of interaction and overlap of the interdependent tasks. An independent and unco-ordinated approach is not appropriate in the reciprocal interdependency of the construction process, yet construction is one of the most fragmented sectors to call itself an industry. In particular, the divide between construction and design is almost unique in the delivery of a product. This is further compounded by the difficulty in identifying a single customer for the finished product – developers, owners, users, taxpayers and shareholders are all likely to input their brief as stakeholders.

A further concept Walker introduces is sentient groups, a term given relevant context by identifying the strong commitment of individuals in the industry to commit themselves to professional or to disciplinary groups [3]. This sentience within groups triggers strong loyalties and causes conflict at the task interface between two sentient groups, such as architect and contractor. The group norms and educational traditions create differences in problem-solving approaches and priorities. These need to be resolved by finding common ground, but can create conflict. Known contractual conditions help commonality but they often fall short. Here, an integrated approach has so much importance because it involves the application of leadership. Integrated leadership does not negate specialisation, but looks to develop creative means of getting people to work together with maximum synergy to intertwine processes that help to cut out waste. It could more accurately be described as managing a change process. The preceding chapters have indicated current and best practice, but have also discussed the emerging processes and the ongoing nature of integration. At this stage, it is worth looking at the case study back ground of an integrated project (Case study 15.1).

Integration

Prime contracting procurement provides an output specification to the PC who mobilises the design and construction team to deliver the estate. The principles of prime contracting are to establish an integrated project team which levers in the direct involvement of the supply chain in strategic decision making and to ensure a high standard of management by the PC to create conditions that respect the workforce. Some of the mechanisms suggested by DE for achieving long-term VFM are incentivised payments, improved supply chain management, continuous improvement and partnering [4]. Other principles that contractors take on board are a three-stage selection process including negotiation, a single point of contact, clear allocation of risk, through life costing, open book accounting, prompt payment of suppliers and the early appointment of a disputes' review board to provide timely interventions to manage disputes. The contract has standard core provisions and will be tailored to suit specific project circumstances. It is connected to the Ministry of Defence programme called Smart Procurement of which Building Down Barriers has been a successful pilot run for its application to construction.

Case study 15.1 Andover North Site Redevelopment (ANSR)

Figure 15.1 The ANSR external view and inside of office facility [5].

Andover North Site Redevelopment (Figure 15.1 illustrates the completed building) is one of the first tranche of *capital* prime contracts rolled out by DE, post 2000, to improve performance in DE procurement as an alternative to PFI.

The site is the HQ for defence logistics on a brown field site and consists of the development of a technical building, offices for 780 people, a separate training facility, a crèche, sports facilities, a mess with accommodation, a gatehouse and 34 acres of landscaping. It is worth £40m and was started in January 2000 and it was completed in October 2002 with an additional 6-year maintenance (compliance) period. For details see the Appendix [5].

Procurement

ANSR is procured on a prime contract and this involves a core team including the key designers and the three cluster leaders co-ordinating the supply chain. The core team tendered as a joint consortium and were chosen on a balanced scorecard basis and were chosen on a three-stage selection process. Factors taken into account when selecting a PC are project management ability, technical competence, financial standing, supply chain management, willingness to share risk and understanding of the MOD culture [4].

In practice, integration and differentiation take place on a project simultaneously. Simplistically different skills and roles are differentiated in the Organisational Breakdown Structure (OBS) and responsibilities are assigned to associated task completion through the WBS. At the same time those tasks are integrated by the sequencing of the tasks in a programme or a network so that a logical relationship is agreed and programme slots are allocated. If there is some common working on the task so that one task affects another in terms of quality or health and safety then more detailed integration is required. As the process gets more complicated it may be necessary to integrate construction methodology and design and this is the rationale for procurement systems that move away from the traditional approach and bring in the contractor at an early stage. The capital cost and the operating costs are integrated to assess the balance for maximum value. This is the rationale for WLC so that the alternative design options may be properly evaluated and decisions made properly. It will also affect the selection process for contractors as the lowest tender for capital cost may not be the best value. Emerging out of this is the integration of construction and facilities management and the letting of a DBFO contract on a two-stage tender so that a competitive first stage leads to a negotiation stage where value is squeezed out on the basis of sharing the savings.

Leadership and team work

Further integration is a factor of managing people and getting the team performing by getting some synergy. This requires effective leadership, motivation and team-building exercises that make the team aware of its strengths and weaknesses and the generic roles that are required. Integration may also mean resolving conflict and having mechanisms in place that encourage 'win–win' situations and using principled negotiation techniques that support this philosophy. Trust has been raised in the context of supply chain management and collaborative methods, such as partnering, recognise the need to develop deeper, preferably long term relationships, that allow straight talking and cut waste by not attaching blame, but concentrate on trying to sort out the problem. There are several areas for building trust such as the use of joint ventures, transparent accounting and the abolition of retention. In the case study, the PC and core team agreed to go with a fixed margin for all core members and established themselves as a virtual joint venture.

Leadership flourishes in the integrated approach because it encourages a wider range of leadership taking away the bottleneck of channelling all outputs through one person. It requires a clear sharing out of responsibility amongst cluster leaders. In the case study, a core team of cluster leaders have the democratic right to challenge the PC who holds the largest share, but has less than a controlling 50% share of the combined vote (see Figure 15.3). Empowerment is a buzz word, but allows a widening of ownership and commitment. Figure 15.2 shows the team structure.

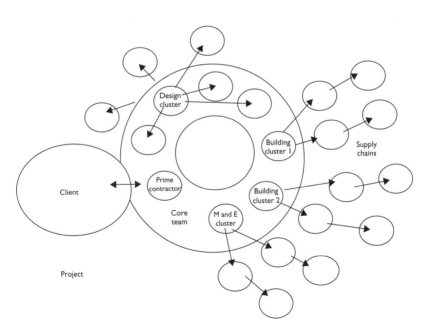

Figure 15.2 Diagram showing leadership through clusters and supply chain.

A team may also choose to develop generic roles on the basis of natural tendencies such as the Belbin or Myers Briggs assessments to profile the team better, but also to make provision for gaps and to understand the way others work either in meetings or in problem-solving situations when inputs are needed. A recognition of ways to increase synergy within the team can be carried out by using the team to create its own motivation.

Integrated decision making is about getting inputs from the whole team and assigning the final decision to the team members who are accountable for the outcomes. This provides mutual support and a no blame culture encourages more experimentation without taking away accountability. One model indicates the importance of generating ideas, evaluating them from a team and a task point of view and then identifying clearly the way forward based upon the evidence.

Risk and value

The early involvement of contractors in the design team, to ensure that there is an input on constructability at the early stages of the project, allows a wider and more integrated consideration of risk and value from all parties that are going to be expected to carry risk. It also allows contractors to gain a better understanding of the values and objectives of the client in managing those risks appropriately. Contractors under traditional forms of procurement are excluded until construction, which is too late for them to make any significant contribution to design, which could increase design effectiveness without abortive costs. There is no incentive at this stage to introduce new products and methodology which may induce VFM. Worse still, contractors may identify health and safety problems with the assumed methodology, reducing value by introducing late changes to the design and abortive costs. In the case study, Figure 15.3 shows how the power was shared among the cluster leaders and the risk apportioned and passed down the supply chain. The percentages shown indicate the derivative weightings of risk sharing and profit sharing.

Innovation

Innovation emerges out of an integrated approach, because it requires a collaborative no-blame approach and the allowance for experimentation. Time is also required because innovation is iterative like design and needs time for research, reflection and testing. Innovation usually starts off as a creative brainstorming process around a particular idea, so the focus of innovation becomes wider. The next step is to narrow the ideas down by a process of screening. This might happen in a VM workshop and the ideas can be narrowed down by RM. At this stage, an integrated team can decide to give space for rapid prototyping as Maylor [6] calls it. An example is

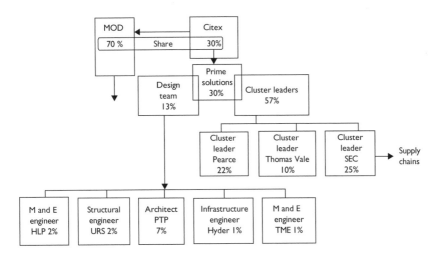

Figure 15.3 Pain gain sharing amongst core team [5].

given in the case study when it was agreed to join the training building to the office block making a key saving in capital costs by reducing external cladding and in FM, by sharing registration and staffing levels. In the case study incentives for innovation emerged from the pain:gain/loss:profit which the client agreed with the PC (Figure 15.3), so that if savings were made then all benefited from a proportional share of the gain agreed at the beginning to represent the degree of risk which each organisation was prepared to take.

The use of an innovation fund which compels management to develop all reasonable ideas received from the workforce through a suggestions box could improve the rate of ideas. These ideas could be patented to the workers and any savings over and above the costs of development shared. This encourages the ownership of ideas and the harnessing of knowledge and creativity in the project.

A different type of innovation is connected with adjusting the project process. An example of this would be instituting weekly payments instead of monthly payments for suppliers to enhance their cash flows, in return for a reduction in price. The idea has the potential to get the subcontractor to give the project priority treatment (less delays), as it improves supplier motivation. It could also improve productivity. It has a cost because the cash flow of the main contractor will be worse, but the contractor is already motivated. An evaluation is needed to see if the productivity improvements and the reduction in price together more than cover than cost. Safeguards need to be in place by monitoring the sustainability of the productivity improvement assumed. Overall it has a good chance, because it is a 'win–win' with a likely profit increase for the contract, a cash flow worry off the mind of the supply chain and better relations. It has a chance of failure if extra quantity surveyors are hauled in, eroding the profit gain through overheads and exasperating relationships by paring supplier profits in the cash back.

In the case study prefabrication workshops were set up on the site for early preparation of mechanical and electrical elements. This was complemented by the use of a building services wholesaler on the site who carried out all the purchasing on behalf of the contractor in return for being the supplier.

Information systems

An integrated information system requires the building of a system which delivers information in a timely and user-friendly way. Information can be delivered to people in different formats to suit each user, or it may be delivered in a common format that all users understand and can receive when and if they need it both to focus their attention and where necessary to understand the 'whole' picture. The latter is an integrated approach, and on larger projects it may well mean the development of a common electronic format with proper access to information portals. Consistency is more easily achieved with a single format. Too much information can be just as confusing as information that is non-existent or incomplete. Late information is disruptive and if response to queries is delayed this is also disruptive. What are the priorities and who decides? Information that is delivered instantaneously can help to reduce late or partial access to information, but can also hide priorities. An integrated system needs to be developed to the extent that it is economic for the benefits gained and will work better if there is face to face communication amongst the team.

Meetings need to be an interactive forum and so should be used to gather information and make joint decisions. Other alternatives such as teleconferences – phone or video may also save time. Co-location is a continuous opportunity for feedback and quick decisions, but may be expensive to organise, so virtual co-location (electronic alternatives), or shorter regular periods of co-location may be an alternative for less formal interaction and teambuilding. In the case study virtual collocation was used by the development of an extranet, which provided full information to parties and was properly managed to prioritise distribution and alert the most relevant parties.

Conflict resolution

Integrated leadership has been discussed as spreading responsibility productively, but integration is required between the leadership of different organisations so that the responsibility for key decisions such as programme, cost and quality control are integrated as a whole for the overall project objectives to avoid conflict. This involves decisions in terms of access to work areas, shared lifting and scaffold and the decision to expend money, for example, scaffold sheeting, to afford weather proofing for several organisations. The quality of one organisation impinges upon the quality, profits and programming of subsequent work. Strategic leadership is active and is needed to plan and predict conflict and make strategic decisions for its mitigation and to exploit opportunities for combined solutions to problems. This has been discussed as feed forward control for conflict management. In the case study, prime contracting makes provision for a conflict panel that has a standing brief to mediate as a third party, but could also have a monitoring role in anticipating non-productive conflict and providing an additional eye for the project manager. The case study also indicates the existence of a peer review team consisting of the PC and directors not directly involved in the project, who examine critically the systems and progress as a way of auditing and giving early warning where there are conflict factors, which threaten the commercial integrity of the project and therefore by implication the success of the project as a whole. The client was invited onto this review team, but declined in order for it to carry out, without inhibition, a necessary process which was seen to protect their own profits.

Integrated decisions

The stage-gate system is a way of integrating the client approvals with the tasks to be carried out by the project manager through the life cycle of the project. The OGC (formerly Treasury procurement) Gateway Review™ shows five gateways for decisions from inception through to completion for application to major project management. These provide a framework which can be related to the RIBA plan of work. This may be harmful to the project moral if these gateways are associated with bottlenecks or stop points in the project and this requires committed leadership on both sides to identify the potential problems and to ensure approval requirements, which are intimately known and planned so that project control and momentum are not rested from the project manager by poor submissions or delayed approvals. For example, planning permission is an unknown final committee decision, but the planning officer and others may be integrated into the team and should be used to reduce uncertainty to ensure better local knowledge and hence submissions.

Integrating personal and project objectives

The respect for people movement is an example of a more integrated approach between the employee and the potential employees and organisations in the industry. More and more main contractors and consultants facilitate production rather than do it. The CBPP report [7] reviewed this initiative, but it is only one of several methods for identifying the wider picture of integrating people into the business and having access to the best talent available for the project. Various initiatives including female recruitment, diversity, equality, better site conditions, safety, health and CSCS cards and career development and life-long learning have been introduced, but there are no clear indications of improvements. The initiatives need to be integrated together to harness new ideas.

The EFQM model is a more general model for excellence used by 800 organisations across Europe mostly as a self-assessment device. It puts a great emphasis on people issues and awards prizes for good performances overall, integrating the context of people management and people satisfaction with the business reality and proving that good business is looking after your workforce.

In the case study, it was a contract requirement to do a 360° peer review, which was carried out by two core team directors for each of the core team managers, including one from your own organisation. These were to be found to be more hard headed than the third-party coaching approach and gave opportunities for the directors to tour the site and access documents so that personal success in delivering the project was directly under the spotlight. This retrieved a direct connection between the performance of the individual and their ability to deliver on the job as well as bringing in a 360° view from an outsider.

Integrated time and cost control

Programme scheduling and control is carried out by planners who have limited communications with the cost managers responsible for budget control and cash flow and the design team. Control measures for time and cost need to be integrated with the production of the supply chain, otherwise they may work against each other. Earned value is often cited as providing a link with the time schedule prediction by comparing the budget spent with the budget that should have been spent at time now and predicting the slippage or advance of the programme. This link is tenuous and assumes that the cost of an activity and the time it takes and its criticality are directly related. This is patently not true and although it may give some approximate guide especially if the cost of materials is stripped out of the cost, more is required to integrate the two. Object oriented systems are beginning to be rolled out where basic cost estimates are being linked to design components (objects) which are related to the way that buildings are constructed so that time allocation may be given when the resources are allocated or normal. For example, it would be possible to use this as a powerful 'what if' tool for VM as well as in change management in assessing the sensitivity of different designs and methodologies with an immediate view of its affect on cost. After letting the contract, market prices could be fed into the software to give early warning and it would aid the use of instantaneous control systems for change control such as the NEC system, for compensation events is an attempt to put this into practice. With or without the presence of such a tool integration of corrective action and the reason for problems should be integrated with the team to meet project objectives on a regular basis. A project process map may be helpful to longer-term projects and strategic partnering or measured-term contracts.

Design quality

Integration of the project design with the urban planning process (Chapter 9) requires more attention to the wider impact of the development on the built environment. These issues have led to the setting up of organisations like the CABE. Aspirations are to better assess this impact such as is currently carried out on large projects in Europe in doing an Environmental Impact Analysis (EIA). However, the emphasis should be to lever in greater value for the community by the provision of attractive and inspirational structures, that not only provide employment for the community, but may supplement the enhancement of other social solutions such as crime reduction (security by design), a sense of place by the use of attractive spaces and landscaping and building up collective community pride as a spin off in its functional provision. These may be imposed legally by the imposing planning agreements' permissions, but perhaps they are better delivered by seeing them as challenge for added value that maintains the client budget constraints, but also enhances their reputation locally. This is not new to an enlightened client, but has generally been associated with a large budget – there is evidence to suggest that this is not the case. The case for a considerate contractor has already been argued for increased partnership with the public during the construction period. Schemes are already in position for this and have often led to other benefits for the contractor and better acceptance of building schemes in general.

Culture

The culture is an integrating tool for bringing people together with the same objective and needs to be led by the client or the project manager. The core project team of consultants and key contractors needs an open, no-blame culture to build confidence to admit mistakes and to take forward a solution to put it right as a team. This culture of openness can be extended to the client so that trust is built in to the contract and the opening up of client partnering withvalue savings built in to the agreed budget to reduce the cost and incentives to share the gain.

Figure 15.4 The pact that was signed by the ANSR team [5].

The closer culture in the case study meant a partnership charter (Figure 15.4) was established to unify objectives. This meant that the project could operate a single bank account for the core team members so that profit margins can be standardised for all and payments made according to the proportionate value of work complete, a single PII liability was negotiated to cover professional liability of all parties. This applied peer pressure on team members to perform well and to give each other a hand to perform well if profit was to be maximised and not eroded. The major challenge for developing this type of culture is to pass this down the supply network so that the weakest link is also performing well. SCM is not well developed in construction because of short-term relationships and the traditional attitudes that have meant that suppliers down the chain have gained special relationships without the chance for 'back to back' value-inducing contracts. There is also a requirement for public organisations to regularly widen their partnerships and to market test or to share out work between an approved list. In the case study the second and some of the third level of supply chain were bought into the training regime by the cluster leaders.

Health and safety, quality and the environment

There is a close relationship between a continuous improvement, quality approach and health and safety and good business practice. Many project managers point out that health and safety is good business because a safe site, where there is less perception of danger and a sense that management is 'looking after its workforce', often produces more productive work and so the investment in safe systems reaps a reward. The relationship between better welfare, a clean site and health and safety is important and pays back the welfare measures. A radical approach to improving conditions may even help make the industry attractive to more good quality managers. In the case study and in many other projects, awards were available for individual ideas for improving health and safety. If this was connected with quality and the development of behavioural approach then it might be feasible to make quality or 'first time right' awards and to develop a culture of worker collaboration and continuous improvement. This could be connected to the *reduction* of supervision levels and the simplifying of the QA paperwork to save money.

Legislation such as the Construction Design and Management Regulations have encouraged a more integrated approach in the approach to health and safety, by bringing together designers, contractors and clients in assessing health and safety risk. It is recognised that the blame for many accidents is interconnected and a poor design consideration can lead to unnecessary lifting and injury problems, due to poor management and organisation on site, indirectly caused by tough competition and client choice of lowest price tenders from 'work hungry' and poorly organised contractors with poor resourcing. It is also recognised that the safe design and use of products is interconnected and needs to be integrated by better communications for safe handling and use.

Customer satisfaction

Customer satisfaction is the starting theme for this book, as an integrating factor in the delivery of projects and sustainable profitability for the contractor and the consultant. Relationship marketing has been discussed as a way of developing an understanding of the clients' business and involving all levels of the workforce in building relationships with the client as a starting point for the integration of objectives. An integrated approach would include the setting up of an internal and external customer system so that, there is a culture of continuous improvement. This takes the strain off project management. It is important to recognise the differentiation principle to give a unique service to the external customer in accordance with their particular aims and objectives. Customer accounts have a single key contact over a period of time and a dedicated core team may be required by the client. However, this will need to be balanced with the need to offer variety for the supply chain and to maintain the challenge, buzz and variety that many people join the industry to attain. Maiser's principle of customer satisfaction is to meet specification by slightly more than agreed to induce delight, get the customer loyalty and refer also to other potential customers.

The integrating principle is fundamental to success and examples of best practice represented by the case studies are inspiring, though there is plenty of room for further improvements. The potential for a wider application outside the UK and Europe is obvious even though best practice cases have not been mentioned here plenty of them exist.

References

1 Lawrence P.C. and Lorsch J.W. (1967) 'Organisation and environment: managing differentiation and integration. Graduate School of Business Administration'. In Walker A. (2002). *Project management in Construction*. Harvard University, Boston. Blackwell Publishing, Oxford.
2 Walker A. (2002) *Management of Construction Projects*. 4th Edition. Blackwell Publishing, Oxford.
3 Miller E.J. and Rice A.K. (1967) Tavistock Institute.
4 Defence Estates (2001) *Prime Contracting Initiative: Spearheading Innovative Procurement Initiatives*. Briefing Pack. PCI IPT Communications.
5 Brown A. (2002) Andover North Site Profile: Bucknall Austin Solutions Team. Strategic Forum for Construction website http://www.strategicforum.org.uk/sfctoolkit2/help/andover.doc. The case study material is derived from the write up on Andover North Site Redevelopment.
6 Maylor H. (2003) *Project Management*. 4th Edition. Pearson Education, Edinburgh.
7 Rethinking Construction Respect for People Working Group (2003) *Framework for Action*. May. 2nd Edition.

Appendix

The Andover North Site Redevelopment (ANSR) Project

A. Brown

Prime contracting embraces the main drivers in Smarter Procurement and the Bucknall Austin Prime Solutions (BA) model pushes the boundaries further with truly innovative thinking and approach.

The Andover North Site Redevelopment (ANSR) Project [1] was the First Prime Contract to be let by DE and this project encapsulates the smarter thinking that we developed in our 'Prime' model (Figure A.1).

The project is unique in that it is Consultant led, BA assembled and managed a 'Virtual Organisation' for the works.

The BA Core Team comprised:

Bucknall Austin Prime Solutions	Prime contractor, Management functions
Pearce Construction & Thomas Vale	Building Cluster Leaders
SEC	M&E Cluster Leader
Percy Thomas	Architect, Design Cluster Leader
URS Thorburn Colquhoun	Structural Engineering
Hoare Lea & Halcrow	M&E design
Hyder	Infrastructure design

Headlines include:

- Team Charter agreed by all (including Client).
- Core team have equity shares and participate in the pain/gain mechanism with DE.
- Single declared profit for the whole team.
- Single Project PII insurance (including fitness for purpose).
- No claims.
- H&S and sustainability at the top of our agenda.
- 'Sharing' resources/specialisms within the team.
- Back-to-back bespoke contracts down the supply chain.
- Core Team Project Board with monthly Director's meeting to deal with issues.

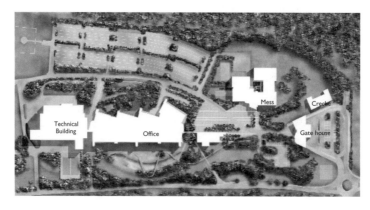

Figure A.1 ANSR is a £40m contract with 6.5-year Compliance (maintenance) period. It provides a HQ for the Defence Logistics Organisation comprising Office for 800 staff, Gatehouse, Creche, Technical Facility, Mess and landscaping (34 acres).

- Project Bank Account – payments from MoD into account with whole team paid simultaneously by electronic transfer direct by bank.
- Virtual co-location of whole team using electronic collaboration tool.
- Prime Specific Supply Chain pre-qualification.
- Target Cost with GMP.
- Open book accounting and approach.
- Incentivisation throughout supply chain.
- Awards scheme for individuals on site.

The given headlines were set out in the Integrated Project Agreement which was signed by team.

The result – a 'one for all/all for one' culture, whereby everyone was focused on the project, rather than themselves and all had a vested interest in its successful outcome.

The project generated innovation at all levels within the supply chain including:

- Prefabrication workshop erected on site to fabricate ductwork, pipework etc. in controlled, safe environment.
- Electrical wholesaler on site, all components ordered 'live' and delivered 'just-in-time' on open book basis.
- Low energy buildings – office design follows on from Abbeywood principles.
- Building structure, orientation, design assists thermal control of internal environment.
- Heat pumps in lieu of chillers.

- Intelligent lighting.
- Flexible layouts – low churn costs.
- Excellent BREEAM rating.
- Fabric roof to streets.
- Enclosed steelwork left unpainted.
- Site levels optimised – minimising material off-site.
- Re-use of topsoil, minimising imported material.
- 'Balanced scorecard' approach to Component Option – encompassed Through Life Costs, programme, risk, H&S, Environment, aesthetic The ANSR Project completed on time and cost in 2 October to a high quality, representing a real milestone for this new way of working in the UK.

IIP

The Bucknall Austin Prime Solutions (BA) team incorporates like-minded organisations who either have IIP accreditation or are working towards accreditation.

Customer satisfaction

Andover was the first Prime Contract let by DE and the whole emphasis of Prime is based upon a totally open approach. This meant that all issues on the Project were visible to the client and they played an active role in managing them with the design, construction and maintenance teams. Indeed the Client and BA team worked together as an integrated unit for the total delivery of the Project. The Client team have also adopted an innovative and proactive approach to new working practices.

The need for formal letters was kept to an absolute minimum with the use of the Information Channel (hosted by BIW) and electronic mail being the main focus for the transfer of information, clarification and comments.

Customer satisfaction was assured by constant liaison and review. Being practically co-located with the client enabled the team to discuss and resolve any issues as they arose, preventing them escalating into a serious problem or contractual dispute.

Directors from the core team (who were not directly involved in the project) undertook quarterly peer reviews. These reviews encompassed design and delivery and the Clients' view on our performance. The peer reviewers made recommendations for changes that were incorporated prior to the next review.

Integrated supply chain

One of the main drivers with prime contracting is developing and working with an integrated Supply Chain. The approach developed for Andover provided for integration at all levels.

The project encompassed 5 buildings plus 34 acres of external works and our Core team, which was the highest level of the supply chain, included two Building Contractors and a Services Contractor. All three worked together in partnership with C.H. Pearce undertaking two of the largest buildings and Thomas Vale's the other three plus the site wide external works. SEC, again in partnership rather than as subcontractor, undertook all the services works across the five buildings together with the site infrastructure.

The three contractors already had their own well-developed supply chains which we tailored, so as to suit the needs of the project. All Supply Chain members chosen for Andover were pre-qualified, irrespective of whether they were existing and new. These pre-qualification processes include completion of a seven-page questionnaire, briefing meetings and workshops plus final interview. The reason for this was to ensure that as well as possessing the right technical skills, the organisations could and would embrace the culture of Partnering and Prime Contracting.

The Supply Chain were actively engaged in the design development and value-engineering workshops held on Andover, to ensure that their buildability knowledge and technical expertise was used in the decision-making process in order to Optimise Through Life VFM. Many of the key suppliers were actively involved at tender and preferred bidder stages as key decisions were made during this period.

Sustainability

The Strategic Brief set out the MoDs policy on environmental impact and sustainability. Our policy was to use this as a benchmark and strive for greater goals. We employed a specialist sustainability consultant from U.R.S. Thorburn Colquhoun to advise and audit the works, working alongside the Prime Solutions team. Not only were our designs independently reviewed in terms of functionality and efficiency, but also the WLC of materials selected were considered in the design selection.

The requirement was to achieve a very good BREEAM rating and we are on course to achieve an excellent rating. Our philosophy not only embraces the Capital build, but also seeks to address sustainability through the life of the facilities. The particular measures adopted are as follows:

- Recycling the demolition material and incorporated them in the works.
- Designing out the need to remove excavated material for site.
- We used natural materials, from proven sustainable sources in the construction.
- Low energy intelligent light fittings.
- Dual flush toilet systems including training occupiers in efficient usage.
- Heat pumps/air handling unit (AHU) Recuperators to recycle and extract free heat from extract air.

- Using a collaboration tool for the storage and dissemination of the design and correspondence reducing the number of paper copies held by the team.
- Paper recycling on site.
- Waste materials were segregated for recycling during construction.
- Incorporation of virtually all existing trees on the site within the final design.
- Using low labour intensive building methodology.
- Integrated transport – providing new bus stop and turning area for local buses.
- On-site cycle and pedestrian routes linked to local transport infrastructure.
- Education/Training of occupiers for the economic use of water and energy.
- During occupation, the management and operation of the buildings will be audited at regular intervals to check on the performance of the facilities and how they are utilised to ensure optimum operational efficiency is maintained.
- Energy usage will be continuously monitored and audited from data provided from the Building Energy Management System (BMS) and incoming service supplies. This will show the pattern of use in practice and highlight any variances from the predicted use, allowing fine-tuning of the systems and controls.

These regular audits will allow the Client and PC to best manage energy and reliability maintenance and maintenance of the structural fabric.

Partnering

The BA team comprises not only like-minded organisations but also like-minded individuals within those organisations, who practice what they preach. We believe Partnering and the further developed concept of prime contracting is fundamentally important to improving industry performance and giving better value to our clients.

Partnering is the cornerstone of our approach on Andover, although BA are the appointed PC, we partner with all our core team. The project has its own Project Board, made up of director representatives of each Core Team organisation, which meets monthly to review and to resolve key project issues – all decisions are made democratically. No single organisation, including BA, holds a casting vote.

There is shared responsibility across the Core team irrespective of where a problem originates, each member has a shareholding in the 'Virtual Organisation' and liability (and benefit) is apportioned in accordance with this shareholding.

As well as developing a bespoke suite of back-to-back contracts right down the Supply Chain, we also developed a Partnering Agreement called 'The Integrated Project Agreement' which all parties signed. This is far more than a charter, it sets out how we will partner together, it defines the responsibilities of each organisation to another, the sharing mechanism and sharelines, it empowers the Project Board and defines its authority and also includes an internal dispute escalation procedure (which we did not use!)

The partnering approach also and most importantly extends to the Client team. The management of both the Client and PC form an integrated team which is co-located on site. Decisions about the project are made jointly and where disagreements occur both sides have worked proactively together to resolve them without the need for recourse to more formal contractual remedies.

The BA team partnership has been successful on Andover that the organisations are currently working to create a legal Joint Venture out of the Virtual Company.

Innovation

Because prime contracting is so new and indeed Andover was the first project to be awarded there was no precedent for many of the issues which arose, which in turn generated considerable innovation.

The Integrated Project Agreement and our approach to sharing project risks and rewards, as discussed earlier, are good examples of such innovation.

The Project Bank Account is another example of innovation which was developed for Andover but has far broader applications. It was developed to address the age-old problem of Contractors withholding monies paid in interim payments and due to their Supply Chain for their own financial gain.

Its operation is simple but effective and foolproof. A payment is agreed with the client in the normal way including a breakdown of payments due to the Supply Chain. This information is passed to the bank in advance of the client making payment directly into the account, when the money is deposited, the bank make payments to the whole supply chain simultaneously in accordance with the payment breakdown.

Single Project Professional Indemnity Insurance (PII), we worked closely with the insurance industry and DE to establish a PII policy that gave cover to the whole team (designers and design within the supply chain) that also covered the Fitness for Purpose requirements within the Contract.

Incentivisation of the Supply Chain, we developed a system with our clients' approval whereby the whole Supply Chain are incentivised (by sharing in the benefits) to identify good ideas which save cost or improve VFM. We also have a site-based award scheme to reward safe working practices, site cleanliness, good working practices and the like.

Combined Profit for BA team, the project is set up on an Open Book basis and the headline profit quoted in the tender is a combined weighted figure for the whole team. This again avoids the problem of profit on profit, on profit up the supply chain which often happens.

Web-based Project Collaboration Tool, we operated a collaboration tool on the project from before contract award right through to completion. This created a virtually co-located team and provides a controlled medium for depositing and exchanging Project information. It proves an audit trail for any document, drawing, report or the like and cuts down considerably the need for paper copies.

We are currently developing this system further for future projects to become a complete management system, by combining it with a Knowledge Management system, reporting protocols, templates and self-generated management reports.

Prime Knowledge Management System, we have process mapped the whole Prime process converting it to an intranet-based electronic matrix with hyperlinks to a network of supporting documents and the like. This enables every task to be together with its mandatory and advisory practices defined and a selection of proforma templates to be used. This will sit on the updated Collaboration Tool.

Reference

1 Brown A. (2002) Andover North Site Profile: Bucknall Austin Solutions Team. Strategic Forum for Construction website http://www.strategicforum.org.uk/sfctoolkit2/help/andover.doc

Glossary

Business plan is a client-based proposal for formalising the development of a new part of their business or facility. It consists of some basic objectives and feasibility assessments and may be developed as the project planning gets underway. On receiving director outline approval, it moves the project from conception to inception, which is the arbitrary beginning of the project life cycle as defined in this book.

Client is the commissioning sponsor of the facility and is the main business decision maker in the project team. The client is the key customer, but not the only one. In contract terms the client is often called the employer, although they may have an agent which separates them from any decision making. The client will have the responsibility for paying the bills.

Construction manager is manager of the construction and commercial interests of the contractor. In construction management procurement the construction manager provides construction advise and procurement management direct to the client on a fee basis.

Consultants are payable by fee as advisers to the client and members of the project team. An executive project manager is a consultant.

Contractor/subcontractor/specialist contractor Contractor or main contractor generally refers to the co-ordinating role of construction activities in which there are sub or specialist roles. In the modern context of supply chain or supply nets there is still a hierarchy, but co-ordination, may be operated in clusters, accountable direct to the project manager, client or construction manager.

Design team is responsible for all aspects of the design. It consists of specialists in specific areas of design such as a building services engineer (mechanical, electrical and other services), structural engineer (building structures), civil engineer (ground engineering and civil structures), acoustic engineer, landscape architect etc. They may work individually for the client, indirectly for the lead designer (later), design and build contractor or PFI provider.

Facilities management This is the provision of non-core services to the clients facility. It is the planning, organisation and managing of a facility on a day-by-day basis to maintain the physical assets. A facility is a building or an infrastructure.

Integration or integrated team is the concept of seamless working between the client and their consultants and contractor supply chain. It assumes a moving away from an interface management in support of a virtual organisation, where individual organisational objectives are subjugated (integrated) to project objectives. It should lead to an early appointment of the whole team to take part in the development process.

Lead designer (LD) co-ordinates the design team effort. In traditional procurement (see later), they have access to the client, receive the brief at inception stage, develop the design, arrange to tender for construction and exercise a quality checking role. During construction they are the key communication channel. This gives them a project management role as well as design role in the absence of an executive project manager. LD is usually the architect in building projects or the engineer in civil projects.

Planning supervisor is a co-ordinator for ensuring a suitably integrated health and safety approach in the pre tender health and safety planning and post construction information handed over to the facilities team. They play a statutory role that is complimentary to the LD.

Procurement specifically applies to the method of tendering and contracting between the client and their suppliers. The different methods vary from traditional separate procurement of design and construction to turnkey arrangements with a single point of contract. The choice determines the balance of risk allocated between the client and members of the project team. The project manager needs to fit procurement type to client needs.

Programme management is the management of several related parallel or sequential projects. When concepts such as strategic partnering and supply chain management are discussed, there is a need to look at issues which are 'inter' projects as value is often built on carrying forward knowledge and experience between projects. Strictly these are business strategies, but they are recognised as being project driven as they are based on business with a single client. An example of programme management would be a developer who has an overall programme to develop a site with commitments to several users, yet they have let out the design to several architects and contractors to cover infrastructure and various buildings or building types.

Project is defined in BS6079–1 Guide to Project Management as 'A unique set of co-ordinated activities with definite starting and finishing points, undertaken by an individual or organisation to meet *specific objectives* within defined schedule, cost and performance parameters'. (BSI 6079 (2000) Guide to Project Management.)

Project co-ordination (table 1.1 CIOB) The non-executive functions which make up the project either after appointment of the consultants or after the appointment of the contractor. (Chartered Institute of Building (2002) *Code of Practice for Construction and Property*. CIOB. 3rd Edition. Blackwells Publishing, Oxford. p. 6.)

Project definition is the development of the project brief and scope up to the planning application stage, so that the risks have been identified and the VFM has been optimised to suit business needs. (Chapter 3 develops this theme.)

Project execution plan (PEP) is a master plan that develops the details of the strategic stage of the life cycle and provides a baseline schedule in areas such as cost and programme, together with strategic control documents for, organisation, health and safety, systems and policies. It also recognises the need for continuous development of the plan to meet the dynamic conditions of the project.

Project management (CIOB 2002) is 'the overall planning, co-ordination and control of a project from inception to completion, aimed at meeting a client's requirements in order to produce a functionally and financially viable project that will be completed on time, within authorised cost and to the required quality standards'.

Project manager (Annex A BS6079)   The individual with the responsibility for managing a project to specific objectives. An executive project manager manages the project on behalf of the client from inception to completion.

Project sponsor (Annex A BS6079 and CIOB 2002)   The individual or body for whom the project is undertaken and who is the primary risk taker. The individual representing the sponsoring body and to whom the project manager reports.

Project team CIOB (2002: 6)   This includes those who have a professional role such as design, funding, the contractors and subcontractors led by the client representative who may also be called the project manager. It does not include the project sponsor.

Quantity surveyor is a specialist in the financial and contractual areas of project management. Utilised by the client side for preparation of bills of quantities, auditing payment and providing project financial advise. Utilised by the contractor for cost planning, procurement and preparing payment claims.

Risk management (RM) is a proactive approach to the identification, assessment, allocation and mitigation of risk factors to rationalise and optimise contingency arrangements.

Stakeholders of the project are those who have the potential to influence the course of a project. The most direct influences are likely to come from the direct parties to the contract and the project team, with influence from the statutory permissions that are required on behalf of the community. The shareholders have a much less direct influence in projects.

Sustainable construction is a term to describe environmentally attractive solutions to the delivery and maintenance of facilities, but as its name implies should consider the impact of the building on scarce resources, the community and globally.

Traditional procurement means the separation of design and construction. Design is awarded on a fee basis and construction is awarded on a priced 'fixed' price basis to a main contractor, using design and specification documents prepared by others. Specialist contracts are sub awarded by the contractor to their supply chain. There is limited whole life project management, but key client contact is through the LD. Standard contracts are well known.

User is the ultimate occupier/operator of the facility and may have nothing to do with its commissioning.

Value engineering is the tactical application of VM in the later design stages relates design supply and construction.

Value management (VM) or analysis is the strategic analysis of functional requirements and the optimisation of provision within the physical asset provision and its maintenance and running throughout its life cycle. It should be specifically separate from cost saving which has the sense of cutting quality or provision and is a reactive activity.

Index

Note: Page numbers in bold indicate principle reference and cs indicate case study.